Dynamic Patterns

Dynamic Patterns

The Self-Organization of Brain and Behavior

J. A. Scott Kelso

1995

A Bradford Book
The MIT Press
Cambridge, Massachusetts
London, England

This book was set in Palatino by Asco Trade Typesetting Ltd., Hong Kong and was printed and bound in the United States of America.

Library of Congress Cataloging-in-Publication Data

Kelso, J. A. Scott.
  Dynamic Patterns : the self-organization of brain and behavior / J. A. Scott Kelso.
     p.   cm.
  "A Bradford book."
  Includes bibliographical references and index.
  ISBN 0-262-11200-0
  1. Neuropsychology. 2. Self-organizing systems. I. Title.
QP360.K454   1995
612.8—dc20
                                                        94-32105
                                                          CIP

To Betty, Kate, and Alex for their love and a little tale

Contents

Foreword

The human brain is the most complex system we know of. The study of behavior may serve as a window into understanding basic brain activities. While in contemporary neurophysiology and molecular biology, brain activity is traced back to the actions of its individual neurons and even to its molecules such as neurotransmitters, Scott Kelso's book promotes quite a different perspective, namely the study of the outcome of cooperative behavior of many neurons and many other cells. This book fascinates me. I am delighted to see how beautifully it presents the results of a combination of experiments, ingeniously devised and performed by Scott Kelso, and the application of concepts of synergetics, an interdisciplinary field of research that I founded and developed.

Every page of the book provides us with delightful reading and opens new perspectives again and again. It also clearly elucidates historical developments. I mention as examples the presentation of early work by A. S. F. Leyton and C. S. Sherrington on the great variability of brain activities, or of the prophetic words by Aharon Katchalsky and co-authors on brain activity in terms of waves, oscillation, and sudden transitions.

I am sure that this book will not only find a broad and highly interested readership, but will also become a landmark in the field of behavioral sciences and brain research, possibly comparable to Schrödinger's famous book *What Is Life?*

Hermann Haken

Preface

What could be more important than understanding human behavior? Every-one of us has a brain composed of billions of cells that somehow, by virtue of their coordination, give rise to how we think, act, decide, remember, perceive, learn, and develop. How do the individual parts like nerve cells, muscles, and joints of the body interact in such a way that *patterns* of behavior emerge? What are the organizing principles and how are they embodied in the mecha-nisms that generate behavior? Since the time of Newton, physicists have sought to establish the laws and principles that govern the physical world. Consideration of more complex phenomena such as those observed in living things has, until recently, been excluded from physical science. Even though we claim to know the laws that govern the behavior of matter, such laws don't tell us anything at all about how or why we reach for a cup of coffee or walk down the street.

This book is a small step toward filling the gap between the known laws of how matter behaves and how human beings behave. Just as classical physics derived its macroscopic laws from observations about the motion of planets and terrestial bodies, I will describe how it is possible to construct some of the laws that appear to govern (or at least describe and occasionally predict) the dynamic behavioral patterns produced by animals and people. These are used as a foundation on which to build a deeper understanding of phenomena such as perceiving, intending, anticipating, learning, adapting to the environment —and as a window into the brain itself.

The centerpiece of the book is a theory of coordination, how things are put together to produce recognizable functions. The things themselves might be made of matter, such as neurons, muscles, parts of the body, or they might be mental "things" such as perceptions and ideas. Coordination, I argue, is a fundamental feature of life. Imagine a living system composed of components that ignored each other and did not interact with themselves or the environ-ment. Such a system would possess neither structure nor function. In this book I address the basic nature of this interaction, how it occurs and why it is the way it is. The core thesis is that human behavior—from neurons to mind—is governed by the generic processes of self-organization. Self-organization re-fers to the spontaneous formation of patterns and pattern change in open, nonequilibrium systems. I argue that regardless of the level of description one chooses to study (and this I would stress is always a personal choice), the same basic pattern-forming principles are in evidence.

For a system to function its components must be (at least transiently) coupled or linked in some fashion. Such a coupling should guarantee properties that are critical to (complex) living things: multifunctionality, stability, and the ability to flexibly change according to environmental or internal demands. In a number of cases it turns out to be possible to specify exactly what form this coupling takes. Significantly, it captures the coordinative relations between the parts, regardless of the parts themselves (parts of the body, parts of the nervous system, parts of society) and the medium over which the interaction between the parts takes place.

It is difficult to define a discipline or even a field that would neatly encompass the contents of this book. This may be part of its charm or its demise. I have called the overall approach *dynamic patterns*. The term contrasts with more statically conceived notions of structure, and is intended to capture what is, in effect, the functional organization of living things. In biological systems, dynamic patterns of behavior can be produced by a diversity of mechanisms. Similarly, the same kinds of dynamic patterns can be observed at different levels of organization. Thus, both dynamic patterns and pattern dynamics are somewhat independent of the stuff that realizes them *and* the level at which they are observed.

Theoretically, the dynamic pattern approach is founded on and greatly inspired by Haken's theory of nonequilibrium phase transitions that are at the heart of pattern formation in nature. As the term *dynamic patterns* suggests, pattern variables, once identified, are expressed in the mathematical language of nonlinear dynamical systems, including stochastic aspects. Experimentally, the main examples come from psychology, not of the exciting kind like sex, dreams, madness, or genius, but of the very basic kind common to us all, voluntary movement, perception, and learning. Even for such commonplace biological activities, relevant variables are not known but have to be discovered. This book provides a strategy for finding functionally relevant pattern variables and their equations of motion, how patterns evolve and change in time.

Although I have tried to relate my discussion to other approaches, this book is very much a personal view. By far the strongest prejudice I have is that there are deep, unifying principles underlying human behavior and the brain. Abstract generalizations, of course, must be accompanied by unwavering devotion to detailed fact. My selection of experimental examples is not only biased, but the examples themselves are sufficiently restrictive that their generality might be questioned. Nevertheless, they convey what I want them to, illustrating not only how complicated behavior may arise from simple laws, but also how complex systems like the human brain can generate both simple and complicated behavioral patterns.

Is this book accessible to the nonspecialist? I hope so. I wrote it hoping to engage a broader audience. Dynamic pattern language might be a bit arcane, but once you are familiar with it, it becomes second nature. Because the same concepts and terminology are applied at different levels of description, the jargon is restricted more than usual. Indeed, perhaps the greatest appeal of *Dynamic Patterns* is that it provides a conceptual framework and vocabulary that is just as natural for the mind as it is for the processes going on in the brain or behavior in general. Without such commensurate description, I argue, it is impossible to see the connections among neural, mental, and behavioral events.

The Polish physicist Jacob Bronowski once remarked that, compared with the physical sciences, the distance between fact and theory in the biological and psychological sciences is small. Often what passes for theory in these fields is a redescription of the facts, albeit in a different language. The latter usually *represents* the facts and is simply not abstract enough for understanding how the brains and minds of people behave. In *Dynamic Patterns*, it turns out that rather abstract variables are required to describe the collective or co-ordinative behavior of complex living things. Ironically enough, the dynamics of these abstract pattern variables exhibit features such as stability and insta-bility, crises and intermittency, attraction and repulsion, that are remarkably close to everyday life. The fact that this language of behavior also has a precise mathematical meaning enhances, I think, intuition. Nevertheless, the challenge I posed myself in this book was to explain the self-organizing processes underlying behavioral pattern formation and change in everyday "dynamical" language using only a minimum of necessary formulae.

A great deal is still to be learned, but it is my hope that this book will provide a springboard for future developments. That was, in fact, my prime purpose in writing it. After twenty-some years of doing mostly experimental work, it seemed time to pause and make a statement even as the research rushes on. My sense is that a lot of people are poised to plunge into the multidisciplinary field that surrounds *Dynamic Patterns*. Perhaps this book will be the fluctuation they need.

Acknowledgments

This book is something my friends and colleagues have been trying to get me to do for a long time. I hope they are not disappointed with the result. Certainly, their scientific contributions occupy a large part of this book, as I trust will be clear from reading it.

I have been fortunate to have had many collaborators and associates over the years. I owe a great debt to my early mentors in science at the University of Wisconsin-Madison, especially George Stelmach and Bill Epstein. The latter has visited me regularly in Florida over the years, and has been a constant source of intellectual stimulation and encouragement, as well as a close friend. As a young assistant professor at the University of Iowa in the mid-1970s, I benefited greatly from close interactions with students, one of whom, David Goodman, is developing the present approach with his own students. Gerry Zimmerman, then a speech scientist at the University of Iowa and now a lawyer, introduced me to the problems of speech production. Little did I know that I would pursue these issues in great detail at the Haskins Laboratories in New Haven, Connecticut. There, Michael Turvey profoundly influenced my thinking on perception and action. Although we have not always agreed, I hold him and his views in great respect, and value our longstanding friendship dearly. I am very grateful to my students and postdoctoral associates at Haskins, John Scholz, Eric Bateson, Bruce Kay, Kevin Munhall, and Eliot Saltzman for the work we did together. Looking back, it was difficult research, but great fun.

Of course, the last ten years at the Center for Complex Systems at Florida Atlantic University has been a wonderfully fulfilling experience. From two rather dilapidated trailers out on a runway north of campus in 1985, the Center has become one of the focal points of the University with over forty faculty, students and staff. As this book goes to press, the Center is poised to offer the first PhD degree in Complex Systems and Brain Sciences in the United States. That this has been possible in such a short time is due to the support given by the FAU administration to multidisciplinary research, the hard work of the participants involved (for which I thank them), and the research and training funds provided by our sponsors, principally the National Institute of Mental Health, the U.S. Office of Naval Research, and the National

Science Foundation. For their support over the years which enabled much of the research described here to be done, I am deeply grateful.

On the theoretical side, my research at the Center over the last ten years has been aided beyond measure by collaborations with a number of young mathematicians and physicists, including Mingzhou Ding, Tim Kiemel, Gonzalo DeGuzman, and, especially, Armin Fuchs, and Gregor Schöner. Their insights combined with skills that I do not have have been a great help to me and the research program in general. On the empirical side, at the Center I have been fortunate to observe and participate in the development of a new breed of theoretically oriented and computationally sophisticated scientist. I count the following students and associates in this category: John Jeka (my first PhD student at the Center), John Buchanan, David Engstrom, Tom Holroyd, Liz Horvath, Pam Case, Phil Gleason, Paul Treffner, Collin Brown, Sylvie Athénes, Gene Wallenstein, Joanna Rączaszek, and Pier Zanone. I thank them for joining this adventure and most of all for working with me. They have taught me a lot.

The Center has had many visitors over the years, some of whom helped shape the perspective drawn in this book. Among these I would like to single out three. Steve Wallace, an old graduate student friend at Wisconsin took a sabbatical leave to work with me at the Center in the late 1980s and has continued to pursue the dynamic pattern approach in innovative fashion with his own students ever since. Similarly, a good friend Esther Thelen has become one of the chief proponents of the approach in the field of developmental psychology. I am especially grateful to her and Mark Latash for their constructive criticism of an earlier draft of this book. Finally, Jim Lackner has visited my wife, Betty, and me regularly since we came to Florida. His astute scientific mind, combined with a sharp wit, has been a joy to us over the years, and kept me going when times were tough.

I taught a course to graduate students and postdoctoral fellows on the subject matter of this book in the spring of 1994. I thought it went quite well. One assignment was to write a critique of each chapter and pull no punches. These critiques prevented me from making some major gaffes, and I am very grateful to those who took the assignment seriously. In particular, the comments of Paul Treffner, Tom Holroyd, Joanna Rączaszek, John Buchanan, Pam Case, Helen Harton, Phil Gleason, and Armin Fuchs were especially helpful. These people are not to be blamed for the errors or misinterpretations that remain.

I had a lot of help with the production of this book for which I am extremely grateful. My loyal and skillful assistant of many years, Betty Harvey, typed all the drafts and revisions. To write creatively I must work by hand, and Betty transformed these pages into beautiful, formated, amazingly error-free typescript. And always, I might add, with her usual friendly spirit. My students rag me about still writing by hand, to which I reply that Charles Dickens wrote *A Christmas Carol* in six weeks long before the age of word processors. John Buchanan, one of my senior graduate students, collated and sometimes made the figures in a timely fashion which helped a great deal. His unflappable nature smoothed the process and soothed me. For a number of the computer graphics and model simulations I owe a debt of gratitude to Armin Fuchs. Nothing was too much trouble to Armin or Tom Holroyd (who also produced some of the graphics).

My wife, Betty Tuller, discussed, read, and criticized every word of this book. Without her intellectual and emotional support this book would never have been written. She was and is the backboard for many of the ideas discussed herein. For seventeen years we have worked, played, and loved together. This book is dedicated to her and our children.

It is seldom that a single event can play a dominant role in one's scientific life, but such is the case for the subject matter of this book. I refer to an invitation from Hermann Haken, rendered to me over twelve years ago, to come to Stuttgart to work with him on modeling some of my experiments. Our collaboration continues to be the intellectual highlight of my life. As I relate in the text itself, Haken's theory is the backbone for much, if not all of the research described herein. His notion of self-organization, that all open systems in nature are driven to organize themselves into patterns, is the very core of this book. The key concepts of synergetics, among which are order parameters, control parameters, instability, and slaving, figure prominently throughout. Moreover, it is the recently confirmed predictions of Haken's theory that support one of the central theses in this book, that the brain is *fundamentally* a pattern-forming, self-organized, dynamical system poised on the brink of instability. By operating near instability, the brain is able to switch flexibly and quickly among a large repertoire of spatiotemporal patterns. It is, I like to say, a "twinkling" system, creating and annihilating patterns according to the demands placed on it. When the brain switches, it undergoes a nonequilibrium phase transition, which according to Haken's theory, is the basic mechanism of self-organization in nature. For not only your genius, Hermann, but also your friendship and generosity extended to me over many years, I am forever grateful.

Dynamic Patterns

1   How Nature Handles Complexity

Nature uses only the longest threads to weave her pattern, so each small piece of the fabric reveals the organization of the entire tapestry.
—R. P. Feynman

People are called creative or crazy when they see a likeness between things that were thought not to be alike before. Creativity, however, entails more than seeing a likeness or drawing a metaphor. Abstracting something new is what matters. The examples that I will talk about in this chapter and the concepts that emerge from them initially will seem very strange and far removed from what we are really trying to understand, the behavior of humans. But I hope to show later on that the concepts and tools introduced here actually provide the foundation for a theory of how mind, brain, and behavior are related. At the heart of this theory is how patterns are formed in complex systems. I view the brain not as a box with compartments that contain sadness, joy, color, texture, and all the other "objects" and categories that one might think of. Instead, I envisage it as a constantly shifting dynamic system; more like the flow of a river in which patterns emerge and disappear, than a static landscape. "Meaningful patterns," as Sherrington said over fifty years ago, "but never abiding ones."[1] This is an entirely different image from the brain as a computer with stored contents or subroutines to be called up by a program. In nature's pattern-forming systems, contents aren't contained anywhere but are revealed only by the dynamics. Form and content are thus inextricably connected and can't ever be separated.[2]

Like a river whose eddies, vortices, and turbulent structures do not exist independent of the flow itself, so it is with the brain. Mental things, symbols and the like, do not sit outside the brain as programmable entities, but are created by the never ceasing dynamical activity of the brain. The mistake made by many cognitive scientists is to view symbolic contents as static, timeless entities that are independent of their origins. Symbols, like the vortices of the river, may be *stable* structures or patterns that persist for a long time, but they are not timeless and unchanging.

One of the goals of this chapter is to show how patterns in general emerge in a self-organized fashion, without any agent-like entity ordering the elements, telling them when and where to go. I intend to show that principles of self-organization lie behind all structure or pattern formation, and, later, that the brain itself is an active, dynamic, self-organizing system. This paradigm of understanding how cooperation among the parts of a system generates patterns opens up an entirely new (or at least different) research program for the behavioral and brain sciences.

What do I mean by understanding? I obviously don't mean some privileged level of analysis where the fundamental units of brain function reside. Or that some molecule constitutes an elementary unit of cognition. Or even, as some would have it, that behavior itself is composed of some fundamental unit, like a reflex or a program, that must be studied and further decomposed, perhaps, into some even more elementary constituents. My point is that we will not be bound to the properties of matter itself, or draw our laws strictly from them. In the theory that I will propose no single level of analysis has priority over any other. The psychologist studying overt behavior doesn't have to look over her shoulder at the neuroscientist studying the insides of the nervous system, or the neuroscientist at the chemist, or the chemist at the physicist, and so on.

Understanding will be sought in terms of essential variables that characterize patterns of behavior regardless of what elements are involved in producing the patterns or at what level these patterns are studied or observed. To quote Napoleon Bonaparte, "They may say what they like, everything is organized matter." I will try to avoid dualities such as top-down versus bottom-up and macro versus micro because I think they can be misleading. What is macro at one level, after all, can be micro for another. Molecules, for example, are too macroscopic for a particle physicist to study, and too microscopic for the typical physiologist. Understanding will come by deploying a particular *strategy* to investigate complex systems such as human beings, and the reductionism will be to *principles* (or what I call *generic* mechanisms), that apply across different levels of investigation.

This emphasis is on self-organized pattern formation precisely because we will see that the same principles are at work even though the stuff that is producing the patterns may be very different indeed. Emergent properties are a significant feature of all complex systems in nature, and our goal will be to identify some of the key processes involved in emergence. Details will be shown to be important for understanding how cooperation among the components of a system creates new patterns in a self-organized fashion. Once we accept this more abstract, pattern level of analysis, we start to see common underlying processes and principles of self-organization that transcend animate and inanimate nature.[3] And we start to see how enormously complex behavior can arise from a few simple (but nonlinear) rules. At first blush, all this must appear rather strange. You mean, the reader asks, that some deep relationship might exist between, say, the changing weather pattern and me changing my mind? Sounds totally crazy! But I would say the real question is not whether it's crazy at all, but whether it's crazy (or creative) enough.

WHAT IS A PATTERN?

Pattern (*pat'* ern), n. 1. A natural or chance marking, configuration or design; patterns of frost on the window. 2. *A combination of qualities, acts, tendencies etc. forming a consistent or characteristic arrangement:* the behavior patterns of teenagers.[4]

Everybody knows the world is made up of processes from which patterns emerge, but we seldom give pause to what this means. Maybe it's enough to know a pattern when we see it. Some patterns appeal to us more than others: the beautiful snow crystal, a butterfly's wings, the hive of a honeybee. Pattern is at the heart of great science and art: praise be the one who discovers or creates some underlying unity or truth in the world that surrounds us. Human beings are exceptional at detecting patterns even when the patterns are embedded in a field of randomness and disorder, as we will see later (chapter 7). I often tell my students that the genius of Galileo was that he could abstract the relation (form a pattern in his mind) between a ball rolling down an inclined plane and the motions of the planets. Even the greatest intellectual achievements of human beings have to do with patterns.

When we talk about patterns, we step back from things themselves and concentrate on the *relations* among things.[5] My main concern will be with the behavioral patterns or modes of organized behavior produced by living things on different levels of description. Although I won't try to explain the behavior patterns of teenagers (at least just yet), I will extricate elementary concepts and principles that apply to *all* of pattern formation, relatively independent of who or what the component elements are. And since patterns continually form and change, I will introduce some of the main concepts and tools that scientists use to describe their evolution.

KINDS OF PATTERNS

Most of an organism most of the time is developing from one pattern to another not from homogeneity into a pattern.
—A. M. Turing

The brilliant Englishman Alan Turing, whose name is often associated with the development of the digital computer (and breaking the secret code used by the Nazis in WWII) had a lesser-known but profound interest in how patterns form in nature.[6] I put his words up front because they capture the essence of the kinds of patterns on which I wish to concentrate. First, Turing's reference to organisms places an emphasis on the patterns produced in animate nature (see box). It will turn out, however, that the basic pattern-forming principles apply to both animate and inanimate systems.

The caveat, of course, is that an inanimate system, say, a fluid or a chemical reaction, must be *far from thermal equilibrium*. Enormously interesting and

THE TWO SIDES OF TURING

I am reminded of a famous scientist who was always arguing that the brain is not a Turing machine. When I noted during one of his lectures that there was another Turing, he was quite adamant: "No, No, there's only one Turing. You know, the Turing of the Turing machine!" After writing some equations on the blackboard describing chemical patterns, he paused and stared at me. "Ah, I see what you mean," he said. The equations that he had written on the board were the very ones Turing used to describe "the chemical basis of morphogenesis" in a remote paper published in 1952. My "two Turings" refer to the genius who, on the one hand, made significant contributions to programmable computers, and on the other, showed how patterns in nature can emerge without any programmer at all. Only quite recently have the patterns predicted by Turing been observed experimentally, and Turing's theory still figures quite prominently in developmental biology.[7]

important structures studied by physicists, such as ferromagnetism, or superconductivity in which matter changes its microscopic structure as the temperature is lowered, are not our primary interest. And even though we might find ice crystals beautiful to look at, they are too orderly and rigid states of matter to be of relevance here. Systems in thermodynamic equilibrium are as dead as anything can be.

Only systems that are pumped or energized from the outside (or, like living systems who happen to possess metabolic machinery, from the inside and the outside) are capable of producing the kinds of patterns and structures that interest us. These are called open, nonequilibrium systems: open in the sense that they can interact with their environment, exchanging energy, matter or information with their surrounds; and nonequilibrium, in the sense that without such sources they cannot maintain their structure or function.

In the last twenty years or so, tremendous progress has been made in understanding how patterns form in open, nonequilibrium systems, especially by my friend and colleague Hermann Haken.[8] Thanks to him and others,[9] ordinary matter, under certain conditions, has been shown to exhibit lifelike properties. I will draw especially on Haken's work because it provides the theoretical foundation for some of the later developments in neuroscience and psychology that are the primary focus of this book. My aim is to show how principles of pattern formation may be adapted for and tailored to special features of living things, how they perceive and act, learn and remember.

Another reason why Turing's quotation captures what we are after is that he emphasizes that even primitive organisms are not structureless things composed of homogeneously distributed elements. They always have some kind of intrinsic organization. Although this seems rather obvious, it will turn out that profound consequences result from recognizing that organisms are not tabulae rasae. Many scientists who study learning, even at the cellular level, fail to appreciate this point, which I will develop in chapter 6. Parenthetically, the important role of intrinsic organizational processes resonates strongly with early Gestalt theory in psychology, which sought natural principles of autonomous order formation in perception and brain function (see chapter 7). As we will see, entirely new concepts of self-organizing dynamical systems are necessary to replace the older physical ideas available to the Gestaltists.

Relatedly, *becoming* is a process of change, and this change often takes the form of an order-order transition from one pattern to another. The famous physicist Erwin Schrödinger, who earned the Nobel Prize for his contributions to quantum mechanics (the Schrödinger wave equation), also stressed the principle of order-order transitions for understanding life.[10] He considered the physicists' emphasis on disorder-order transitions as completely irrelevant to the emergence of life processes. In physics, different aggregate states of matter—solid, liquid, gaseous—are called *phases*, and the transitions between them are called *phase transitions*. When vapor changes to liquid and eventually to ice, this is an example of progressive change from disorder to order. We see immediately that life processes have nothing to do with this kind of phase transition, and that entirely different principles having to do with nonequilibrium phase transitions are required.

Finally, Turing's reference to developing and evolving patterns stresses the dynamic aspect of pattern formation and even the continuously changing nature of the patterns themselves. For this reason, the principles we seek are for dynamic patterns, not static ones. All structures in animate nature are actually dynamic. We tend to think of some of them as static, but this is not really the case. Here I will adopt the view that structures and behaviors are both dynamic patterns separated only by the time scales on which they live.

PRINCIPLES OF DYNAMIC PATTERN FORMATION

Any principle of pattern formation has to handle two problems. The first is how a given pattern is constructed from a very large number of material components. We might call this *the problem of complexity of substance*. The brain is an example par excellence of a compositionally complex system. It has approximately 10^{10} neurons, each of which can have up to 10^4 connections with other neurons and 50 plus neurotransmitters (chemicals that are necessary for the neurons to work). Second, any principle of pattern formation must address not just how one pattern but many patterns are produced to accommodate different circumstances.

Biological structures like ourselves, for example, are multifunctional: we can use the same set of anatomical components for different behavioral functions as in eating and speaking, or different components to perform the same function (try writing your name with a pencil attached to your big toe). We might call this second problem *the problem of pattern complexity*. Moreover, how a given pattern persists under various environmental conditions (its *stability*) and how it adjusts to changing internal or external conditions (its *adaptability*) have to be accounted for. The processes that govern how a pattern is selected from myriad possibilities must be incorporated in any set of organizational principles for living things. These processes often involve *cooperation* and *competition*, and a subtle interplay between the two.

Dynamic Instabilities

To explain the mechanisms underlying pattern formation, let's use the familiar example of a fluid heated from below and cooled from above. First, a word of caution. No one is saying that the brain (or living things in general) is simply a fluid composed of homogeneous elements. Far from it. Rather, the fluid as used here illustrates some of the ways nature handles complex nonequilibrium systems containing many degrees of freedom. In particular, it allows us to illustrate the key concepts of pattern formation and change that will provide a foundation for understanding the emergence of biological order and change. Like all great physical experiments, the beauty of the fluid example—called the Rayleigh-Bénard instability—is that even though it is performed in the laboratory, it opens the view to the bigger picture, such as weather patterns, cloud formation, and turbulence in fluids. This idea of choosing (some might say inventing) an experimental model system that adequately captures some of the essential features of the bigger (real world) problem will be a crucial theme throughout this book.

I often do a version of the Rayleigh-Bénard experiment with my daughter when I fry an egg for her on a Saturday morning. She gets a charge out of the patterns that form in the oil as it's heated. Here's how it goes (figure 1.1). Take a little cooking oil, put it in a pan, and heat it from below. If the temperature

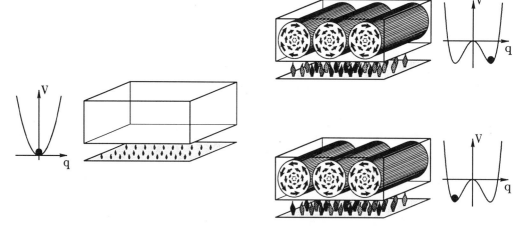

Figure 1.1 (*Left*). A layer of liquid heated weakly from below displays no macromotion. The liquid is in a rest state, as indicated by the ball resting in the minimum of the potential well. (*Right*). At a critical value of the temperature gradient the liquid displays macroscopic rolling motions. The ball can assume one of two possibilities, rolls rotating in one direction or the other.

difference between the top and bottom of the oil layer is small, there will be no large-scale motion of the liquid. Notice the liquid contains *very many molecules*, and the heat is dissipated among them as a random micromotion that we cannot see. We call this heat conduction.

Now already we see this is an *open system*, activated by the application of a temperature gradient that drives the motion. As this driving influence increases, an amazing event called an *instability* occurs. The liquid begins to move as a coordinated whole, no longer randomly but in an orderly, rolling motion. The reason for the onset of this motion is that the cooler liquid at the top of the oil layer is more dense and tends to fall, whereas the warmer and less dense oil at the bottom tends to rise. The resulting convection rolls are what physicists call a *collective* or *cooperative* effect, which arises without any external instructions. The temperature gradient is called a *control parameter* in the language of dynamical systems. Note that the control parameter does not prescribe or contain the code for the emerging pattern. It simply leads the system through the variety of possible patterns or states.

Rolling motions are not the only possibilities. In an open container, surface tension can also affect the flow. The net effect of this force is to minimize the surface area of the fluid, causing tesselation of the surface and the formation of hexagon cells (figure 1.2 and plate 1). In the center of each hexagon, liquid rises, spreads out over the surface, and sinks at the perimeter where the

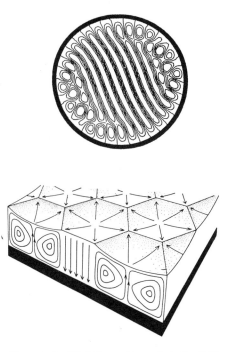

Figure 1.2 (*Top*) Convection rolls in the fluid as viewed from above. (*Bottom*) Tesselation of the liquid surface with hexagonal cells. Where the tension is greatest the surface becomes puckered, reducing surface area (Reprinted with permission from reference 13. Copyright © (1980) by Scientific American, Inc. All rights reserved.

How Nature Handles Complexity

hexagons join. An important point is that two quite distinct mechanisms— one to do with buoyancy and the other surface tension—can give rise to the same dynamic pattern. Thus, the mechanism-pattern relation is not necessarily one to one. Similarly, the same mechanism can give rise to different patterns. This will turn out to be a dominant characteristic of biological systems as well, which we will have to explain.

Self-Organization

Such spontaneous pattern formation is exactly what we mean by *self-organization*: the system organizes itself, but there is no "self," no agent inside the system doing the organizing. A single Bénard preparation might contain something on the order of 10^{20} molecules, each of which is subject to random disordered motion. But once the rolling motion starts, even in parts of the fluid, all these molecules begin to behave in a coherent fashion. The system is no longer merely a haphazard collection of randomly moving molecules: billions of molecules cooperate to create dynamic patterns synchronized in time, and extending over large distances in space many orders of magnitude larger than the molecular interaction. It's like a wave in Yankee stadium: the individual fans communicate and cluster together in groups to create a nearly synchronized pattern that spreads throughout the stadium.

In the Rayleigh-Bénard system the amplitude of the convection rolls plays the role of an *order parameter* or *collective variable*: all parts of the liquid no longer behave independently but are sucked into an ordered, coordinated pattern that can be described precisely using the order parameter concept. Where does this order parameter concept come from? It really comes out of what mathematicians call linear stability analysis (in our example, analysis of the equation of motion that describes the oil's behavior, which we won't go into here). The basic idea is that the initially random starting pattern can be considered as a superposition of a whole bunch of different vibratory modes described by this equation. For a given temperature gradient some (most) of these modes will be damped out. But others will increase in velocity, and the mode or pattern with the biggest rate of increase will dominate. This principle of the most unstable solution, called the *slaving principle* of synergetics,[11] serves as a *selection mechanism*: from random initial conditions a specific form of motion is favored. It's this coherent pattern that is described by the order parameter, and it is the order parameter dynamics that characterizes how patterns form and evolve in time. One of our main questions will be, what are the relevant collective variables and collective variable dynamics for complex systems such as the nervous system?

Collective Variables and Circular Causality

We should clarify one further important aspect of the collective variable or order parameter notion. In synergetics, the order parameter is created by the

Excuse me your Honor, but I believe
I was here first.

Figure 1.3 The chicken and egg problem according to Farcus. (Reprinted with permission from Universal Press Syndicate)

cooperation of the individual parts of the system, here the fluid molecules. Conversely, it governs or constrains the behavior of the individual parts. This is a strange kind of *circular causality* (which is the chicken and which is the egg?), but we will see that it is typical of all self-organizing systems (figure 1.3). What we have here is one of the main conceptual differences between the circularly causal underpinnings of pattern formation in nonequilibrium systems and the linear causality that underlies most of modern physiology and psychology, with its inputs and outputs, stimuli and responses.

Some might argue that the concept of feedback closes the loop, as it were, between input and output. This works fine in simple systems that have only two parts to be joined, each of which affects the other. But add a few more parts interlaced together and very quickly it becomes impossible to treat the system in terms of feedback circuits. In such complex systems, as W. Ross Ashby elegantly pointed out years ago, the concept of feedback is inadequate.[12] What is more important is to realize that richly interconnected systems may exhibit both simple and complex behavioral patterns. Returning to the Bénard example, there is no reference state with which feedback can be compared and no place where comparison operations are performed. Indeed, *nonequilibrium steady states* emerge from the nonlinear interactions among the system's components, but there are no feedback-regulated set points or reference values as in a thermostat. Hence in the present case the questions of who sets the reference value, who programs the computer, who programs the programmer, and so on do not even arise. The child who breaks open a toy to find out how it works would be very disappointed if that toy was self-organized. Self-organizing systems have no deus ex machina, no ghost in the machine ordering the parts.

How Nature Handles Complexity

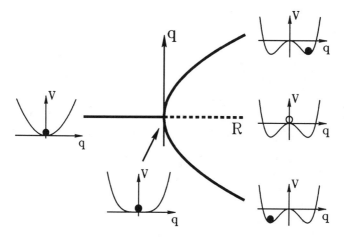

Figure 1.4 The corresponding bifurcation diagram and potential landscape V for the fluid behavior shown in figure 1.1. R is the control parameter and q represents the state of the fluid. Solid (open) balls represent stable (unstable) states of the potential. Similarly, thick lines are stable, dashed line is unstable.

Symmetry Breaking Fluctuations

A few more surprises are in store for us with this simple physical example. If we look at the arrangement of the convection rolls we see that their motion can be either in one direction or the other (see figure 1.1). Once this direction of motion begins in a particular cell, it will not change. But how does the fluid decide which way to go? The answer is Lady Luck herself. The symmetry of left- and right-handed motion is broken by an accidental fluctuation or perturbation. Of course, no agent-like entity inside the system decides which direction the fluid must go. Certainly, a decision appears to be made, but no decision maker tells the fluid what to do.

Bifurcations

Naturally, if one had a perfect system it would be equally probable that the rolls would go in one direction or the other. A good way to visualize this is the *bifurcation diagram* (figure 1.4). Below the critical control parameter value, R, for the onset of coherent motion, the only possible macroscopic state is the state of rest. The little ball representing the state of the fluid, q, is in the minimum of the potential, $V(q)$. At the instability point (shown by the arrow), a bifurcation (or branching) from this state of rest occurs: two rolling motions whose rotation speed is equal but opposite emerge spontaneously. Only one, of course, is realized in an experiment. Graphically, the little ball ends up in one well or the other. Notice the previously stable state (rest) is now unstable, shown by the dashed line in the bifurcation diagram and the open ball at the maximum of the potential.

This is a classic example of a *pitchfork bifurcation* (looks exactly as it sounds) between the state of rest and the state of macroscopic motion in one direction or the other. A stable solution of the dynamics becomes unstable, and a transition to bistability occurs. Bifurcations will be a persistent theme running throughout this book for a number of reasons. Here I'll just mention two. First, bifurcations constitute a mechanism for change and flexibility. Needless to say this isn't the smooth, inexorable gradual change advocated, for example, by Charles Darwin for the process of evolution. In self-organizing systems, a small change, say, in the temperature gradient applied to a fluid can produce a huge collective effect. Second, the bifurcation pictured in figure 1.4 nicely shows how open systems often have several options (here just two) for the same environmental conditions. Chance decides which solution nature adopts. Fluctuations are thus crucial to understanding how patterns are formed. They are always probing the stability of the system, allowing it to discover new and different ways to solve problems. Of course, fluctuations are intrinsic to all natural systems, not just a source of noise to be damped out. In certain cases that I'll describe later, they may actually help *amplify* weak background signals, a phenomenon that physicists and engineers refer to as stochastic resonance (see chapter 7).

Although our bifurcation picture only shows the first convection instability, more and more complex patterns—a whole hierarchy of instabilities—may arise if the temperature gradient is further increased. New patterns are created again and again in ever increasing complexity, until eventually the behavior becomes irregular and even turbulent. The components of the system (the fluid molecules) have almost too many options to adopt, and behavior never settles down. The analogies to society, the economy, and our mental health, although potentially misleading, are almost too powerful to overlook.

Origins of Instabilities: Competition versus Cooperation

The story is not quite over. A little more has to be said about the *origins* of instability. These are, I think, quite universal and will be important when we come to matters of the nervous system and behavior. As I said before, convective motion in the Bénard experiment is an example of a collective effect, which means cooperation among the many parts of the system. Lurking underneath this cooperativity, however, is competition. The competition, in this case, is between how forcefully we drive the system (buoyancy forces in the experiment are directly related to the temperature difference) and how well the system holds itself together (governed by two *dissipative* forces, viscosity and thermal conduction). The ratio of driving forces to dissipative forces is called the Rayleigh number (R in figure 1.4). It is dimensionless, since force quantities occupy the numerator and the denominator of the ratio, and these cancel, leaving just a number.

Subsequent instabilities depend on a second dimensionless number called the Prandtl number, which is the ratio of viscosity (the stickiness of the fluid or the strength of intermolecular attraction) to thermal conductivity. But the message in each case is the same: once one force overcomes the other, instabilities arise, leading to rich and eventually irregular dynamic behaviors. Later on we will see in other kinds of systems, including the brain, that self-organization may arise due to competition not among conventional forces but among different sources of *information*. For now, the word "information" substitutes for the kinds of influences (external and internal) that affect the behavior of living things. An important point that will emerge is that information is *meaningful* and *specific* to the dynamic patterns that biological systems produce.

A further, somewhat related point has to be made before leaving our example and considering what this all means. Rayleigh and Prandtl numbers are dimensionless variables that deal with force fields. They tell us about how instabilities develop, and lead to pattern formation in physical and chemical systems. But it appears that we're left out something quite important. What role, if any, does the environment play? Now this has nothing to do with force in any conventional physical sense, but it does have to do with *geometry*.

In the Bénard example, the geometry takes the form of a third relevant dimensionless quantity called the *aspect ratio*, which relates a typical lateral (horizontal) dimension to the depth of the fluid. But unlike the temperature gradient that expresses the balance of forces and flows in an irreversible process, the aspect ratio is an environmental manipulation that measures the specific influence of lateral boundaries on convective patterns.[13] Oil in a circular dish, for example, may give rise to the beautiful hexagonal honeycomb that is observed when the cell structure of the Bénard instability is viewed from above (plate 1). It's just this that excites my daughter on Saturday mornings when we fry an egg.

Turing Instabilities

Another class of dynamic pattern is analogous in many ways to the Bénard system, but based instead on *reaction-diffusion* processes. Reaction and diffusion turn out to be enormously important in chemical pattern formation, but the reason for discussing them here is that they may also play a role in *morphogenesis*, how biological pattern and form originates. Our friend Turing shows up again, because his pioneering paper "The chemical basis of morphogenesis," provided major insights into the problem. The story is told that when he was twelve years old, Turing entertained himself one summer by extracting iodine from seaweed. Some years later he won the science prize at his school for a mathematical analysis of an iodine clock reaction. Given this apparent preoccupation with iodine, it's fortuitous that Turing patterns were first unambiguously established in an *iodine reaction*, first by Patrick de Kepper's group in Bordeaux, France (1990), and shortly afterward by Harry Swinney's group in Austin, Texas.[14]

In his 1952 paper Turing showed that a spatially homogeneous distribution of chemicals (i.e., no pattern) could give rise by a so-called *Turing* (or diffusion-driven) *instability* to spatial patterns of different kinds.[15] Turing suggested that this reaction-diffusion mechanism might explain such patterns as whorled leaves and the tentacle patterns on Hydra. But by far the most eye-catching application was (and is) the dappling patterns on the coats of animals such as the palomino horse, the giraffe, and the leopard.

How does the Turing instability work? Usually, diffusion is considered a stabilizing process rather than a source of destabilization. However, an analogy along the lines suggested by mathematical biologist James Murray, who has analyzed the Turing mechanism extensively, may help visualize the essential aspects of this novel idea.[16] Consider a forest fire in which a flame front starts to propagate. The fire itself is a source of *activation*; the firefighters are pilots in helicopters spraying fire retardant, a source of *inhibition*. If the firefighters can spray the trees in advance of the flame front, the trees will be prevented from burning. If they cannot, then the fire will spread and the entire area will become charred. The intuitive idea is that the diffusion coefficient of the inhibitor must be much faster than the diffusion coefficient of the activator. That is to say, when the helicopter pilots see the flame front coming, they move ahead of it quickly, thus preventing the fire spreading into the area they've just sprayed. The charred area then is confined to a finite domain whose spatial pattern depends largely on the diffusion rates of activation (the fire) and inhibition (the firefighters). This basic description applies even if the fire begins in random places, rather than in a single location. Indeed, we can imagine the kind of dappling (spatial inhomogeneity) that would occur if the same scene were repeated around each locally distributed fire.

It turns out that one of the main mechanisms for realizing Turing patterns in the laboratory lies in this difference between the effective diffusion coefficients of the reactants. Istvan Lengyel and Irving Epstein showed that the inhibitor substance in the chemical reaction must move quickly relative to the activator.[17] When it does, their model predicts exactly what was observed in experiments: a static pattern of dots containing high concentrations of iodine (plate 2). Almost every animal coat pattern, and probably the spots on tropical fishes, can be generated by this single diffusion-driven process. Instead of chemicals reacting in a gel, biological development is viewed as directed by the concentrations in cells of certain kinds of morphogens (hence morphogenesis). As in our analogy, the basic mechanism is generic: local activation (the fire) plus long-range inhibition (the firefighters).

Even though the mechanisms of reaction-diffusion and convection are different, the basic principles underlying pattern formation are the same. As a parameter governing the reaction is changed, one of the collective modes characterizing the pattern becomes unstable. All the others modes are stable. The principle of the least stable solution, Haken's slaving principle, runs the show again, and the process is entirely self-organized.[18]

Of course, many other factors are involved in biological development that we have not considered here, such as the geometry and scale of the embryonic cells themselves, and the timing of activation and inhibition. Murray provides an in-depth analysis of many of these factors, especially geometry and scale.[19] So how, one might ask, does the gene, the unit of inheritance, fit into this picture? Speaking generally, one idea is that genes don't program biological development as if they contained a set of instructions for building an organism. If we calculated the information necessary to construct an organism, the figure is far larger than could ever be stored in DNA. Rather, it looks more as if the role of genes is to act as *specific* constraints on laws of self-organization. I realize this point is a bit vague at present, but the notion of *information as meaningful and specific to self-organized dynamical processes* will become a major part of the story later on (see especially, chapter 5). For now, I leave the reader with an anecdote attributed to Turing himself. When asked if his theory could explain the stripes on a zebra, he allegedly replied that the stripes were easy but the horse part was tricky.

Some Other Dynamic Patterns

Dynamic patterns, as I've emphasized, are not at all limited to inanimate matter. But for now, only a couple of examples. One that I find a little ironic concerns the molecular biologists' bread and butter, *Escherichia coli*, the common intestinal bacteria. A big splash was made in the *New York Times* science section (Sept. 1992) of the need for new principles to understand the fact that even single *E. coli* cells aggregate into tight balls to protect themselves when they are exposed to antibiotics or noxious stimuli. Lo and behold, they form all kinds of fantastic patterns that can't be predicted from the properties of single cells alone!

Dynamically patterned organization of unicellular organisms has actually been observed since the 1950s.[20] For example, one of the most extensively used models for cell patterning in development is the cellular slime mold *Dictyostelium discoideium*. When you remove the food supply of this little beast, the individual cells aggregate into a multicellular slug that can move about quite independently. When the slug stops moving, a fruiting body forms with densely packed spores. This eventually bursts apart, the amoebae are distributed, and the process begins all over again. Just as the molecules in the fluid interact with each other, so the single amoebae communicate by secreting an attractant whose concentration grows as more and more cells come into contact with it. Wave patterns form in a self-organized fashion entirely analogous to the Bénard instability.

A second example, perhaps closer to home, concerns how the visual cortex forms.[21] This depends mostly on the spontaneous activity of the nervous system even without experience of visual stimuli. The big question is how organized regions in cortex, such as ocular dominance columns, can form based only on spontaneous neural activity. It turns out that an instability mechanism, in the generic sense, also causes formation of ocular dominance patterns. From an initially random state in which all neurons in visual cortex receive roughly equal signals from the left and right eyes, symmetry is broken. Randomness is disrupted because neighboring cells are more likely to fire together. Like amoebae coupled by an attracting medium, when cells fire together, their connectivity is enhanced. Although initially each individual cortical neuron has an equal chance of connecting to the left or right eye, this balanced state is unstable. But all it takes is for one eye to be slightly more successful than the other in activating a single cortical neuron, and before you know it a preference for that eye is set up. Once again, the final dominance pattern is the one that dominates all the others that are mixed together to form the initially random state. The way you predict which ocular dominance column is going to form is to perturb the system a little bit, and observe at the moment of instability which pattern grows the fastest. This is the one that wins the "battle" between the two eyes.

Note again that the cortices of animals and people are not solely static structures, but are sculpted by dynamical processes. Dynamics gives meaning to geometrical forms while also being constrained by them. In short, the creation of structured forms such as ocular dominance columns in the cortex is *activity dependent*. As I said at the beginning of this chapter, the classic dichotomy between structure and function fades, and we begin to sense the intimate relation between them. Ultimately, all we are left with is dynamics, self-sustaining and persisting on several space-time scales, at all levels from the single cell up.

THE MESSAGES OF SELF-ORGANIZED PATTERNS

Nature is not economical of structures—only of principles.
—Abdus Salam

People and brains are not fluids or soap bubbles, as one well-known psychologist insisted after one of my lectures a few years ago. Brains obviously are highly interconnected entities, not continuous media. Life is not a bowl of cherries ... or lime jello, for that matter. So what can we learn about people and brains from the classic example of the Bénard instability? Obviously, the point here is that it's the *organizational concepts* that matter, because they pertain to how complex structures or patterns can emerge and sustain themselves without any detailed instructions whatever. Such behavior is essential to the nature of things, and is not imposed from the outside. The Bénard example nicely illustrates the interplay between chance (fluctuations) and necessity (order) in effecting *selection* in a self-organized fashion. We will have other more pertinent examples later. Indeed, synergetic concepts shed an entirely new light on neo-Darwinism, which tends to ignore principles of self-organization in nonlinear dynamical systems.

We have seen how it is possible for structure and pattern to come "for free": under certain conditions the organization of matter undergoes abrupt change, almost without warning. Later on we will see that identifiable signatures or fingerprints can be detected in *advance* of upcoming change. For now, let's try to extract some of the main conceptual themes of pattern formation thus far, with the aim of using them to motivate our studies of brain and behavior. By summarizing the situation, my aim is to consolidate some of the key ideas that will turn up again and again throughout this book.

Elementary Concepts of and Conditions for Self-Organization

1. Patterns arise spontaneously as the result of large numbers of interacting components. If there aren't enough components or they are prevented from interacting, you won't see patterns emerge or evolve. The nature of the interactions must be nonlinear. This constitutes a major break with Sir Isaac Newton, who said in Definition II of the *Principia*: "The motion of the whole is the sum of the motion of all the parts." For us, the motion of the whole is not only greater than, but *different* than the sum of the motions of the parts, due to nonlinear interactions among the parts or between the parts and the environment.

2. The system must be *dissipative* and far from (thermal) equilibrium. Due to nonlinear interactions in the system, heat or energy doesn't diffuse uniformly but is concentrated into structural flows that transport the heat (dissipate it) more efficiently. As a result of dissipation, many of the system's degrees of freedom are suppressed and only a few contribute to the behavior. Intuitively, dissipation is equivalent to a kind of attraction that may take several forms (see 7 below).

3. *Relevant* degrees of freedom, those characterizing emerging patterns in complex systems, are called collective variables or order parameters in synergetics. An order parameter is created by the coordination between the parts, but in turn influences the behavior of the parts. This is what we mean by *circular causality*, which, incidentally, is not the same as tautology.

4. Order parameters are found near nonequilibrium phase transitions, where loss of stability gives rise to new or different patterns and/or switching between patterns. Order parameters may exist far from transitions as well, but it is difficult to identify them.

5. Fluctuations are continuously probing the system, allowing it to feel its stability and providing an opportunity to discover new patterns. Fluctuations are positive sources of noise, not just something to be rid of.

6. Parameters that lead the system through different patterns, but that (unlike order parameters) are not typically dependent on the patterns themselves, are called control parameters. Such control parameters may be quite *unspecific* in nature; that is, in no sense do they act as a code or a prescription for the emerging patterns. On the other hand, we will see that in biological systems

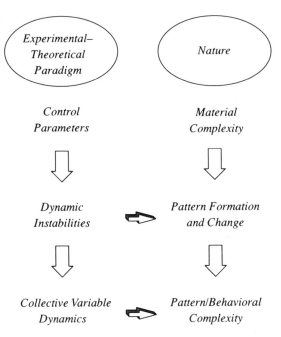

Figure 1.5 Summary of the big picture, from material complexity to behavioral complexity.

boundary conditions may be quite *specific* in an informational sense, especially when a particular pattern is required (cf. chapters 5 and 6).

7. The order parameter dynamics, the equation describing the coordinated motion of the system, may have simple (fixed point, limit cycle) or complicated solutions including deterministic chaos and stochastic (random) aspects thereby giving rise to enormous behavioral complexity.

Thus the main picture so far (summarized in figure 1.5) is that the complexity of matter or substance with all its microscopic constituents, gives rise through the dynamical mechanism of nonequilibrium phase transitions to simpler (lower-dimensional) order parameter dynamics that in turn are capable of generating enormous behavioral complexity. Nobel laureate Murray Gell-Mann is credited with saying, "Surface complexity arises out of deep simplicity." But actually we will see that the opposite is also true, namely, that *surface simplicity emerges out of deep complexity.*[22] That message will ring out loud and clear when we consider in later chapters the experimental evidence in fields dealing, for example, with the coordinated behavior of people and brains.

The Tripartite Scheme

I want to say a little bit more about the generality (some might say universality) of the theoretical framework summarized here, and about the kinds of

How Nature Handles Complexity

rules or laws that constitute the order parameter dynamics. The central idea is that understanding at any level of organization starts with the knowledge of basically three things: the parameters acting on the system (which are sometimes equated with the term *boundary conditions*), the interacting elements themselves (a set of primitives), and the emerging patterns or modes (cooperativities) to which they give rise.

The minimum requirements for characterizing a coherent or cooperative process rest with what I call the *tripartite scheme*. Harking back to my earlier comments on the relative status of macro and micro, we can appreciate that this is not a rigid picture. Especially in biological systems, constraints and borders are constantly being created and dissolved. Cooperativity at one level of organization may act as a parametric boundary condition on a lower level. Conversely, the former may act as an elementary or component process at higher levels of organization. Mind you, no value is meant to be assigned to the words "higher" and "lower" as if one level is better or more fundamental than the other. I simply mean the levels above and below where the cooperativity (pattern, structure) is observed. The present scheme is thus level independent and hierarchical, but not, remembering circular causality, unidirectional.

Simple Laws for Behavioral Complexity

Now a few words about order parameter dynamics and behavioral complexity. And a word of caution as well. In complex systems, as I've emphasized, we have to find the order parameters and control parameters on a chosen level of description. Then we have to identify the order parameter dynamics. There's no free lunch here. It's extremely unlikely that a single master equation set will appear out of the blue. Rather, this book deals with events level by level, looking within and across levels for similarities. The details are always going to differ, but if the conceptual framework sketched here is on the right track, nature will be seen to employ the same *strategy* at all levels. That is, the creation and stability of new patterns and the annihilation of old ones will be governed by the same basic mechanisms and principles.

What will the equations for the order parameter(s) on any given level, look like? For present purposes, it's more important to focus on what they might do. By definition, the order parameter dynamics are of much lower dimension than the system they describe. Just how low depends on the particular system, but theory is well developed for only a small number of relevant degrees of freedom. The approach, as I've said, works best in dissipative systems in which most of the degrees of freedom are heavily damped.

For illustration purposes only, let's take the two-dimensional Hénon map with two control parameters. Maps are an important class of dynamical systems that describe the evolution of a variable x in discrete time. Given the value of x at time zero, the map gives x_1 (x at time 1). Given x_1 the map gives x_2 and so forth. The Hénon map was introduced as a simplified model of a set

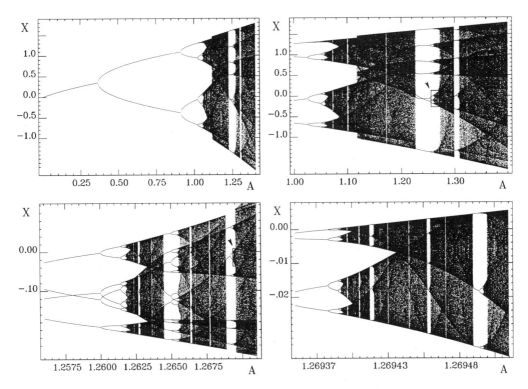

Figure 1.6 Bifurcation diagram of Hénon map $x_{n+1} = A - x_n^2 + By_n$; $y_{n+1} = x_n$. Boxes marked by arrows indicate successive blowups of parameter regions of A, illustrating the self-similar property of the map.

of differential equations formulated by Lorenz to describe fluid flow.[23] Here, we simply want to use it to illustrate some key points about the kinds of behavior that can emerge from very simple, but nonlinear rules.[24]

In figure 1.6 we plot the behavior of just one of the variables, x, of this two-dimensional (x, y) system, as a function of one of the control parameters, A. The other control parameter, B, is fixed at a constant value. As in figure 1.4, this is called a *bifurcation diagram*, but is rather more complicated than the simple pitchfork shown there. Rather, we see that one solution splits into two at a certain parameter value, and splits again and again as the parameter is changed. The *trajectory* or *orbit* of the map (the sequence of values x_0, x_1, x_2, etc.) goes through a cascade of bifurcations leading to deterministic chaos.

This, so-called universal bifurcation sequence has been analyzed extensively (there are actually a number of different routes to chaos), and has been heavily popularized in part because it produces such a compelling and beautiful picture. The goal here, however, is to try to get across what such pictures mean and how they affect the way scientists (especially those interested in brain and behavior) approach problems. For now, our reasoning is mostly by analogy. In the next chapter, analogy will be replaced by a tight theory-experiment relation.

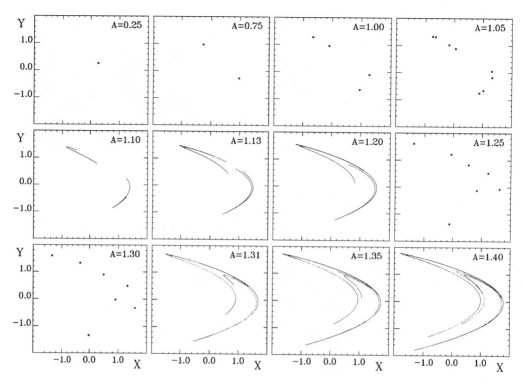

Figure 1.7 Construction of Hénon attractor by successive bifurcations as parameter A is changed (B is fixed at 0.3). See text.

One reason for presenting the bifurcation diagram shown in figure 1.6 is that it is representative of a simple deterministic (but nonlinear) equation that displays both *simple* and *complicated* behaviors depending on parameter values. "Simple" behavior here means a stable fixed point or periodic (limit cycle) orbit that appears for low values of the parameter, A. This means that all initial values of x (initial conditions) converge on an *attractor*. An example of a fixed point attractor is a damped pendulum: regardless of how far you displace its bob it will wind down and eventually come to a halt. An example of a limit cycle attractor is a grandfather clock that oscillates with a certain frequency and amplitude determined only by its parameters, not the initial conditions. If perturbed, trajectories outside the limit cycle spiral inward, while trajectories inside spiral outward toward the limit cycle.

In figure 1.7 one fixed point is stable until, at a certain parameter value ($A = 0.75$), it flips between two points, bouncing back and forth in a period 2 orbit. This remains stable for a while until it flips to a period 4 orbit at a new parameter value ($A = 1.00$) and so forth. Things start to get complicated fast, and eventually (at $A = 1.31$ in figure 1.7) the strange Hénon attractor appears. Note that just as this system is *sensitive to parameters*, a small change producing a qualitatively different behavior, so too it is quite *insensitive* to parameters in other parameter ranges. There, one can observe the *same behav-*

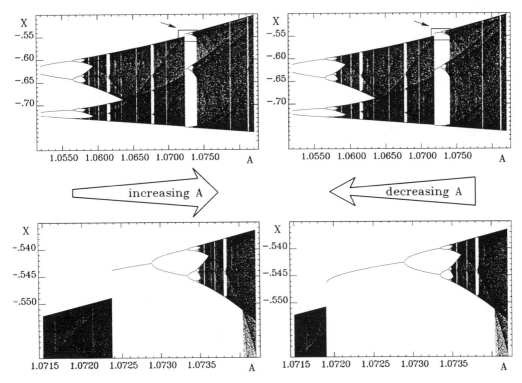

Figure 1.8 Bifurcation diagram of Hénon map illustrates several attractors (multistability) and hysteresis.

ior for different parameter values: nothing much happens no matter how we change the parameter, what scientists call a null effect.

We've already hinted at some of the attributes of "complicated" behavior in the Hénon map. Scientists are still finding new features in this simple map (I'll mention one important new one below), and the list here is not intended to be inclusive. Examples of behavioral complexity are the cascading bifurcations that occur as parameter *A* is varied; multistability, where several solutions coexist for the same parameter value; and bands of chaotic behavior, always interspersed with stable periodic windows.

One of the most famous of these windows is the period 3 orbit. Period 3 refers to an area inside the chaotic region where the behavior of the variable, *x*, bounces among only three values. Boxes in figure 1.8 (top) marked by arrows show the top branch of period 3. Notice that when parameter *A* is *increasing* the onset of the period 3 occurs at a larger value than when *A* is *decreasing*. Such shifts indicate the presence of numerous attractors and an important phenomenon called *hysteresis*. Hysteresis means that when a system parameter changes direction, the behavior may stay where it was, delaying its return to some previous state. This is another way of saying that several behavioral states may actually *coexist* for the same parameter value; which states you see depends on the *direction* of parameter change. Where you come from affects where you stay or where you go, just as history would have it.

How Nature Handles Complexity

At certain well known parameter values ($A = 1.4$, $B = 0.3$) the dynamics of the Hénon map are strange and chaotic, meaning that nearby initial conditions become uncorrelated. Moreover, by zooming in on specific parts of the attractor one can observe its self-similar, fractal structure. In figure 1.6, consecutive pictures are blow-ups of the behavior inside the small highlighted boxes. This self-similarity is a purely geometrical property of the Hénon map itself: the structure of the attractor is repeated successively as one changes the scale of observation. As I've emphasized, geometry and dynamics go together like bread and butter.

Another feature that we would love to have in any biological system is the ability to create and annihilate patterns simultaneously according to task requirements. Many nonlinear dynamical systems embody this kind of property, the Hénon map being a prime example. The reasons why creation-annihilation occurs are quite technical, but have been spelled out by Silvina Dawson and colleagues.[25] Intuitively, they show that you need at least two independent critical points[26] where both contact-making and contact-breaking bifurcations are possible (figure 1.9, top). At the contact-making critical point, creation of a new form occurs. At the contact-breaking critical point, annihilation occurs. A nice way to see this beautiful flexibility property of low-dimensional dynamic systems is to magnify parameter regions of the map where one sees *reversals* of period-doubling cascades (figure 1.9, bottom). It turns out that a surprisingly large range of parameter values exists where such creation-annihilation processes are manifest. One merely has to zoom in to see them.

What insights, if any, have been gained by this discussion? And what are the methodological implications? Obviously, simple and complicated behaviors have been shown to emerge from the same dynamical system. Surface simplicity and surface complexity are both possible outcomes. We don't have to posit a different mechanism for each qualitatively different behavior. Mainstream science tends to make this mistake all the time, and it leads to a huge proliferation of models. I think that this is due at least in part to a one-cause-one-effect mentality and consequent failure to explore the full range of parameter values in a given experimental system. If one only probes a few parameter values and if one sees something different in each case, one is inclined to offer separate explanations. But as we will see again and again in this book, and as illustrated in the example of the Hénon map, how you move through parameter space determines what you see. And some of what you see, of course, is very fancy indeed.

The mechanisms underlying such behavioral complexity are *generic*, but nontrivial. What one always finds at the heart of the evolution of complex behavior are *dynamic instabilities*, bifurcations of different kinds that have to be identified. Complex systems, as we will see, seem to live near these instabilities where they can express the kind of flexibility and adaptability that are fundamental to living things. In principle, at least, we can understand that simple, nonlinear laws are capable of generating enormous diversity of pattern, seemingly for free. The blood, sweat, and tears enter, however (or the

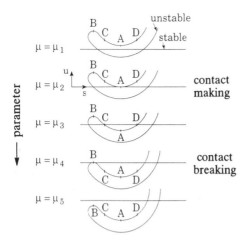

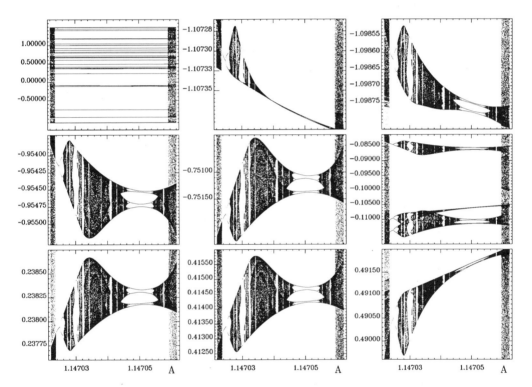

Figure 1.9 (*Top*) Contact making and contact breaking. Pieces of the stable (s) and unstable (u) manifolds of a saddle point are shown for five values of the control parameter, μ. Contact making occurs at $\mu = \mu_2$ at A; contact breaking occurs at $\mu = \mu_4$ at B. (Adapted from reference 25.) (*Bottom*) Reversals of period doubling cascades in the bifurcation diagram of the Hénon map for a fixed range of parameter A ($B = 0.3$). Notice at a gross scale (top left) the y-axis variable x shows no change. However, at finer scales, orbits are created and destroyed, the inevitable consequence of cascade reversals.

How Nature Handles Complexity

fun starts), when we try to discover what these laws are. Keeping Shakespeare in mind:

There are more things in heaven and earth, Horatio, than are dreamt of in your philosophy.
Hamlet, I, v

NEW LAWS TO BE EXPECTED IN THE ORGANISM

When Schrödinger speculated, in his lectures to the Dublin intelligentsia in the early 1940s, that new laws are to be expected in the organism, the world was at war, and concepts of self-organization in nonlinear dynamic systems far from equilibrium were unheard of.[27] The very idea that living systems might involve other laws of physics, hitherto unknown, was quite radical. To Schrödinger's credit, he anticipated that these other laws would require an extension of the laws of physics as established to that date. The modern theory of pattern formation and self-organization in nonequilibrium systems, as exemplified in Haken's synergetics, embraces Schrödinger's order based on order proposal for the organization of living matter. In fact, these new concepts apply over a wide range of complex, dynamic behaviors from disorder to order all the way through to deterministic chaos. They are, so to speak, the foundation on which to build and generalize our understanding of some of the more exotic phenomena that I treat in this book.

Thus, my answer to the question, is life based on the laws of physics? is yes, with the proviso that we accept that the laws of physics are not fixed in stone, but are open to elaboration. It makes no sense to talk about *the* laws of physics as if the workings of our minds and bodies are controlled by well known fundamental laws. As I stressed earlier, it will be just as fundamental to discover the new laws and principles that govern the complex behavior of living things at the many levels they can be observed. We should not expect to construct complexity from simplicity (e.g., by some extrapolation of the properties of elementary particles). At each level of complexity, entirely new properties appear, the understanding of which will require new concepts and methods.[28] Of course, this is not to deny that ordinary matter obeys quantum mechanics, but, again, it is the principles of (self)-*organized* matter at the scale of living things that we are after here. If this sounds like beating a dead horse, I apologize. But as we will see, not everyone (including some of the high priests of physics) agrees.

MATTERS OF MIND AND MATTER

Mind must be a sort of dynamical pattern, not so much founded in a neurological substrate as floating above it, independent of it.
—R. P. Feynman

What has been suggested thus far is that over and over again nature uses the same principles of self-organization to produce dynamic patterns on all scales.

The precise patterns that form may differ from one scale of observation to another, but the basic pattern forming principles are the same. What, then, of the physical basis of mind? Might it too be elucidated by these new concepts of pattern formation? Here again, when eminent physicists and biologists come to consider novel (nay, exotic) properties of living things such as consciousness and creativity, they completely ignore (or are ignorant of) theories of cooperative phenomena far from equilibrium.

In *The Emperor's New Mind*, for example, Roger Penrose looks to ties between the known laws of quantum mechanics and special relativity for clues to human consciousness.[29] According to Penrose, it will be possible to elucidate consciousness once this linkage has been established and formulated into a new theory called correct quantum gravity (CQG). For Penrose, consciousness, together with other attributes such as inspiration, insight, and originality, is nonalgorithmic and hence noncomputable. Fair enough. The human brain is thought to exploit quantum parallelism, that simultaneously available different alternatives coexist at a quantum level. When Penrose or any other mathematical physicist experiences an "aha" insight, this cannot, he thinks, be due to some complicated computation, but rather to direct contact with Plato's world of mathematical concepts.

Underlying the unity of consciousness (and flashes of genius), it seems, is the unity of Platonic and physical worlds. The precision of quantum mechanics and relativity (admittedly SUPERB theories in Penrose's capital classification scheme) provides "an almost abstract mathematical existence for actual physical reality." I resonate to the theme that concrete reality may, on the one hand, be represented in abstract and mathematical ways, and on the other, that mathematical concepts may achieve a kind of concrete reality. This is what I mean when I propose that the linkage across levels of description in complex systems is by virtue of *shared dynamical principles*. But in his parlance, I would place Penrose's physics of mind in the TENTATIVE or, to be mischievous, MISGUIDED category of theories.

In his fourfold classification of physical theories as SUPERB, USEFUL, TENTATIVE, and MISGUIDED, Penrose doesn't even mention theoretical developments that have occurred in the field of nonequilibrium phase transitions that are at the core of pattern formation in open, physical, chemical, and biochemical systems. The brain is a complicated object, yet few, he thinks, would claim that the *physical* principles underlying its behavior are unknown. Penrose begs to differ and opts for quantum gravity. To the contrary, I believe that understanding mental *processes* requires new concepts, arising at scales far removed from the quantum level. Psychology, I am saying, is not just applied physics, or, for that matter applied biology.

The sought-after oneness or globality of thought emerges, in my view, as a collective, self-organized property of the nervous system coupled, as it is, to the environment. Unlike Penrose, I do not think "where is the seat of consciousness?" to be a particularly useful scientific question. Consciousness, in my opinion, is unlikely to be seated in any single place in the nervous system considered by Penrose, such as the upper brain stem or the cerebral cortex. Following concepts of self-organization, global features of brain function are far more likely to be bound up in the coordination or cooperativity *between* places. In fact, exciting evidence already exists that such coordination occurs in the brain at a variety of different levels (see chapters 7, 8, and 9). Whether or not coordination is related to *consciousness*, however, is an entirely different matter.

The theory of self-organized pattern formation in nonequilibrium systems also puts creativity and "aha" experiences in a new light. The breakup of old ideas and the sudden creation of something new is hypothesized to take the form of a phase transition. Like Saul on the road to Damascus, a fluctuation drives the system into a new state that may be so globally stable it can last forever. An exciting possibility that we will consider is that there may be ways to see these transitions in the human brain.

What kind of a thing, then, *is* the brain? And what is the connection between brain and mind? What's the thread holding the fabric together? Let's put our cards on the table. You will not read here that the brain is a computer that manipulates symbols. Neither the brain nor the neurons that compose it *compute*, although they may combine signals in ways that seem analogous to addition, multiplication, division, and subtraction. The nervous system may act *as if* it were performing Boolean functions. But computation is just a metaphor for how the brain works. Thus, the question of how neurons compute and the principles of neural computation might be important for engineers, but they are irrelevant here.

People can be calculating, but the brain does not calculate. *People* can program computers, but the brain doesn't program anything, not even the movement of my little finger. A computer is a complex structure that exhibits complicated behavior. The brain too, is a complex structure that, at least transiently and in particular contexts, lives on a low-dimensional parameter space where enormously complex behavior is possible. The brain of the neural network field, the brain of artificial intelligence research and its cognates, needs no new physics. The brain inside your head does.

The thesis here is that the human brain is *fundamentally* a pattern-forming, self-organized system governed by nonlinear dynamical laws. Rather than compute, our brain "dwells" (at least for short times) in metastable states: it is poised on the brink of instability where it can switch flexibly and quickly. By living near criticality, the brain is able to anticipate the future, not simply react to the present. All this involves the new physics of self-organization in which, incidentally, no single level is any more or less fundamental than any other.

THE MIND REVEALED? OR, WHAT THIS BOOK'S ABOUT

A short time ago the headline "The mind revealed?" appeared in *Science*, the principal publication of the American Association for the Advancement of Science.[30] The main facts behind the hype were remarkably simple (in a sense, probably far too simple). When you move a bar across the eye of a cat, neurons in primary visual cortex, often located quite far apart, display frequency locking and phase locking, just as if the neurons were soldiers marching perfectly in step to a pipe band. Neuroscientist Charles Gray and colleagues also found that the correlation between neuronal activities was much greater to a *single* moving bar than it was if the same two neurons were stimulated by two shorter, separate moving bars.[31] The interpretation of these facts is what produced the headline. Sir Francis Crick, the Nobel laureate, and others[32] interpreted these phase-locked oscillations as the way the brain links features of an object into a single coherent percept. Since an object seen is often heard, touched, and smelled, such a binding mechanism seems necessary to provide perceptual unity. At any time, the relevant neurons for coding features such as color, motion, and orientation must cooperate to form some sort of global activity. It is this global activity that, in Crick's opinion, is the neural correlate of consciousness. The notion is intuitively appealing in light of our physical examples of self-organization, in which global ordering can emerge from local interactions. "Local" in the nervous system, however, refers to *connectivity* among cells that may themselves be widely separated in space.

Nothwithstanding that Gray et al.'s initial observations were on anesthetized cats (they have since been reproduced in awake animals), at present there seems to be no special or compelling reason why consciousness (as opposed to some other poorly defined property of living things) should require phase and frequency synchronization. Rather, what I find fascinating about these results is that they might reflect a *ubiquitous tendency* of nature to coordinate things.

A central idea that we will explore is that nature doesn't care what these things are. What matters is the self-organization, the way things cooperate to form coherent patterns. Self-organization provides a paradigm for behavior and cognition, as well as the structure and function of the nervous system. In contrast to a computer, which requires particular programs to produce particular results, the tendency for self-organization is intrinsic to natural systems under certain conditions. Of course, defining the necessary and sufficient conditions is one of the most important challenges we face.

Many of the ideas and results discussed in this chapter have not for the most part made contact with mainstream cognitive science, neuroscience, and its progeny, cognitive neuroscience (and vice versa, one might add). Science is certainly a highly fractionated enterprise these days, so perhaps it's not so realistic to expect connections to be made among quite specialized fields. But there's more to it than that, I think, and it's exemplified in the following story in which a well-known science writer, Roger Lewin, confronted well-known philosopher of mind-cum-neurobiologist Patricia Churchland:

"Is it reasonable to think of the human brain as a complex dynamical system?" I asked. "It's obviously true," she replied quickly. "But so what? Then what is your research program?... What research do you do?"[33]

The very idea that the brain follows principles of self-organization in which global states might be characterized in terms of collective variables that follow a dynamic capable of generating enormous human behavioral complexity seems anathema to the materialistic reductionist.

In this book I will report some results of a research program that I and others have conducted over the last twenty years that seeks to elucidate principles of self-organization for biological systems. Behavior will be the starting point of our analysis. Is behavior itself self-organized? What are the collective variables or order parameters, control parameters, bifurcations, and so on? How do we understand the influence of the environment, intentions, and learning in terms of self-organized dynamical systems? How can information be conceived as meaningful and specific to dynamical processes? How can dynamics be formulated so that it is commensurate with information?

Behavior, in particular, the coordinated actions of animals and people, might seem a strange place to begin our understanding of the brain or mind. But in actual fact, behavior supplies the main route (more a guiding light) into our analysis of mental and brain events. How can one understand the function of the brain without knowing what it's for? How can one make meaningful correspondences between events in different domains without a sophisticated analysis of each?

The technological wizardry and fact finding of modern neuroscience has led to a situation in which behavior is largely ignored.[34] On the other hand, the main thrust of many recent treatments of complexity and self-organization is with mathematical and computer models. Little contact with experiment is made, and interplay between theory and experiment, so crucial to the development of science, is lacking. Behavior as a source of insight into principles of self-organization and, importantly, their necessary extension to accommodate the special properties of living things that I shall describe, is virtually ignored. Yet, to my mind, it is just as important to establish that functional behaviors are self-organized as it is to establish that principles of self-organization also govern the operation of the human brain. In fact, the claim on the floor is that both overt behavior and brain behavior, properly construed, obey the same principles.

2   Self-Organization of Behavior: The Basic Picture

Im Anfang war die Tat! *("In the beginning was the deed.")*
—J. W. von Goethe

SOME HISTORICAL REMARKS ABOUT THE SCIENCE OF PSYCHOLOGY

Science always requires a language, and the science of psychology is no exception. But psychology is in a tricky position, scientifically speaking. The reason is that, according to Webster's dictionary, it must do double duty as the science of mind and of behavior. Even if we leave the brain out of psychology, which seems a bit counterproductive, the language of mind (percepts, images, thoughts, feelings, etc.) is very different from that of behavior.

It's a strange definition of psychology that contains the fundamental problem of psychology: what is the relation between mind and behavior? Even when we bring the brain back in, the best we seem to be able to do is *correlate* its physicochemical and physiological activities with different aspects of experience. Strangely enough, in the contemporary cognitive and brain sciences, any empirically adequate language of description for the two different domains of brain and cognition seems to suffice. Perhaps this is why correlations between the two are often less than compelling: after all, *something* must be going on in the nervous system when we perceive and act, think, learn, and make choices. That we should find some kind of correlation is hardly surprising.

From my point of view, not only does the presence of correlations fall far short of explanation, but also we seem to be correlating apples and oranges. Thus, with few exceptions, the science of psychology, broadly conceived here to include the cognitive and brain sciences, tacitly assumes that the physical and the mental are independent, irreconcilable categories. To abandon such an assumption must surely seem reckless. For me, however, the greatest drawback to understanding the mind-body problem is the very absence of a common vocabulary and theoretical framework within which to couch mental, brain, and behavioral events. Without commensurate description, how is it possible to see the interconnections? And, without a common conceptual language to reconcile the mental and the physical, how can psychology be called a science?

In this chapter I will show, through the use of a specific example and its detailed analysis, that the concepts and language introduced in chapter 1 are precisely what psychology needs. But first, as a context within which to embed these new ideas and results, let's briefly consider some of the shortcomings of other approaches.

Behaviorism

When we look at the history of psychology it's easy enough to speculate why the dictionary includes both mind and behavior in the definition: scientific psychology is a litany attesting to the continual tension between the two. B. F. Skinner, for example, appropriated the term *behaviorism* for a science of behavior, yet limited his analysis of behavior to the consequences produced on the environment.[1] An astonishing fact about behaviorism was that it did not actually deal with behavior or action, but only the results or *outcomes* of individual acts. The central concept of Skinner's behaviorism, the *operant*, captured nothing about how behavioral actions were organized spatially and temporally. Put another way, behaviorism acknowledged that pigeons can press a lever and rats can run a maze, but it didn't care a hoot about how the lever was pressed or the maze run. It treated the organism as a dimensionless point and ignored the form of behavior produced. Of course, a person may select a particular action based on the consequences of such action, but this tells us nothing about the coordination of action per se. Yet we know that it is this coordination that breaks down in various brain disorders such as Parkinson's disease and Huntington's chorea. And we know that it is important when people speak or walk or play the piano. One of the greatest drawbacks of modern robotic devices is that they lack this flexible coordinative ability.

Perhaps it is because all of us, from early childhood, are so used to coordinating our bodies that the science of mind and behavior (and certainly behaviorism) virtually ignored the problem of coordinated action. By analogy, all of us are familiar with falling, but it took us thousands of years to come up with the notion of *gravitation*.[2] People, as the Gestalt psychologist Wolfgang Köhler noted years ago, tend not to ask questions about phenomena with which they are thoroughly familiar. Such, it seems, has been the lot of coordination as far as most of psychology is concerned. All of us know, in a way, *what* coordination is, but little is known about *how* or *why* it is the way it is. One is reminded of the story about a tourist from the dogstar Sirius who described the most miraculous machine he had ever seen:

A remarkable machine unlike any other I had seen before was rushing toward me.... It apparently did not have any wheels but nevertheless moved forward with an amazing speed. As I was able to see, its most important part was a pair of powerful elastic rods each one consisting of several segments.... Each rod moved along a complex curved arch and suddenly made a soft contact with the ground. Then it looked as if lightning ran along the rod from the top to the bottom, the rod straightened and lifted off the ground with a powerful, resilient push and rushed upwards again.... As I was told, the machine consisted of more than two hundred engines of different size and power, each one playing its own particular role. The controlling center is on the top of the machine, where electrical devices are located that automatically adjust and harmonize the work of the hundreds of motors.[3]

Maybe we should be more like this tourist to see coordinated action as the miracle it is.

Ethology

The ultimate aim of the field of *ethology*, the naturalistic study of behavior, was to describe actions in terms of patterns of muscle activity.[4] Yet the overwhelming amount of detail combined with difficulties in recognizing and classifying relevant chunks of behavior has proved to be an enormous barrier to understanding. This is an oversimplification, of course, and there are notable exceptions, such as the seminal work of Erich von Holst in the 1930s that we will discuss later. Recently, ethologists such as Ilan Golani[5] advocated using a movement notation scheme devised by Eshkol and Wachman (E-W) to choreograph dance sequences.

Like a musical script, the E-W system provides a permanent record that allows for the reconstruction of behavioral actions. Without going into all the details, it treats the body as a set of limb segments, the movements of which are described relative to an imaginary sphere centered, say, at the carrying joint. This sphere, like your friendly globe of the world, is marked by coordinates analogous to lines of longitude and latitude. Thus the values of the two coordinates can be used to specify the position of the limb. A nice feature of the idea is that the coordinates can be defined with reference to a sphere centered on the joint of a particular limb, a partner involved in the movement or some fixed reference point in the environment, thus affording a description of movement in terms of an individual actor, one actor relative to another, or, indeed, one actors' body relative to some outside event, such as orchestral music (figure 2.1).

The E-W scheme has been used to describe everything from the social behavior of jackals and Tasmanian devils (small, ferocious, carnivorous marsupials that inhabit Van Diemen's land), to the ritualized behavior of geese and aggressive interactions between Australian magpies. One of the most impressive applications is by Golani and John Fentress,[6] who examined the ontogeny of facial grooming in mice. They were able to show how grooming develops from a small set of simple, stereotyped movements into the rich and precise repertoire of adult mice.

More recently, Golani in collaboration with David Wolgin and Philip Teitelbaum[7] used the scheme to analyze recovery of function after lesions to the rat's brain, as well as the behavioral effect of drugs. Their results suggested

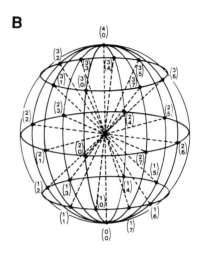

Figure 2.1 (A) The Eshkol-Wachman (E-W) scheme showing a sphere centered at the shoulder joint. The path of the elbow is traced on the surface of the sphere. (B) The sphere, like a geographer's globe, is marked by coordinate lines analogous to latitude and longitude. Positions on the surface of the sphere are specified by two coordinates. The E-W notation allows movement of a limb to be described by its initial position, its final position, and the trajectory from one to the other.

similarities between the way an adult animal *recovers* from brain damage and the way a young intact animal *develops* certain exploratory behaviors, a version of the ontogeny recapitulates phylogeny theme.

From misguided or nonexistent descriptions of behavior in terms of outcomes or results of action, the E-W system substitutes a formal movement notation scheme for the description of behavior. Although it allows for objective and accurate description, it is not at all motivated by theoretical considerations or even the context within which action occurs. Any other ingenious measurement system could do the job just as well, if not better. Nevertheless, the E-W system seems the best tool that ethologists have at the moment to describe naturally occurring behavior.

My view is that accurate description is not enough for a science of behavior, whether of brains or people. Necessary perhaps, but not sufficient. I doubt very much that naturally occurring behaviors are the place to find laws of behavioral and neurological organization. Rather, most naturalistic behavior is simply too complicated to yield fundamental principles. The latter, after all, are hidden from us and it takes, I believe, either special strategies or pure serendipity (of the Archimedes in the bathtub kind) to reveal them. Relatedly, description and explanation are obviously not the same. Explanation demands theory and a coupling of theory to experiment. No matter how refined a formal description of behavior is (e.g., dance notation), there is no guarantee (indeed it seems highly unlikely) that a purely formal approach will provide any deep insights into the organization of behavior. In fairness, Skinner had a theory of behavior, but by ignoring behavior, he threw the baby out with the bathwater. The ethologists refined the measurement of behavior, but the returns, in terms of theoretical insights or understanding, have been modest to say the least.

Cognitive Psychology

So much for psychology as the science of *behavior*. What about psychology as the science of mind? Noam Chomsky, the MIT linguist and activist, is known to have disliked the term "behavioral science." For him it suggested a far from subtle shift of emphasis toward behavioral evidence itself and away from the abstract mental structures that such evidence might illuminate. Chomsky's concern for human language as a subject of study in its own right, his concept of linguistic competence, an internalized system of rules that determines both the phonetic structure of a sentence and its semantic content, bolstered the emergence of modern cognitive science.[8] Chomsky's abstraction away from conditions of language use ("performance") to the study of formal rules of structures, and his generally mentalistic, antibehavioristic stance, were aimed at shifting the science of psychology away from behavior and back to mind.

Of course, these days the characterization of mental life is dictated by a machine metaphor: the brain is viewed by many as a sophisticated computer whose software is the mind. Laymen and scientists alike are prone to describing almost any activity as involving "information processing." There are detractors, however. The philosopher John Searle recently argued that the brain, as an organ, does not process information by some imaginary computational rule-following any more than the gut does![9] Certainly one can *model* some of the functions of the brain on a computer as we do, say, with the weather, but that should not make us believe that the brain, any more than the weather, is a computer. Yet many, in my view, take the machine metaphor far too literally.

In one of my main fields of research, the control and coordination of movement, the computer metaphor has predominated for years.[10] It's easy to see why. Actions must be precisely ordered spatially and temporally. Order, it seems obvious, must be *imposed* somehow on the motor elements.[11] But how? The machine perspective says that order originates from a central program that elicits instructions to select the correct muscles and contract and relax them at the right time. "Just so," as Rudyard Kipling might say. Another artifact familiar to proponents of the machine perspective is the servomechanism. This is good when you want to regulate some property (e.g., limb position, temperature) using feedback. A template or reference level compares the feedback it receives with its own value, and based on this comparison, emits orders to an output device to eliminate any error.

Programs, reference levels, and set points feature heavily in explanations of intelligent behavior, but where do they come from? How, for example, does a given reference signal attain its constancy? If a reference signal at one level is simply the output of another servomechanism at a higher level, this leads to what philosophers term an "infinite regress," or, as Daniel Dennett would say, "a loan on intelligence" that somehow has to be repaid.[12] Any time we posit an entity such as a reference level or a program and endow it with content, we mortgage scientific understanding. The loan can be repaid only when these "phantom users" are vanquished. From the present point of view, it is best not to use artifactual constructs at all. Computers and servomechanisms are not natural systems but artifacts whose characteristics are not especially relevant to understanding living things. Supplanting artifactual machine views of mind and action with the language of dynamical systems and the concepts of self-organization may be easier said than done, but that is the journey we embark on here.

In short, Chomsky and others before and after him tore apart content and process. Chomsky's nearly entire emphasis on the competence part of his performance-competence distinction of linguistic behavior is now pursued to the extreme by program theorists who see the brain as the programmer and the body as a mere slave. The thesis here, however, is that psychology might be better off if it tried to explain the richness of behavior of living things in terms of self-organization, which does not require science to take out a loan on intelligence.

In self-organizing systems, contents and representations emerge from the systemic tendency of open, nonequilibrium systems to form patterns. As we noted in chapter 1, and as will become more and more apparent as we proceed, a lot of action—quite fancy, complicated behavior—can emerge from some relatively primitive arrangements given the presence of nonlinearities. That is, intelligent behavior may arise without intelligent agents—a priori programs and reference levels—that act intelligently.

We will need neither the formal measurement schemes of ethology nor the formal machine vocabulary of cognitive science. Instead, we will emphasize the necessary and sufficient conditions for the emergence of dynamic patterns in a complex system, like an animal with a nervous system immersed in a contextually rich environment.

Gestalt Psychology

Before leaving psychology (which we never really do, since this book is largely about a science of psychology), I should mention two approaches to which I am far more sympathetic than those discussed thus far. I will say more about them later when we consider perceiving as a self-organized process, but mention them here for the sake of closure, even though they are only loosely related to the central topic of this chapter. Both views, nevertheless, are intimately related to the idea of self-organization, but in ways that in my view are quite complementary. Both are antagonistic to the machine stance.

I refer first to the Gestalt theorist Wolfgang Köhler,[13] who viewed psychological processes as the dynamic outcome of external constraints provided by environmental stimulation and internal constraints of brain structure and function. No programmable machine metaphor for Köhler. Instead, macroscopically organized brain states were deemed the relevant stuff of mental life. The latter cannot be observed at the micro level of individual neurons, nor can they be derived by exclusive scrutiny of the microscopic elements. According to Gestalt theory, only at the molar level of description will correspondences be found between mental life and brain states.

Gestalt psychologists of the 1930s and 1940s insisted on the primacy of the language of physics, albeit extended appropriately to include organizational and dynamic aspects of mind. The perceptual process, for example, had to be understood as a result of autonomous creation of order in the perceptual system itself. Köhler's field-theoretical model of perception viewed the brain not as a complex network of many different interacting neurons working together, but as a homogeneous conductor akin to a container full of water. This view was, I'm afraid, hopelessly wrong. What was *not* wrong in my opinion was Köhler's emphasis on order formation, his adherence to the methodology of natural science, and his insistence that physical or physiological explanation be paired with the reality of phenomenal experience.

Scholars such as William Epstein and Gary Hatfield in the US, and Michael Stadler and Peter Kruse in Germany recently reappraised the Gestalt program.[14] Epstein and Hatfield quite correctly, I think, note that neither the technological nor the conceptual tools available to Köhler and his school were up to the task they set themselves. Brain imaging techniques were nonexistent, and the physics of open, nonequilibrium systems had not yet appeared on stage. The German scientists, although perhaps not entirely unbiased (forgiveably so), argued that Gestalt theory anticipated some of the concepts of complex, nonlinear systems presented in chapter 1. Fairness dictates, however, that we recognize that the latter were in no way inspired by Gestalt theory. Nevertheless, I am quite sure that the founders of Gestalt theory would be positively disposed to efforts to establish that brain and overt behavior follow natural laws of self-organization.

Ecological Psychology

Of course, it's not only the nervous system of people and animals that is potentially subject to laws of self-organization. Consider, as the perceptual psychologist James Gibson did, how we drive an automobile.[15] For Gibson

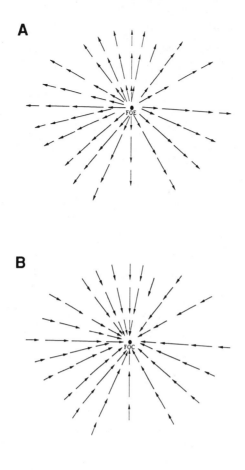

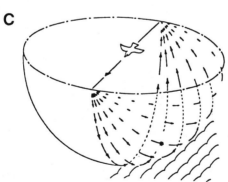

Figure 2.2 (A) Optic flow relative to the focus of expansion (FOE). (B) Optic flow relative to the focus of contraction (FOC). (C) Optic flow for a bird flying in a straight line. (From reference 15.)

and his followers, including Michael Turvey, Peter Kugler, and Robert Shaw, an essential construct is the *optical flowfield* that specifies properties about the car's motion in relation to the environment. This flow, like a fluid, spreads out as an object is approaching us or we are moving toward a surface (figure 2.2).

From the rate of divergence of optical flow it is possible to detect a simple parameter, tau (τ), that specifies time to contact.[16] How we slow down or speed up the car is determined by how we move in relation to the optic flowfield. Moving forward on a straight line produces radial expansion of the flowfield, moving backward radial contraction. Ask yourself how gannets (the large seabirds that plunge dive in such places as the Firth of Forth in bonny Scotland) or your regular housefly "know" when to close their wings or extend their legs as they approach a surface. In all these and many other cases such as long jumping, catching a fly ball, running over rough terrain, and driving a car, timing is controlled in a direct fashion by using τ.

Of course, this is a much longer and more detailed story than I want to pursue here (see chapter 7). Gibson's essential point is that information must be meaningful and specific to the control and coordination requirements of action. Rather than grounding perceptual theory on brain states or as computational rules that generate three-dimensional forms from two-dimensional images on the retina, the Gibsonian program asks how structured energy distributions are lawfully related to the environments and actions of animals. Note that the flowfield and the information it contains are *independent* of the particular visual system that occupies the moving point of information. That's why the gannet, the fly, Carl Lewis (the Olympic long jumper), and Juan Fangio (for many years the world's top race car driver) all use the same macroscopic optical quantity to guide their action.

Perhaps then, this teasing quote from Gibson's 1979 book (published after his death), *The Ecological Approach to Visual Perception*, is not entirely out of place, setting the stage for what is to come:

The rules that govern behavior are not like laws enforced by an authority or decisions made by a commander; behavior is regular without being regulated. The question is how this can be. (p. 225)

What? No deus ex machina? No skeleton in the elevator? No élan vital? No entelechy? How can this be?

ARE ACTIONS SELF-ORGANIZED? IF SO, HOW?

Here's the basic two-pronged problem. The human body is a complex system in at least two senses. On the one hand, it contains roughly 10^2 joints, 10^3 muscles, 10^3 cell types, and 10^{14} neurons and neuronal connections. As Otto Rössler once said, finding a low dimension within the dynamics of such a high-dimensional system is almost a miracle.[17] On the other hand, the human body is multifunctional and behaviorally complex. When I speak and chew, for example, I use the same set of anatomical components, albeit in different ways, to accomplish two different functions. Sometimes, against the wishes of my sainted mother, I do both at the same time. Next time you watch a film, observe the rapidly flowing and shifting scene of sound and motion. Where does one event begin and another end? Where are the boundaries separating the flow of events? When the voice lowers, when the eyes look askance, when the face flushes, when the head turns aside, when the hands fidget, what does all this have to do with what is being said?[18] Referring back to the first chapter, the joint challenges of *compositional complexity* and *pattern complexity* seem to confront us again with a vengeance.

The great Russian physiologist Nikolai Bernstein (1896–1966) proposed an early solution to these problems.[19] He lived in the Soviet Union during the time that Pavlov's views were considered the only ideologically correct explanation of higher brain functions. But he was dead against the notion that the function of the brain could be understood in terms of combinations of conditioned reflexes. One of his chief insights was to define the problem of coordinated action as a problem of mastering the many redundant degrees of freedom in a movement; that is, of reducing the number of independent variables to be controlled. For Bernstein, the large number of *potential* degrees of freedom *precluded* the possibility that each is controlled individually at every point in time. How, he asked himself, does coordination arise in a system with so many degrees of freedom? How do we take a multivariable system and control it with just one or a few parameters?

The Synergy Concept

The resolution to this problem offered by the Bernstein school contained two related parts. The first was to propose that the individual variables are organized into larger groupings called linkages or synergies. During a movement, the internal degrees of freedom are not controlled directly but are constrained to relate among themselves in a relatively fixed and autonomous fashion. Imagine driving a car or a truck that had a separate steering mechanism for each wheel instead of a single steering mechanism for all the wheels. Tough, to say the least! Joining the components into a collective unit, however, allows the collective to be controlled as if it had fewer degrees of freedom than make up its parts, thus greatly simplifying control. Peter Greene, a computer scientist-mathematician who did much to promote the work of the Bernstein school in the US in the early 1970s, likened this idea to an army general saying, "Take hill eight," with the many subordinate layers of the military (subsystems) carrying out the executive command.[20]

For the Russian scientists, later led by the eminent mathematician Israel Gelfand, synergies constituted a dictionary of movements in which the efforts of the muscles were the letters of the language, and synergies combined these letters into words, the number of which was much less than the number of combinations of letters. For Gelfand and colleagues, the language of synergies was not just the external language of movements but also the internal language of the nervous system during the control of movement.[21]

The second, absolutely crucial aspect of the synergy concept is that it was hypothesized to be *function* or *task specific*. The notion of synergy is actually an old one, but earlier ideas associated synergies with *reflexes*. The reflex, in fact, was considered the basic building block of behavior. As one of its greatest advocates, Charles Sherrington, said in the early 1900s "simple reflexes are ever combined into greater unitary harmonies, actions which in their sequence one upon another constitute in their continuity what might be termed the 'behavior' of the individual as a whole."[22] Shades of Isaac Newton. For Bernstein, the reflex didn't contribute to the solution to the coordination problem. Instead, it was part of the problem. The reflex was just another piece that had to be somehow glued together with other pieces; fancy words like "great unitary harmonies" weren't much of a glue.

Testing the Synergy Hypothesis

Bernstein's hypothesis was not about hard-wired anatomical units; rather, synergies were proposed to be functional units, flexibly and temporarily assembled in a task-specific fashion. How might this hypothesis be tested? A stringent test would be to perturb the synergy by challenging only one of its members. If the organization is really a synergy, then all the other functionally related members should readjust immediately and spontaneously to preserve the functional goal.

A good deal of research has gone into identifying these functional synergies in such tasks as maintaining upright posture, walking, and grasping an object.[23] The experimental examples I like best involve coordinated movements of both arms and the production of speech. I like them because, in the first case, they are so simple you can do them yourself sitting in an armchair. In the case of speech, they are technically very difficult to do but are well worth the effort. Both experiments helped put the synergy hypothesis on solid ground.

Pick two targets, one for each hand, that are a different distance away. The task is to reach for them when given a "go" signal. It is well known that the movement time for a single limb depends on the distance the limb has to move and the precision requirements of the target. What happens when the two hands must move very different distances to targets whose precision requirements also differ? This question came up in a graduate seminar I taught in 1977. I well remember one of my students, Dan Southard, using two pieces of curtain material as targets and showing that the limbs reach both targets practically simultaneously. In other words, the brain coordinates both limbs as a *single* functional unit. This is revealed only when you do high-speed film analysis of the movements,[24] which in those days was extremely time consuming. I use the word "functional" because obviously the nervous system does not *have* to control both limbs as a single unit. Only under the task requirement to do two things at once does it create a functional synergy out of its myriad participating elements.

Speech, of course, is a complex system par excellence. The production of a single syllable requires the interaction among a large number of neuromuscular elements spatially distributed at respiratory, laryngeal, and oral levels, all of which operate on very different time scales. We breathe in and out roughly once every 4 seconds, the larynx vibrates at a fundamental frequency of about 100 times a second, and the fastest we can move our tongues voluntarily is about 10 repetitions a second. Yet somehow despite (or maybe because of) these complications the sound emerges as a distinctive and well-formed pattern. For a baby to say "ba" requires the precise coordination of approximately thirty-six muscles. The brain must have some way to compress all this information into something relevant.

Betty Tuller, Carol Fowler, Eric Bateson, and I considered speech to be a prime candidate for testing the synergy hypothesis.[25] But how to test it? First we had to construct a device to perturb an important speech articulator. We chose the jaw, in part because it moves up and down (and sometimes sideways) for all kinds of sounds, and earlier work by John Folkins and James Abbs showed that it was possible to perturb the jaw and obtain interesting results.[26] The very fact of "pipe-block" speech suggests that some kind of synergizing process is going on. Freeze the jaw's motion by putting a pipe in your mouth, and you still don't disrupt speech. (Pipe and cigar smokers do it all the time: a functional analogue to the stiff upper lip). However, one might argue that a lot of learning goes into producing speech with a pipe in your mouth. The strongest test of the synergy hypothesis would be to perturb a person's jaw suddenly during speech and see if other remote members of the putative synergy spontaneously compensated *the very first time* the perturbation was applied. Remember, the synergy concept refers to a functionally, not mechanically, linked assembly of parts. Any remote responses observed should be specifically related to the speech sound actually being produced.

We used infrared light sensors to transduce movements of the lips and jaw. Fine wire electrodes were inserted into speech muscles such as the genioglossus, a major tongue muscle, to monitor electromyographic (EMG) activity. With the help of Milt Lazanski, an orthodontist, I designed a special jaw prosthesis made of titanium for the two subjects. Gaps for missing teeth due to old rugby injuries enhanced the ability to set the prosthesis firmly into the subject's mouth.

The results were stunning. When we suddenly halted the jaw for a few milliseconds as it was raising toward the final [b] in [b æ b] (rhymes with lab), the upper and lower lips compensated immediately so as to produce the [b] but no compensation was observed in the tongue. Conversely, when we applied the same jaw perturbation during the final [z] in the utterance [b æ z], rapid and increased *tongue* muscle activity was observed exactly appropriate for achieving the tongue-palate configuration for a fricative sound, but no active lip compensation.

In short, the form of cooperation we observed in the speech ensemble was not rigid and sterotypic; rather, it was flexible, fast, and adapted precisely to accomplish the task. The many components of the articulatory apparatus always cooperated in such a way as to preserve the speaker's intent. The functional synergy, as it were, revealed.

FROM SYNERGIES TO SYNERGETICS

Synergies correspond to some kind of collective organization that is neurally based. They simplify *control*, or, as Bernstein would have it, they render control of a complex multivariable system possible. But how are synergies formed? What principles govern their assembly? Bernstein saw coordination as the *organization* of the control of the motor apparatus. Just as the theoretical concepts of self-organized pattern-formation in open systems were not available to the Gestaltists, so it was with the Bernsteinians. Nevertheless, both groups looked to the future possibilities of "antientropic processes" in open systems to account for autonomous order formation in perception and action. In his last book, *The Coordination and Regulation of Movements* (Oxford: Pergamon, 1969), Bernstein foresaw the end of the honeymoon between the sciences of cybernetics (servomechanisms and the like) and physiology, and Köhler intuited that the general systemic tendency toward equilibrium of inanimate matter (linear thermodynamics) was not really applicable to the organism.

Michael Turvey, long an advocate of Bernstein's approach to motor control and Gibson's ecological approach to visual perception, summed up the research conducted in the context of Bernstein's formulation of the degrees of freedom problem as the *first major round* of theorizing and experimentation on coordination.[27]. The second major round revolves around some of the questions that I have raised above. In a sense, round 2 really started with a pair of papers published in 1980 by Peter Kugler, Turvey, and me. More concretely, it began with a rather vague and unsubstantiated claim, namely, that a functional synergy is a *dissipative structure* "that expresses a (marginally) stable steady state maintained by a flux of energy, that is, by metabolic processes that degrade more free energy than the drift toward equilibrium."[28]

A somewhat controversial Nobel Prize (aren't they all) had just been awarded in 1977 to Ilya Prigogine for his theory that, as a system is driven away from thermodynamic equilibrium, it may become unstable and then evolve new, coherent, dissipative structures. As I mentioned earlier, Hermann Haken introduced the term synergetics in the late 60s to describe an entire interdisciplinary field dealing with cooperative phenomena far from equilibrium. (Synergetics, by the way, is not a cult, but rather Haken's theory of how pattern-formation phenomena that arise in different contexts and disciplines are related, e.g., in the laser, chemical reactions, and fluid dynamics.)

If I may digress just a bit, the approaches of Haken and Prigogine (and for that matter, Rene Thom's catastrophe theory) are actually very different, even though they've often been bundled together in popular treatments. Prigogine's original theory was heavily weighted toward equations of a thermodynamic character that describe the behavior of ensemble averaged macroscopic quantities. Haken's work, from its very beginning, always included an essential role for, and explicit treatment of, microscopically generated fluctuations. This is also where it deviates from Thom's completely deterministic theory.[29] Fluctuations, as we will see, turn out to be quite crucial, both conceptually and methodologically, to our understanding of self-organization in living things.

Kugler and Turvey[30] stress the relationship between the stability and reproducibility of oscillatory movements and dissipative structures. One of their major goals is "to explain the *characteristic quantities* (emphasis theirs) of a rhythmic behavior—for example, its *period, amplitude* and *energy* per cycle (emphasis mine) [which] cannot be rationalized by neural considerations alone" (p. 4). In a series of experiments in which subjects oscillated a pair of hand-held pendulums whose length and mass could be independently varied, they and their colleagues discovered a number of fascinating relationships between these "characteristic quantities." For example, the pendulum period, over variations in the masses and lengths of pendulums, was proportional to mass to the 0.06 power and to length to the 0.47 power. They were able to match these empirical results with a model in which the characteristic frequency was that of the free, undamped motion of the pendulum with a spring attached a short distance away from the pendulum's axis of rotation. The spring represents muscles and tendons that elastically store and release mechanical energy. In their words, the wrist-pendulum system is a "macroscopic mechanical abstraction ... in which ... the only forces at work in the abstraction are the gravitational force (F_g) and an elastic force (F_k)" (p. 178).

A rather amazing feature of this macroscopic mechanical abstraction is that it helps explain certain features of quadruped locomotion, specifically the limb frequencies of animals (large and small) moving about the Serengeti plains. When plotted against limb length or mass, the stepping frequency, from Thompson's gazelle to the black rhinoceros, falls on three straight lines, one for each locomotory mode. Kugler and Turvey's pendulum-with-spring model, which represents the joint effects of gravity's tendency to return the limb to its equilibrium position and the spring's stiffness or restoring torque, fits the data remarkably well. In fact, for each animal cruising across the Serengeti they found that the ratio of spring torque to gravity's restoring torque was unity for walking, six in trotting, and nine in cantering. As Turvey elegantly points out, there is universality to the design of locomotion, a particular exploitation of nature's laws.[31]

But which laws are we talking about? Everything that has been empirically established thus far in hand-held pendulum studies appears to be consistent with Newtonian physics. The limbs of an animal (or person) behave like an inverted pendulum coupled to a spring. Colin Pennycuick, who collected the Serengeti locomotion data, uses them to support the following claim:

[A]t the intermediate scales of biology, Newtonian physics still works as well as it ever did. The reason is that Isaac Newton was himself a medium-sized animal, and naturally discovered laws that work best over the range of scales which he could perceive directly.... Biology occupies that range of scale in which Newtonian mechanics can account for physical processes to a level of precision appreciably higher than that to which most biologists are accustomed. In biology, if not in physics, Newton still rules.[32]

From the point of view of self-organization in complex systems in which dynamic instabilities play a central role, nothing could be farther from the truth. In the context of *coordination* in living systems, appropriate observables are not usually provided by Newtonian mechanics but have to be discovered. The Kugler-Turvey research program stressed scaling laws among (averaged) physical quantities such as mass, length, and frequency of oscillatory movement, and produced some important results. But such quantities tell us nothing about how the limbs are coordinated (e.g., in a walk as opposed to a trot or gallop), or the principles of neuromotor organization through which such coordinative modes spontaneously arise, stabilize, and change. Entirely new quantities are necessary to capture the coordination of living things as a self-organized phenomenon. Dynamic instabilities, long at the core of pattern formation in open nonequilibrium systems, provide a way to find them. In summary, if coordinated action is based on functional synergies and if functional synergies are indeed self-organized, most if not all of the criterial features of self-organized, synergetic systems—multistability, bifurcations, symmetry breaking fluctuations, etcetera—should be found in behavior itself. How, then, do we go from a potentially fruitful analogy to experiments at the bench? How can behavior be understood as a consequence of self-organizing processes? Obviously we have to find an experimental model system that may give some of these ideas a concrete and precise meaning. Such a model system should be simple and accessible, yet still retain the essentials of the coordination problem. A tall order indeed.

REQUIREMENTS OF A THEORY OF SELF-ORGANIZED BEHAVIOR

Theory is a good thing but a good experiment lasts forever.
—Peter Leonidovich Kapitsa

The front page of the January 1993 issue of the American Psychological Association's *Monitor* contained the following headline:"Chaos, chaos everywhere is what the theorists think." According to the article, psychologists picked up on the chaos idea in the early 1980s and "have been applying it with a vengeance ... to both hard core scientific aspects of psychology and clinical psychology, including both family and individual therapy."

I am not going to comment on the rhetoric surrounding the buzzword chaos and how it provides a more holistic view of human life, except to say, chaos of what? What are the relevant variables that are supposed to exhibit chaotic dynamics? What are the control parameters? And how do we find them in complex living systems where many variables can be measured, but not all are relevant? Certainly, if we are so inclined we can use the word chaos to explain everything, but how do we find the nonlinear equations of motion, whether continuous or discrete, in the first place? What is the x in nonlinear equations of the type $\dot{x} = f(x, \lambda)$, the derivative of a variable x with respect to time is a function of x and a parameter, λ? What are the attractors? What does the bifurcation diagram look like? Are these concepts and mathematical tools even relevant? How does one establish them, even in a single case?

All the hype about chaos and fractals tends to sweep these questions under the rug while everyone admires the nice pictures. Don't get me wrong, I like chaos and fractals. Some of my best friends do this stuff. I also like numerical simulation and computer graphics—couldn't do without them, in fact. They allow you to see inside a mathematical theory. But, as a scientist, I want to know what these pictures represent; I especially want to know that the mathematical equations represent (some small portion of) reality. There has to be some connection between mathematical formulae and the phenomena we are trying to understand. Without this connection, as the popular song goes, we're "p___ing in the wind." Establishing a connection between theory and experiment is one of the canons of science that the "chaos, chaos everywhere" crowd seems to ignore.

Once Again—Dynamic Instabilities

Unlike the fluid patterns and chemical reactions described in chapter 1, or Haken's famous laser example where the microscopic level of molecules or atoms is well-defined, in biology and psychology the path from the microscopic dynamics (e.g., the brain with 10^{14} neurons and neuronal connections) to collective order parameters for macroscopic behavior is not readily accessible to theoretical analysis. So how might the spontaneous formation of pattern—self-organization—be studied? What kind of dynamical law gives rise to the self-organization of behavior? The answers to these questions are rooted in the notion of *instability of motion*.

What's so special about instabilities? First, they provide a special entry point because they allow a clear distinction between one pattern of behavior and another. Instabilities demarcate behavioral patterns, thereby enabling us to identify the dimension on which pattern change occurs, the so-called collective variable or order parameter concept of synergetics. As I mentioned earlier, very many observables may, in principle, contribute to a description of behavior even if observation is restricted to a single level. If we study a system only in the linear range of its operation where change is smooth, it's difficult if not impossible to determine which variables are essential and which are not. Most scientists know about nonlinearity and usually try to avoid it. Here we actually exploit qualitative change, a nonlinear instability, to identify collective variables, the implication being that because these variables change abruptly, it is likely that they are also the key variables when the system operates in the linear range.

Second, instabilities open a path into theoretical modeling of the collective variable dynamics. In other words, they help us find the equations of motion. The idea is to map observed patterns onto attractors of the collective variable. Instabilities, as we have seen, are created by control parameters that move the system through its collective states. Candidate control parameters have to be found, and instabilities offer a way to find them. Collective variables and control parameters are the yin and yang of the entire approach, separate but intimately related. You don't really know you have a control parameter unless its variation causes qualitative change; qualitative change is necessary to identify collective variables unambiguously.

Third, instabilities provide a means to evaluate predictions about the nonlinear, collective variable dynamics near crisis or critical points. Two predicted features of synergetics concern *critical fluctuations* and *critical slowing down*. In the former, values of collective variables undergo large fluctuations as instability is approached. Fluctuation enhancement, in fact, may be said to *anticipate* an upcoming pattern change. Critical slowing down refers to the ability of the system to recover from a perturbation as it nears a critical point. This recovery process takes longer and longer the closer the system is to a critical state. Measurement of the time it takes to return to some observed state—local relaxation time—is an important index of *stability* and its loss when patterns spontaneously form.

Finally, on a more conceptual level, instabilities are hypothesized to be one of the generic mechanisms for flexible switching among multiple attractive states; that is, for entering and exiting patterns of behavior. Thus, although transitions may be realized or instantiated in a multitude of ways on many different levels, the generic mechanism of instability is universal to all of them.

To summarize briefly, we have tried to rationalize instabilities on both methodological and conceptual grounds as a fundamental mechanism underlying self-organization. All that remains now is to establish their existence in human brain and behavior, specifically, in the experimental laboratory.

The Phase Transition Story

I must admit that how the next sequence of events unfolded is still a bit of a mystery to me. It's a strange mixture of intuition and serendipity. Or, as Louis Pasteur was purported to say, luck favors a prepared mind. The background is this. I was aware through reading Haken's work that when macroscopic patterns of behavior change qualitatively, the dynamics of the entire system may be dominated by one or a few order parameters: when rolling motion starts in Bénard cells there is an enormous compression of information. I was aware also of Schrödinger's order-order transition principle as his proposed new physical principle of biological organization. So some rather vague form was circulating in my mind. But how to create an experimental way to study these ideas so that they might no longer be vague but mathematically exact? To *want* the rules of behavior to be self-organized is one thing, but finding a means to realize one's desires is another issue entirely.

One intriguing idea was that gait transitions—when an animal shifts from, say, a trot to a gallop—might be analogous to the simplest form of self-organization known in physics, namely, the nonequilibrium phase transitions analyzed by Haken. Unfortunately, no one had studied gait transitions in this way, and it was quite impossible to conceive of doing the experiments at Haskins Laboratories, which is world famous for its research on speech, not animal locomotion. Imagine, then, the following scene.

It is the winter of 1980 and I'm sitting at my desk in my solitary cubicle late at night. Suddenly from the dark recesses of the mind an image from an ad for the Yellow Pages crops up: "Let your fingers do the walking" To my amazement I was able to create a "quadruped" composed of the index and middle fingers of each hand. By alternating the fingers of my hands and synchronizing the middle and index fingers *between* my hands, I was able to generate a "gait" that shifted involuntarily to another "gait" when the overall motion was speeded up. Talk about the spontaneous formation and change of ordered patterns!

On hindsight, the emergence of this idea was itself a kind of phase transition reminiscent of the kind experienced by my favorite sleuth, Philip Trent. As his friend, Dr. Fairman explains: "'What Trent means is to put it quite simply, that a certain concept had planted itself in his subconsciousness, where an association of ideas had taken place which abruptly emerged, quite spontaneously and unsought, in the sphere of consciousness.' The Inspector gazes grimly at the speaker for some moments. Oh! If that's all he means, why couldn't he say so? You *have* relieved my mind. He turned to Trent, 'You had a brain-wave—is that it?'"[33]

The effect was unbelievably compelling, a real party trick, as one reviewer from the journal *Nature* said. I quickly found that the situation could be simplified even further to involve just the two index fingers. Was this a paradigm that perhaps might provide a window into self-organization in biology and behavior, that might take us from a potentially fruitful analogy to experiments at the bench? As my colleague Pier-Giorgio Zanone (whose work with me on learning you'll see more about in chapter 6) is fond of remarking, "I can't believe it!" Neither, frankly, could I. The next step was to establish the reproducibility of the phenomenon in a series of experiments. That's just a first step, of course, but here's the gist of what I did.

A Phase Transition in Human Hand Movements

My original experiments involved rhythmical behavior.[34] There are a lot of good reasons why rhythmical movements are a good place to start. Rhythmical behaviors are ubiquitous in biological systems. Creatures walk, fly, feed, swim, breathe, make love, and so forth. Rhythmical oscillations are archetypes

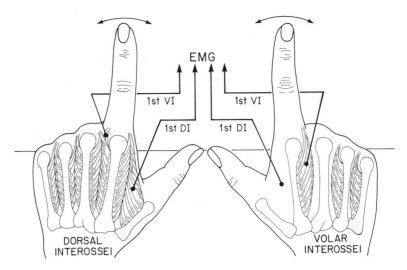

Figure 2.3 One version of the bimanual phase transition paradigm. Subjects move their index fingers rhythmically in the transverse plane with the same frequency for the left and right fingers. The movement is monitored by measuring continuously the position of infrared light-emitting diodes attached to the fingertips. The electromyographic (EMG) activity of the right and left first dorsal interosseus (DI) and the first volar interosseus (VI) muscles are obtained with platinum fine-wire electrodes. (Drawing by C. Carello.)

of time-dependent behavior in nature, just as prevalent in the inanimate world as they are in living organisms. Although they may be quite complicated, we have the deep impression that the principles underlying them should possess a deep simplicity. Ordering or regularity in time is important also for technological devices, including computers.

The task for my subjects, initially colleagues at the lab who wondered what on earth I was up to, was to oscillate their index fingers back and forth with the same frequency of motion in each finger (figure 2.3). Subjects can stably and reproducibly perform two basic patterns, in-phase (homologous muscle groups contracting simultaneously) and antiphase (homologous muscle groups contracting in an alternating fashion). Using a pacing metronome to speed up finger twiddling, oscillation frequency was systematically increased every few seconds from 1.25 cycles per second (Hz) to 3.50 Hz in small steps. Figure 2.4 shows a time series when the subject was instructed to begin moving her fingers in the antiphase mode.

Before going too much further, I should say a bit more about the instructions I gave because they are important. Subjects were required to produce one full cycle of movement with each finger, for each beat of the metronome. Furthermore, if they felt the pattern begin to change, they should not consciously try to prevent it from happening but rather adopt the pattern that was most comfortable under the current conditions. "If the pattern does change," I told them, "don't try to go back to the original pattern but stay in the one that's most comfortable. Above all, try to keep a one-to-one (1 : 1) relationship between your rhythmical motions and the metronome beat." My

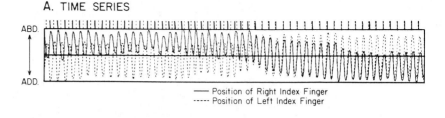

A. TIME SERIES

ABD.

ADD.

—— Position of Right Index Finger
----- Position of Left Index Finger

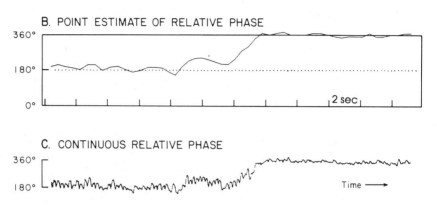

B. POINT ESTIMATE OF RELATIVE PHASE

360°

180°

0°

2 sec

C. CONTINUOUS RELATIVE PHASE

360°

180°

Time ⟶

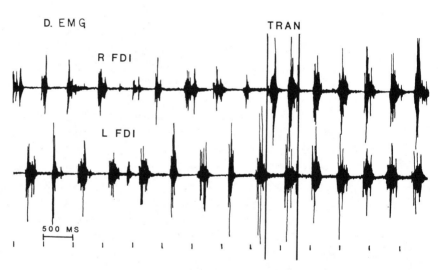

D. EMG

TRAN

R FDI

L FDI

500 MS

Figure 2.4 (A) The time series of left and right finger position shows the transition from antiphase movement to in-phase movement. From left to right the movement frequency, (F), was increased. (B) The point estimate of relative phase (obtained from the relative position of the left finger's peak extension in the right finger's cycle) changes from fluctuating around 180 degrees to fluctuating around 360 degrees. (C) A more refined measure of relative phase is the continuous estimate, obtained from the difference of the individual finger's phases that were calculated from the phase plane $(x, \dot{x})$ trajectory. (D) The EMG record of left and right first DI muscles also shows the change in phasing.

reasons for all this will become clearer later on when we consider the role of volition or intentionality in the self-organization of behavior (chapter 5). For now, it's important to establish the mechanisms underlying involuntary or spontaneous pattern formation and change.

As figure 2.4 clearly reveals, around a certain frequency of movement (the critical region), subjects spontaneously switch from the antiphase parallel motion of the fingers to an in-phase symmetrical pattern. No such switching, however, occurs when the subjects start in the in-phase mode. They stay there throughout the entire frequency range. Thus, while people can produce two stable patterns at low frequencies, only one pattern remains stable as frequency is scaled beyond a critical point.

I devised a way to monitor the transition behavior by calculating the phase relationship between the two fingers. A *point estimate* of relative phase is the latency of one finger with respect to the other finger's cycle time or period, determined from its peak-to-peak displacement. When the latency, t, of one finger, (x_L), is divided by the period, T, of the other (x_R) and multiplied by 360 degrees, we obtain the relative phase in degrees (figure 2.5A). This measure evaluates coordination at only one point in each cycle. I also made a *continuous*

A

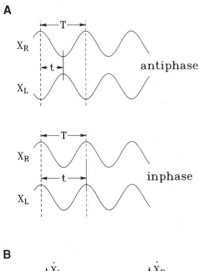

B

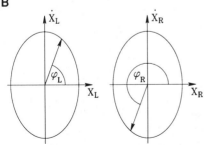

Figure 2.5 (A) Calculation of relative phase as a point estimate from two time series. (B) Calculation of the continuous relative phase from phase plane trajectories (see text).

Self-Organization of Behavior: The Basic Picture

estimate of relative phase by calculating the relative phase at the sampling rate of 200 times a second. Figure 2.5B shows how this was done. The *phase plane trajectory* (a plot of each finger's velocity, $\dot{x}$, versus its position, x) of each finger is shown. Normalizing the finger oscillations to the unit circle, the phases, ϕ_L and ϕ_R, of the fingers are obtained simply from the arctangent, $(\dot{x}/x)$, if x is the normalized position. The continuous relative phase is then just the difference $(\phi_L - \phi_R)$ between these individual phases at every sample. In figure 2.4 we see that the relative phase fluctuates before the transition and stabilizes thereafter. Amazing!

I first formally reported the result at a major meeting of experimental psychologists in the United States, the Psychonomic Society, in 1981. My talk wasn't very well attended. In those days, self-organized phase transitions in psychology were hardly in vogue. A little later, in March 1982, Arnold Mandell and Gene Yates invited me to a conference they were organizing at Ray Kroc's ranch (the late Ray Kroc of McDonald's hamburger fame) in Santa Barbara called "Nonlinearities in Brain Function." With all the hoopla about chaos in the brain—not to speak of other body parts—in the last few years, Mandell and Yates are seldom mentioned. In my opinion, history will reveal them as visionaries, far ahead of their time.

I will never forget the Kroc conference. It was one of the intellectual highlights of my life. Arriving at the ranch after a bus journey through the Santa Barbara hills (during which Mandell, referring to the latest work in nonlinear dynamical systems, told me excitedly, "You ain't seen nothin' yet!"), we quenched our thirst at a huge sideboard containing individual dispensers of every drink imaginable. This Irishman's dream.

But when it came to proposing theoretical models of my phase transition experiments, the well, so to speak, was dry. And this well included some of the top theoretical physicists and applied mathematicians (as well as neurobiologists) in the world. For example, I roomed with a young theoretical physicist from Los Alamos, Doyne Farmer, who was later featured prominently in James Gleick's book *Chaos*. Farmer was and is right at the forefront of the nonlinear dynamics business, and a brilliant teacher to boot. I felt embarrassed to show my little toy model borrowed from catastrophe theory. Little did I know at the time that the picture I'd formed was a reasonably good guess, but hopelessly wrong in detail. It was only an image, a vague analogy, at best. Here are my notes from 1982, word for word, describing this crude picture which was resurrected from my files (figure 2.6):

Think of an asymmetric potential well; choose the initial condition by applying a "force"(?) favoring the left-hand well [the antiphase pattern]. As this potential system is scaled(?) the right hand well [the in-phase pattern] becomes strongly favored, i.e., the depth of the right-hand well relative to the left is increased. When the left-hand well is somewhat flat, the system is particularly influenceable such that any increase in "dissipative noise" will effect a shift into the right-hand well (a favored mode).

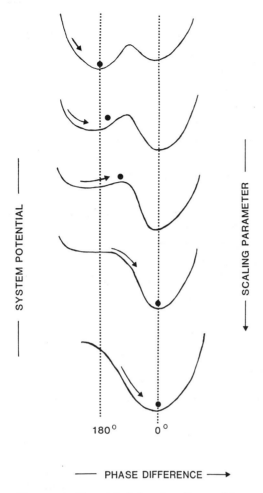

SYSTEM POTENTIAL

SCALING PARAMETER

180°   0°

⟶ PHASE DIFFERENCE ⟶

Figure 2.6   The original phase transition model presented by the author at the Kroc Foundation conference on "Nonlinearities in Brain Function."

Brackets are added to clarify the meaning of left- and right-hand wells. The question marks are in the original notes and reflect my uncertainty about which words to use. More sophisticated alternatives to figure 2.6 were suggested to me at the Kroc meeting—Niemark bifurcations, forced Duffings, and the like. Although I didn't fully understand them at the time, these turned out to be wrong too.

One person who was unable to attend the Kroc conference due to illness was the theoretical physicist Hermann Haken. Although I didn't know it (or him), he was just the person I was looking for.

To cut a long story short, after reading a draft of my experimental paper that I had sent to him for comments, Haken invited me to come to Stuttgart in the summer of 1983 to work with him and his co-workers on a theoretical model of phase transitions in human hand movements. From that point on, a strategy evolved in which perceptual-motor coordination was viewed no longer as a fairly peripheral (to some) topic of study in its own right, but as a window into biological self-organization. The mystery is that none of this would have happened had I not imagined my fingers as walking.

From Phenomena to Theory

To recap, the main features of my experiment were fourfold. First was the presence of only two stable coordination patterns between the hands. Which one was observed was a function of the initial conditions, meaning how subjects were instructed to move their hands at the beginning of the experiment. The fact that humans can stably produce, without a lot of learning (see chapter 6), only two simple coordination patterns between the hands remains for me an absolutely amazing fact. A complex system of muscles, tendons, and joints interacting with a much more complex system composed of literally billions of neurons appears to behave like a pair of coupled oscillators. A truly synergetic effect! Later, Betty Tuller and I showed that even skilled musicians and people who have had the two halves of their brain surgically separated to control epileptic seizures are still strongly attracted to these two basic patterns.[35] That is not to say that other timing patterns are impossible; only that people have a great deal of difficulty producing them. Second was the abrupt transition from one pattern to the other at a critical movement frequency. Third was the result that beyond the transition, only the symmetrical pattern was stable. Fourth, when cycling frequency was reduced, subjects did not spontaneously return to the initially prepared antisymmetrical pattern but stayed in the symmetrical one.

Now that we are in possession of the main facts, the next step is to identify candidate collective variables and control parameters. Since the fingers are moving at a common frequency, one candidate order parameter for coordination might be the relative frequency or frequency-ratio between the time-varying components. But frequency-related measures are inadequate because they refer to events occurring in an individual component, not *between* components. As I emphasized before, to understand coordinated behavior as self-organized, new quantities have to be introduced beyond the ones typical of the individual components. Also, we need a variable that captures not only the observed patterns but transitions between them. Only the *phase relation* appears to fulfill these requirements.

Unlike many other possibilities, it is relative phase that reflects the cooperativity among the components and embodies the kind of circular causality typical of synergetic systems. Thus, on the one hand, the interaction of the subsystems (here the individual finger motions) specifies their phase relation, and on the other, phase specifies the ordering in space and time of the individual subsystems. Also, as figure 2.4 shows, the phase relation changes far more slowly than the variables describing the behavior of the individual components that are oscillating to and fro—another typical feature of the collective variable or order parameter concept. But the most important reason why phase is a suitable order parameter is that it changes abruptly at the transition and is only weakly dependent on the prescribed frequency of movement outside the transition region. Since frequency of movement induces a qualitative change in phase, it may be considered an appropriate control parameter.

The final step is to develop a theoretical model that captures the main qualitative features of the data. If we can do that, quantitative predictions may be expected to follow. But extracting a law of coordination from a set of measurements is not so trivial. The big plus here is that we've done a simple experiment that contains many of the desirable features of biological systems that we want to understand, such as stability, flexibility, switching capability, and so on, yet at the same time prunes away many of the real life complications typical of naturally occurring behaviors. Just as Galileo used an inclined plane (which he could manipulate) to understand the free fall of objects (which he could not), so this phase transition situation allows us to understand how coordinated actions are self-organized. Now the aim is to obtain a precise mathematical description of coordination, stripped down to its essentials.

A Brief Digression

I promised myself as well as the reader that I would limit the number of mathematical equations in this book, relegating them to the technical literature. But the main equation describing coordination is about to appear on center stage, and to develop it I need a few elementary concepts from the field of *dissipative dynamical systems*. A dynamical system is simply an equation or set of equations stipulating the evolution in time of some variable, x. In our case we are interested in the temporal evolution of our hypothesized collective variable, relative phase. How does it change from moment to moment as the control parameter varies?

A dynamical system lives in a *phase space* that contains all the possible states of the system and how these evolve in time. A dissipative dynamical system is one whose phase space volume decreases (dissipates) in time. This means that some places (subsets in the phase space) are more preferred than others. These are called *attractors*: no matter what the initial value of x is, the system converges to the attractor as time flows to infinity. For example, if you stretch a spring or displace a damped pendulum, they will eventually wind down and stop at their equilibrium positions. The attractor in each case is a fixed point or simply *point attractor*.

Some people say that point attractors are boring and nonbiological; others say that the only biological systems that contain point attractors are dead ones. That is sheer nonsense from a theoretic modeling point of view, as it ignores the crucial issue of what fixed points refer to. When I talk about fixed points here it will be in the context of collective variable dynamics of some biological system, not some analogy to mechanical springs or pendula. Other kinds of attractors than fixed points also exist, such as limit cycles and chaotic attractors, but we'll discuss them more fully as they emerge in specific examples later on.

An important concept related to the idea of attractors is the *basin of attraction*. For a given attractor, this refers to the region in phase space in which almost all initial conditions converge to the attractor. Several attractors with different basins of attraction may also exist at the same time, a feature called *multistability*. Multistability, the coexistence of several collective states for the same value of the control parameter, is, of course, an essential property of biological dynamics. When a control parameter changes smoothly, the attractor also usually changes smoothly. However, when the parameter passes through a critical point, a qualitative change in the attractor may take place. This phenomenon, as mentioned before, is called a *bifurcation*, the mathematical term used in dynamic systems theory, or *nonequilibrium phase transition*, the term preferred by physicists because it includes the effects of fluctuations.

Finally, when the *direction* in which the control parameter varies is changed, the system may remain in its current state or switch at a later point, thereby exhibiting *hysteresis*. This means that an overlapping region exists where, depending on the direction of parameter change, the system can be in one of several states. As we have stressed, bifurcations and hysteresis are hallmarks of nonlinearity in complex biological systems.

The Haken-Kelso-Bunz Model

What, we asked ourselves, is the layout of attractor states in our hand movement experiments, and how is that layout altered as a putative control parameter is changed? Answers to these questions rest on the nature of the collective variable relative phase, ϕ, which turns out to possess an amazing *symmetry* in both space and time. A symmetry is simply a transformation that leaves the system the same afterward as it was before. What can systems with symmetry do? Imagine your fingers walking again. If you look in the mirror while you do the antiphase and in-phase movements, you can exchange left or right hands and the phase relation does not change. In other words, a *spatial* symmetry exists. Similarly, the hand motions are periodic, repeating at regular intervals in time. If we shift time by one period forward or backward the relative phase stays the same. Periodicity, in other words, constitutes a *temporal* symmetry. The fact is that the *only phase relations possible* under left-right exchange and a phase shift of 2π are in phase ($\phi = 0$) and antiphase ($\phi = \pm\pi$). It almost suspends belief that these silly hand movement experiments reveal the existence of a spatiotemporal symmetry that governs the way individual components (here the fingers) interact in space and time.

Of course, much of the action will come when we break or lower this spatiotemporal symmetry, which nature does all the time! But for now, the task is to postulate the simplest mathematical function that could accommodate space-time symmetry, bistability, and the observed bifurcation diagram in the walking fingers experiments. Let's call this function, V, now known in the literature after Haken, Kelso, and Bunz as the HKB model.[36] Since V is time symmetric (periodic), we can write

$$V(\phi + 2\pi) = V(\phi).$$

Since V is mirror image or space symmetric (left-right exchange) we can write

$$V(\phi) = V(-\phi).$$

The first condition allows us to express a function, V, as a Fourier series. According to Fourier, any periodic function, and indeed many functions normally encountered in physics, can be made up as the sum of simple harmonic components such as sines and cosines. The second condition eliminates sines from the function, since only cosines are invariant when ϕ is replaced with $-\phi$. Any intrinsic left-right asymmetry, of course, requires the inclusion of sines. For now, to accommodate all our observations we need include only the first two terms of the Fourier series:

⤳ $V = -a\cos\phi - b\cos 2\phi,$

where the minus signs allow us to interpret the function, V, as a landscape with attractor states for positive values of a and b.

The behavior of the system is easy to visualize by identifying ϕ with a black ball moving in an overdamped fashion in the landscape defined by the function, V. By changing the ratio b/a, inversely related to frequency in my experiment, we can travel through the evolving landscape as shown in figure 2.7. When we initially prepare the system in a state illustrated by the black ball in the upper left panel ($\phi = \pm\pi$) and decrease the ratio b/a (equivalent to shortening the period of the rhythmical coordination pattern) we obtain a

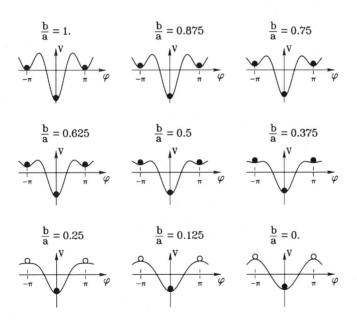

Figure 2.7 The HKB model of coordination. The potential, $V(\phi)$, as the ratio b/a is changed. The little ball illustrates the behavior of the system initially prepared (upper left corner) in the antiphase state. White balls are unstable coordinative states; black balls are stable.

critical value where the ball falls to the lower minimum corresponding to $\phi = 0$.

This means that the hand movements exhibit a transition from the antisymmetrical ($\phi = \pm \pi$) mode into the symmetrical ($\phi = 0$) mode. Notice that when we now reduce the frequency of motion, reversing the direction of parameter change, starting in the lower right portion of figure 2.7 the system will stay in the symmetrical in-phase mode even past the critical point. Theoretically (and experimentally) $\phi = 0$ is the deepest minimum of the function and therefore the most stable coordination pattern. But even more important is that our theory contains the experimentally observed and essentially nonlinear hysteresis effect.

We have built up a theoretical model of the phase transition without any discussion of differential equations. Differential equations arise whenever a law is expressed in terms of variables and their derivatives, or rates of change. Our coordination law may therefore be expressed in terms of the derivative of the collective variable, ϕ, which we denote as $\dot{\phi}$. This is simply the negative derivative of the function, V, with respect to ϕ.

$$\dot{\phi} = -\frac{dV}{d\phi}$$

$$= -a \sin \phi - 2b \sin 2\phi.$$

A beautiful way to intuit this basic coordination law is to plot the derivative of ϕ (called phi dot or $\dot{\phi}$) against ϕ itself for different parameter values. This is called the *vector field* of the relative phase dynamics and is shown in figure 2.8. Note that our coordination law contains stationary patterns or fixed points of ϕ at places where $\dot{\phi}$ is zero and crosses the ϕ-axis. When the slope of $\dot{\phi}$ is negative at the abscissa, the fixed points are *stable* and *attracting*. When the slope is positive, the fixed points are *unstable* and *repelling*. Arrows are drawn in figure 2.8 to indicate the direction of flow. Thick solid lines correspond to stable fixed points; dashed lines represent the unstable fixed points. As one travels from bottom to top in this figure, decreasing the ratio b/a, the stable fixed point at $\phi = \pi$ eventually disappears, leaving only one at $\phi = 0$. The bifurcation, appropriately enough, is called a pitchfork in the jargon of dynamical systems: the stable coordination pattern at $\phi = \pi$ is surrounded by two unstable fixed points delineating its basin of attraction, only to be annihilated at a certain critical point.

Notice again the terribly important fact that there is no one-to-one relation between the parameter value and the coordinative patterns. *Both modes coexist* for the *same* parameter value, necessitating nonlinear models. This is what I meant earlier by the need to formulate a coordination law that is simple, but not too simple. Our elementary coordination law possesses a remarkable symmetry and contains multistability, bifurcation, and hysteresis as primitive behavioral properties.

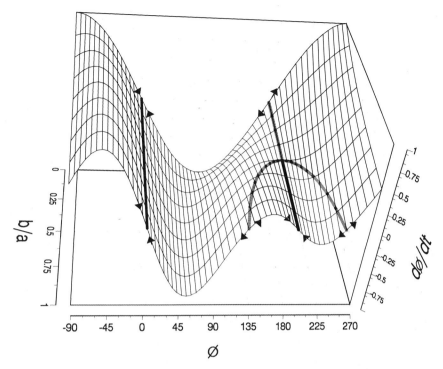

Figure 2.8 The HKB model of coordination expressed as a vector field, arrows indicating the direction of flow (see text for details). Thick solid and lighter dashed lines correspond to attractive and repelling fixed points of the collective variable dynamics. Note the inverse pitchfork bifurcation as the control parameter b/a is decreased.

... And Back Again?

All the main features of my experiments—the presence of only two stable relative phase or attractor states between the hands; transitions from one attractor to another at a critical cycling frequency; the existence of only one attractor state beyond the transition; even hysteresis—have been theoretically modeled, but so what? What makes us think that HKB theory is any more than a compact mathematical formulation? If the experiment is a really crucial one, we still have to prove that our approach has primacy over others, notwithstanding that it can describe our results well.[37] I have to admit that one of the main motivations behind these experiments was to counter the then dominant notion of motor programs, which tries to explain switching (an abrupt shift in spatiotemporal order) by a device or a mechanism that contains "switches." This seems a cheap way to do science, kind of like attributing thunder to the Norse god Thor. I have the same problem with ascribing words such as "schizophrenic," "alcoholic," and "depressive" to genes, but that's another book.

The real power of the synergetic theory of self-organization lies in the central concept of stability, which is important because stability can be lost.

Self-Organization of Behavior: The Basic Picture

That is exactly what happens at nonequilibrium phase transitions where patterns form or change spontaneously with no specific ordering influence from the outside (and no homuncular motor program inside). The hallmark features of such instabilities are, as I mentioned before, a strong enhancement of fluctuations (critical fluctuations) and a large increase in the time it takes the system to relax from a perturbation (critical slowing down). As we will see, our specific theoretical model of hand movements contains these predictions, thus allowing us in principle to transcend mere description.

Critical slowing down is easily intuited from the pictures of the evolving attractor landscape and its corresponding vector field (figures 2.7 and 2.8). In the former, notice how the potential around $\phi = \pm\pi$ deforms, the minimum in question becoming shallower and shallower as the parameter reaches a critical point. If perturbed away from its minimum, the little black ball will relax slowly compared with when the slope around the minimum is steep (top left). Similarly, in figure 2.8 the slope around $\phi = 0$ is greater than the slope near $\phi = \pi$ (180 deg.), which progressively gets shallower and hence less attracting as the parameter approaches criticality. There, the system is poised to change state by just the slightest little nudge.

Critical fluctuations arise because all real systems are subject to random fluctuations of various kinds, such as the environment and the multitude of microscopic components, that produce deviations away from the attractor state. Imagine a soccer team kicking our little black ball entirely at random. When the slope of the hill is steep, the ball can't be kicked very far away from its equilibrium position. When the slope flattens, however, the same magnitude of kick will cause the ball to move much farther away. As a result, near the critical point the attractor state suffers wild critical fluctuations.

I tested these predictions in the bimanual coordination paradigm with the help of John Scholz. I had already noted in the original publications that the phase relation between the limbs became much more variable near the transition, and discussions with Haken encouraged me to look in detail at the fine structure of fluctuations.

Scholz and I measured fluctuations in the two basic coordination patterns as subjects increased movement frequency by calculating the standard deviation from the relative phase time series. Dramatic increases in fluctuations were noted for the antiphase, but not the in-phase pattern before the transition. After the transition, the previously unstable antiphase state (now in-phase) fluctuated at the same low level as the stable in-phase state, a striking confirmation of the prediction.[38]

We tested critical slowing down by applying a little torque pulse to perturb briefly and unexpectedly one of the subject's oscillating fingers. This knocked the fingers away from their established phase relation and allowed us to calculate the time taken to stabilize the phase again at its value before the perturbation. In agreement with theory, we found that as the critical point neared, the relaxation time in the antiphase mode increased while it remained constant or decreased in the in-phase mode. Also we found that perturbations near the critical transition frequency often caused transitions from one mode to the other, exactly what one might expect from a complex dynamical system poised near an instability.[39]

Perhaps I should say a word or two about how to calculate the theoretical model parameters a and b. As we know, the ratio $|b/a|$ in the HKB model corresponds to the control parameter, movement frequency, in the experiment. From another viewpoint, the ratio expresses the relative importance of the phase-attractive states at 0 and $\pm\pi$ (we remind the reader that for $|b/a| > 0.25$, the system is bistable; as the ratio approaches a critical value the antiphase state loses stability and for $|b/a| < 0.25$, the system is monostable at the in-phase state (see again figure 2.7). Therein lies the secret to calculating the parameters of our theoretical model. But to do this, we have to take fluctuations into account explicity. Technically speaking, we have to study the transition behavior by adding a fluctuating force to the HKB model. This means solving the stochastic dynamics of ϕ by transforming it into a Fokker-Planck equation (apologies for the technical jargon). This equation describes the time evolution of the probability distribution for a system described, like the HKB model by a potential, V. Gregor Schöner, an expert in stochastic dynamics, collaborated with Haken and me on this problem.

One outcome of Schöner's analysis was that it allowed us to determine the model parameters, a and b, as well as how strong the noise is in the system, using experimental information on local relaxation time and variability measures of the patterns in the noncritical parameter regime. For example, the relaxation times were predicted as:

$$\tau_{\mathrm{rel},0} = 1/4b + a; \qquad \tau_{\mathrm{rel},\pi} = 1/4b - a,$$

where 0 refers to the in-phase mode and π to the antiphase mode. Similarly, we were able to estimate the noise strength in the system using measures of phase *variability* in the two modes before the onset of the transition.

I should stress that this is not just a parameter-fitting exercise, but rather it allowed us to check the consistency of the entire approach. Moreover, the stochastic theory contained another feature that we were able to examine experimentally, namely, how long the transition should take from antiphase to in-phase, a variable we called the *switching time*. An excellent agreement between the stochastic version of the HKB model and the experimental data was observed, in terms of both the mean switching time and the shape of the distribution of switching times.

Obviously, a lot of mathematical details have been left out of this description. Mathematics, as one wag said, is like sex, better performed in private than in public. However, I did not want to leave the issue of parameter estimation dangling, as if parameters were left freely blowing in the wind (as they sometimes are in theories). A conceptually important result of our analysis is that not just control parameters but fluctuations are instrumental in effecting transitions, probing the stability of coordinated states and pushing the system over the edge from unstable to stable states. Confirmation of theoretical predictions regarding critical fluctuations, critical slowing down, and switching times reveals that the emergence of coordinated behavior may be understood in considerable detail in terms of the physics of nonequilibrium processes. These same effects have now been observed in many other experimental model systems, attesting to the general validity of the theory (see chapters 3 and 4).

On the lighter side, two real world examples of fluctuation phenomena are shown in figure 2.9. One picture, under the headline "Nightmare," shows the

13 weeks of the Dow

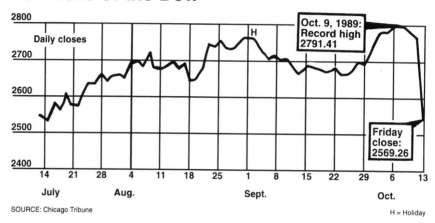

SOURCE: Chicago Tribune

H = Holiday

Janusz Kapusta

Figure 2.9 Two real world examples of fluctuations: (A) economic (Reprinted with permission from Knight-Ridder Graphics Network), and (B) political. (Copyright © 1992 by the New York Times Company. Reprinted by Permission)

fluctuations in the stockmarket before the big drop of nearly 200 points on October 13, 1989. Are they critical? The other ("Switch, Don't Fight, Mr. Perot," *New York Times*, September 18, 1992) shows the independent candidate for U.S. president, Ross Perot, on the brink, trying to decide whether to stay in the race or not. So what pushed him to stay in?

Relating Levels. I. The Components

A key feature of our approach is to characterize coordinated states in terms of the dynamics of collective variables, in this case, with relative phase as an order parameter. Obviously, it is possible to study the system on yet another level of description, namely, that of the individual limb or finger's dynamic

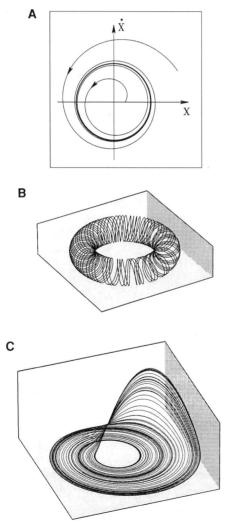

Figure 2.10   (A) Limit cycle attractor. (B) Torus or quasi-periodic attractor. (C) Chaotic attractor.

behavior. Thus we may choose each limb's position, x_i, and velocity, $\dot{x}_i$ ($i = 1, 2$), as collective variables; collective now with respect to the next lower level of description, such as the coordination of neuromuscular activities (see below). The stable and reproducible rhythmic performance of each hand may now be modeled as an attractor in the phase plane, $(x_i, \dot{x}_i)$, in this case a limit cycle. When the hand is on its limit cycle, it oscillates with a certain frequency and amplitude that are functions of parameters only, not of the initial conditions. The stability of this attractor is revealed by the fact that trajectories originating outside the limit cycle spiral inward, whereas trajectories inside spiral outward toward the limit cycle (figure 2.10A).

The stability of the limit cycle and its persistent, self-sustaining character are fundamentally due to a balance between excitation and inhibition (from the nervous system) and dissipation. Dissipation predominates outside the limit cycle, causing the amplitude to decrease; excitation predominates if x_i and $\dot{x}_i$ are small and inside the limit cycle, causing the amplitude to increase.

Bruce Kay, Elliot Saltzman, and I sought kinematic relations that might allow us to identify the form of the limit cycle characterizing each oscillating limb.[41] For example, in studying individual hand movements we found that amplitude of movement decreased monotonically as frequency was experimentally increased. We were able to map this and other observed kinematic relations onto a limit cycle attractor that combined features of the well-known Rayleigh and van der Pol oscillators. In performing this mapping, the concept of stability was once again at the heart of theory. We measured the stability of the attractor using perturbation techniques similar to those described for the coordinated case. Trajectories perturbed away from the limit cycle return more rapidly to strong attractors than weak ones.

Another assumption of limit cycle models is that the oscillation is essentially autonomous. The basic idea is that the phase of an autonomous time-invariant oscillator is marginally stable, unlike that of a driven oscillator that is locked to the driving function. A way to check this is to perturb the hand at different phases of its oscillation and see if and how the phase is reset or shifted. To assess the pattern of phase shift that the rhythm exhibits, we constructed and analyzed *phase-resetting curves* that plot the old phase where the cycle is perturbed against the new phase at which the cycle stabilizes after the perturbation.[42] We found that the oscillation tended to be *phase-advanced* by a perturbation, thereby producing a consistent phase-dependent shift in pattern. From these phase resetting results we concluded that central timing elements in the nervous system responsible for generating rhythms (see chapter 4 and below) are affected by the biomechanical properties (e.g., stiffness, damping) of the limb being controlled.

A final assumption of limit cycle attractors is that they are effectively one dimensional, forming a simple closed curve in phase space. However, real biological systems—and our hand movements are no exception—are not mathematically ideal, perfectly periodic systems. When plotted on the phase plane, a rhythmic movement trajectory appears instead as a band around some average closed curve. But how many degrees of freedom are actually involved? If the variability is due to stochastic noise, it is an infinite number—a daunting prospect. If the band of variability is produced by additional *deterministic sources*, for example, oscillations having frequencies that are incommensurate with the main frequency, the attractor should be *m*-dimensional, one for each oscillatory process. Topologically, such a *quasi-periodic* attractor is defined by a *m*-dimensional torus (T^m). Figure 2.10B shows an example for $m = 2$. Bands of variability on the phase plane may also be produced by *deterministic chaotic processes* that exhibit fractional or fractal dimension[43] (figure 2.10C).

Using a computational method[44] that allowed us to estimate the dimensionality of hand movement trajectories directly, we found evidence for two processes, one at a global scale and one at a small scale. The global process was a low-dimensional limit cycle attractor entirely consistent with earlier kinematic results showing stability in the face of perturbation. The small scale process was essentially infinite-dimensional, that is, stochastic noise. Model simulations on a computer confirmed this result. The main point is that once again, although now at the level of the individual components, we can take a complex dynamic behavior and map or encode it onto a dynamical model whose veracity can be checked experimentally. By such methods (and I know how time consuming they are) it is possible to reach an understanding of the coordination dynamics on different levels.

Relating Levels. II. Coupling

What is the relation between the limit cycle attractors of one hand and the phase entrained coordination dynamics of two hands working together? Coordinative and component levels of description can now be related by coupling the latter to create the former. This is not as trivial as it sounds. The simplest kind of coupling is obviously linear, for example, making the coupling a function of the amplitude differences between the oscillators. This doesn't work, at least if the goal is to derive all the features observed at the coordinative level. It turns out that not only do the oscillatory processes have to be nonlinear, so also does the *coupling*. Minimally, coupling functions have to contain terms that are products of individual oscillator amplitudes and velocities, that is, time derivatives. It is important to emphasize that the coupling is quite unspecific with respect to the patterns of coordination that emerge. That is to say, different coupling functions can give rise to the same coordination patterns.[45] Also, *changes* in coordination can be brought about in a variety of ways.

Figures 2.11a through c show *Lissajous curves* of our nonlinearly coupled nonlinear oscillator model of basic coordination. The different widths and slopes of the tracings reflect the phase difference between the oscillators. In each case, the initial conditions of each oscillator are identical, and all that is done in the computer simulation is to increase the intrinsic frequency of

Self-Organization of Behavior: The Basic Picture

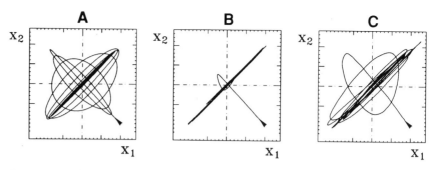

Figure 2.11 Computer simulations of nonlinearly coupled nonlinear oscillators. Arrows indicate the starting antiphase state. The coupling is of the form $\alpha(\dot{x}_1 - \dot{x}_2) + \beta(\dot{x}_1 - \dot{x}_2)(x_1 - x_2)^2$, where x_1 and x_2 refer to the oscillators, and α and β are coupling coefficients. A noise term, $F(t)$, is added to the coupling. (A) Fixed coupling parameters and noise. (B) Increase in coupling strength with noise the same strength as A. (C) Coupling parameters same as A, but noise strength is increased.

oscillation. In figure 2.11a the coupling parameters and the noise level are constant, and a transition occurs as frequency is increased. In figure 2.11b the coupling strength is increased, and the transition occurs almost immediately. A similar result is evident in figure 2.11c, but here the coupling parameters are as in figure 2.11a, and only the noise level parameter is increased.

A nice way to summarize the entire phenomenon is through the torus plots displayed in figure 2.12. The state spaces of the two oscillators ($x_1, \dot{x}_1$ and $x_2, \dot{x}_2$) (figure 2.10A) are plotted against each other. For a rational frequency relationship between the two oscillators, in the present case 1 : 1, the relative phase trajectory is a closed limit cycle and corresponds to a phase- and frequency-locked state. Figure 2.12a shows a stable antiphase limit cycle transiting (figure 2.12b) to a stable in-phase trajectory (figure 2.12c). A flat representation of the torus displays the phase of each oscillator in the interval $(0, 2\pi)$ on horizontal and vertical axes. The constant relative phase between the oscillators is reflected by straight lines.

In the flat representation shown in figure 2.13a the phase relation is π; that is, one oscillator's phase is zero and the other is π. The apparent discontinuity is not real, but is due to the 2π periodicity of the phase. Thus, when you fall off the top edge of the plane in figure 2.13a you reappear at the bottom. Figure 2.13, parts b and c show the transition to the in-phase relationship (zero phase lag between the oscillators).

The conclusions from experiment and theory are inescapable. First, the same behavioral patterns may be obtained from very different kinds and strengths of couplings among the components. In other words, *invariance of function* is guaranteed despite reconfiguration of connections or couplings among component elements. Second, and related, several patterns (here for convenience, two) may be produced by the same set of components and couplings. Such *multifunctionality* is an intrinsic property of the present approach. Third, one can see how difficult it is in complex biological systems to

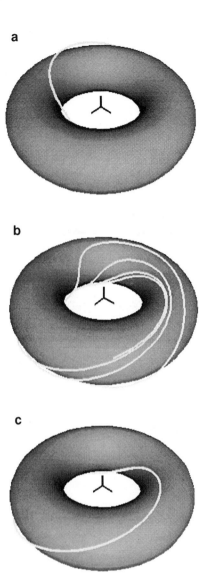

Figure 2.12 Same transition as in figure 2.11 but now displayed on the torus. As coupling is varied a stable antiphase trajectory (a), loses stability (b), and switches to a stable in-phase trajectory (c).

attribute the system's collective or coordinative capabilities to couplings per se. Many mechanisms, both physiological and mathematical, can instantiate or realize the same function. The conspicuous lack of a one-to-one relationship between self-organized coordination patterns and the structures that realize them is a central feature of the present theory, and surely constitutes one of the basic differences between living things and mechanisms or machine.[46]

Self-Organization of Behavior: The Basic Picture

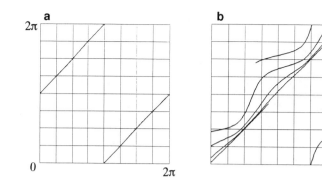
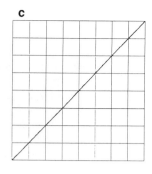

Figure 2.13 Flat representation of the torus displaying the phase of each oscillator on the horizontal and vertical axis. The transition is the same as shown in figure 2.12.

The Tripartite Scheme—Once More with Feeling

It may be possible to carry out this level-independent analysis when we step down to other scales, such as that of neurons and neuronal assemblies (see chapter 8). But for now, let's pull together the main conceptual themes that emerge from the walking fingers example. Figure 2.14 represents the linkages between phenomena and dynamic pattern theory (horizontal mapping) and between levels of description (vertical mapping). Here are the key points to keep in mind.

• A minimum of three levels (the task or goal level as a special kind of boundary constraint, collective variable level, and component level) is required to provide a complete understanding of any *single* level of description.

• Mutability exists among levels. For instance, the component level defined here in terms of nonlinear oscillators may be viewed as a collective variable level for finer-grained descriptions such as the way agonist and antagonist muscles generate kinematic patterns.

• Patterns at all levels are governed by the dynamics of collective variables. In this sense, no single level is any more important or fundamental than any other.

• Boundary constraints, at least in complex biological systems, necessarily mean that the coordination dynamics are context or task dependent. I take this to be another major distinction between the usual conception of physical law (as purely syntactic, nonsemantic statements) and the self-organized, semantically meaningful laws of biological coordination. Order parameters and their dynamics are always *functionally* defined in biological systems. They therefore exist only as meaningful characteristic quantities, unique and specific to tasks.

Reprise

I have demonstrated that simple behavioral patterns and considerable pattern complexity may arise from the process of self-organization, as emergent con-

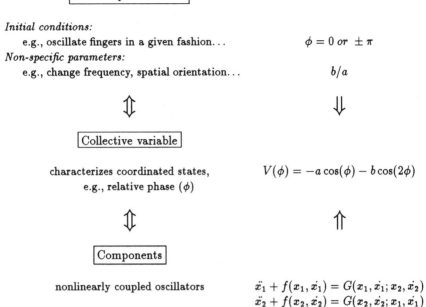

	Phenomena	**Pattern Dynamics**

Boundary constraints

Initial conditions:
 e.g., oscillate fingers in a given fashion... $\phi = 0 \; or \; \pm\pi$
Non-specific parameters:
 e.g., change frequency, spatial orientation... b/a

Collective variable

characterizes coordinated states, $V(\phi) = -a\cos(\phi) - b\cos(2\phi)$
e.g., relative phase (ϕ)

Components

nonlinearly coupled oscillators $\ddot{x}_1 + f(x_1, \dot{x}_1) = G(x_1, \dot{x}_1; x_2, \dot{x}_2)$
           $\ddot{x}_2 + f(x_2, \dot{x}_2) = G(x_2, \dot{x}_2; x_1, \dot{x}_1)$

Figure 2.14 The tripartite scheme applied to understanding coordination: the bimanual coordination example (see text for details).

sequences of nonlinear interactions among active components. Ironically, the very aspect of behavior that scientists, especially psychologists and biologists, usually try to avoid—instability of motion—turns out to be the key generic mechanism of self-organization. The very many neurons, muscles, and joints act together in such a way that the entire system acts as a single coherent unit.

 The discovery and consequent analysis of phase transitions in human hand movements introduces a new paradigm into biology.[47] It appears also that a comparison of nonequilibrium phase transitions in physics and discontinuities in coordinated action goes beyond mere analogy.[48] Of course, in physics, phase transitions remain objects of concentrated research. But the basic events that we have found in voluntary human hand movements, critical fluctuations and critical slowing down, occur over and over again in nonequilibrium systems, and suggest that the same laws and principles are in operation. The step we have tried to take in this chapter, albeit a baby step, is from the identification and descriptive language of functional synergies in action, to synergetics, a theory of how synergies are created, sustained, and dissolved. This is the conceptual and methodological foundation on which I believe a scientific psychology should be built, a science that bridges mental, brain, and behavioral events.

3 Self-Organization of Behavior: First Steps of Generalization

The principal object of research in any department of knowledge is to find the point of view from which the subject appears in its greatest simplicity.
—Josiah Willard Gibbs

HUBRIS TEMPERED?

Does the move from synergies to synergetics constitute a bloodless coup? Are motor programs and the computer metaphor dead? Well, not quite. As with any new paradigm, a little healthy skepticism is to be expected, even among one's supporters. So let's try to shore up the elementary picture drawn in chapter 2 with some further examples before we extend this picture to even more interesting and complex phenomena.

I have argued that laws of self-organized, coordinated behavior are not structure specific or dependent on physicochemical processes per se. Although coordination is always realized and instantiated by physical stuff, the laws themselves are abstract and dynamical. Nevertheless, it's wrong, in my opinion, to view the coordination dynamics—equations of motion for task-specific collective variables or order parameters—as simply mathematical curiosities. The order parameter may seem immaterial, for example, the relative phase, ϕ, between interacting components is a mathematical quantity, but it is no less real for that. As Hermann Haken pointedly remarked, order parameters in biological systems are just as real as our thoughts.[1] However we interpret the order parameter concept, it clearly expresses the coherent relations among the interacting parts of a system. Later on (chapter 8) we'll see that our relative phase order parameter captures the collective action of neurons in the brain itself. Cells that are active about the same time tend to coalesce into cooperating groups, and vice versa.

For now let's proceed to experimental evidence that bolsters the entire idea that behavioral patterns are self-organized. Remember, this means that patterns emerge, stabilize, and change under the influence of *nonspecific* control parameter(s). Such parameters move the system through patterns (coherent collective states) but do not prescribe or encode these states. This is a crucial distinction that will take on special significance in relation to such issues as intentionality, adaptation, and learning, to be considered in later chapters, where the whole question of how to understand selection of *specific* patterns arises.

In this chapter the goal is to show (I hope not ad nauseam) that the same dynamic pattern principles apply to the coordination among different parts of an organism, organisms themselves, and organisms and their environment. The generality extends to different kinds of systems, such as the articulatory system involved in speech, and to different kinds of patterns or functions such as communication among people. I will only discuss a few examples out of the many for which it has been possible to forge a specific linkage between theory and experiment. The bottom line of this chapter and the next is that the same coordinative coupling governs the interaction among different parts of a system and is quite independent of the physical medium (optical, acoustic, haptic) over which the interaction is mediated. Coordination dynamics is not ordinary physics. It deals with the dynamics of informationally meaningful quantities. Coupling in biological systems must reflect functional, not merely mechanical constraints if behavior is to be adaptive and successful.

Our departure point is some recent critical remarks expressed in a friendly spirit by Dutch physicist Wiero Beek. Although he "believes we are searching in the right direction," Professor Beek argues that "in order to turn Kelso's original experiment into a really crucial one ... more proof is needed." In particular, he wonders how a subject's intentions influence actual behavior, and whether critical fluctuations are observed when a horse goes from trotting into galloping. "I never observed it," he says. "Without such conclusive proof, theory runs the risk of being built on quicksand."[2]

We'll consider how intentionality is handled in the present framework in chapter 5. For now, let's have a look at gaits, a topic that intrigued the ancient Greeks. Lest the reader think we stray too far from matters of mind and behavior, the science of psychology, let me remind you that one of the things we all learn to do is walk. The small child's first attempts at walking are a wonder to behold. They consume his conscious self. Only after toilsome learning does walking drop out of consciousness and becomes one of those activities that we all (including most of scientific psychology) take for granted.

ON HARVARD HORSES AND RUSSIAN CATS

Professor Beek's toss of the gauntlet—whether critical fluctuations occur when a horse goes from trotting to galloping—reminds me of the very question that began the scientific analysis of locomotion. But this challenge took the form of a bet, also around a technical scientific point. Legend has it that the wager was for $25,000, a sizeable sum in the late 1800s, and concerned whether the horse, when it trots, ever takes all its feet off the ground. The issue was settled by a famous photographer, Edward Muybridge, who used high-speed photography to show that indeed the horse did. A survey of Muybridge's opus indicates he was interested in photographing other things beyond horse gaits, but I digress.[3]

It is surely not surprising that it is difficult to observe critical fluctuations when a horse changes gait. Any self-respecting horse would want to change gears smoothly without subjecting itself to erratic fluctuations. Of course, that's not quite the point. Critical fluctuations and critical slowing down are predicted signatures of self-organization in nature. Only close to transition points do these signatures show up, and that, as we've seen, usually requires very carefully designed experiments in which great attention is paid to time scales. That critical fluctuations haven't been seen with the naked eye doesn't mean they are not there. Nature doesn't give up her secrets easily: if critical fluctuations were easily observed, the whole notion that actions are governed by motor programs in which no fluctuations are to be expected might never have gained the ascendancy it did.

It is a well-known fact that horses (and other quadrupeds as well as human bipeds) employ a restricted range of speeds in any given gait. If you're

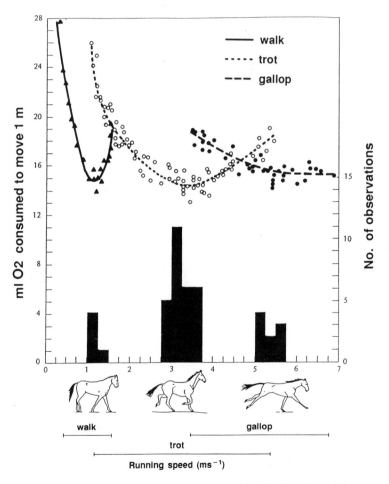

Figure 3.1 Oxygen consumption per meter moved and preferred speed (histograms) of walk, trot, and gallop of ponies. (Adapted from reference 4. Reprinted with permission)

Self-Organization of Behavior: First Steps of Generalization

running a long distance, you want to use a speed that corresponds to minimum energy expenditure.

The histograms in figure 3.1 display the range of speeds that horses select if they are allowed to locomote freely.[4] Note that there are gaps along the abscissa (some speeds aren't selected at all), and that the maxima of the histograms correspond to the *minimum* oxygen consumption per unit distance traveled. The reason you see three curves—one for walk, one for trot, and one for gallop—is because Hoyt and Taylor of Harvard put the horses on a treadmill and actually forced them to walk at speeds at which they might normally trot, trot at speeds they might normally gallop, and so forth.

As revealed in figure 3.1 by the upturn in the curves, it becomes metabolically expensive for the horse to locomote at speeds that deviate from those preferred. The points at which the curves cross is strongly suggestive of a nonlinear system bifurcating into different modes: it becomes too costly for the horse to maintain a given locomotory gait, so breaks into the next most energetically efficient mode occur. The overlap of the parabolic curves suggests the presence of hysteresis, although no data are provided on this point. In fact, all of this is only an interpretation, expressed some years ago,[5] of Hoyt and Taylor's results. The horse locomotion people don't typically think in terms of synergetics and nonlinear dynamical systems. Remember, Newton rules biology!

So what about critical fluctuations? Is the theory on quicksand? Notice in figure 3.1 that the *overlap* of gaits across the range of walking speeds is extremely important functionally: it enables the animal to maintain a stable mode of locomotion in the face of fluctuating demands. Were this overlap feature (multistability) not present, the gait would be extremely unstable. At certain speeds the poor horse would vacillate randomly between different locomotory modes. If one knew what to measure, wild fluctuations would be expected. But, as Beek adamantly points out, he has never seen such wild fluctuations. Nevertheless, one suspects they could be found in (beginning) trotters, before the ponies have learned to stabilize the trot at near galloping speeds.

Since I don't want to wager with the good professor, I'm going to refer him instead to a classic study by Mark Shik and colleagues of the Institute of Biological Physics in Moscow published nearly thirty years ago.[6] Figure 3.2 shows the famous "mesencephalic cat," which was a preparation robbed of voluntary motor output from the cerebral cortex. By electrically stimulating a single site in the cat's midbrain (the mesencephalic region), these Russian scientists were able to induce changes not only in walking velocity, but abrupt gait changes as well. I have taken the liberty of calculating by hand the phase relation between the movement trajectories of the cat's hind limbs. You can see that at a critical stimulation current, a spontaneous transition from trotting to galloping is observed.

But that's not all. In a certain range of stimulus parameters, which the Russian scientists described prophetically as an "unstable region," the animal

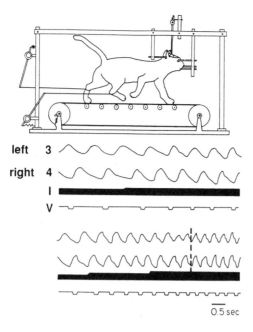

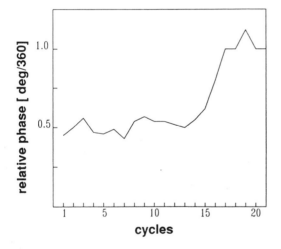

Figure 3.2 (*Top*) The mesencephalic cat. Although the figure doesn't show it, the cat is supported on the treadmill. Recordings of the left and right hind limb movements are shown underneath as a function of the strength of electrical stimulation (I) (Reprinted from reference 6. With kind permission from Elsevier Science Ltd.) (*Bottom*) The relative phase between the hind limbs calculated by the author on the same time series as shown above. Note the shift in gait.

sometimes trotted and sometimes galloped. Had they calculated our hypothesized order parameter, the phase relation among the limbs, they surely would have found the fluctuation enhancement effect predicted by synergetics.

So my answer to Professor Beek is as follows. A horse with an intact nervous system has mechanisms available to stabilize a given gait, even though it may be metabolically costly. Normally, horses avoid potentially unstable regions; they select only a discrete set of speeds from the broad

Self-Organization of Behavior: First Steps of Generalization

range available. In fact, just the ones that minimize energy. Shik's cats, on the other hand, deprived of a cerebral cortex and thus lacking voluntary control, can be taken to "unstable regions" where critical fluctuations can be felt and observed. The bottom line emerging from Shik's cats is that locomotion—the wondrous and numerous spatiotemporal orderings among the limbs—looks very much like a typical example of self-organization under the control of a single nonspecific parameter. (Note for the experts: if stimulation is held constant and the treadmill belt increased, the same dynamical effects can be obtained. The control parameter can be mediated by efferent and/or afferent neural activity.)

About the Harvard horses one can only speculate, but speculate reasonably I think. As the Russians would say, they too solve the motor problem by organizing the necessary interaction among the elements (muscles, joints, limbs). Perhaps, being from Harvard, they are smart enough to avoid unstable regions. Sarcasm aside, any complex motor skill looks natural and smooth when performed by pros. This holds true for skiing, golf, tennis, guitar playing, even juggling. Of course, if one concedes that horses are experts in their locomotory gaits, the implication is that the minima are too deep to show fluctuations. Unstable regions simply may not exist anymore.

COORDINATION AMONG COMPONENTS OF AN ORGANISM

Multilimb Coordination Dynamics

Even though gait shifts occur in quadrupeds and strong hints of dynamical phenomena are certainly present, no experiments to my knowledge have attempted to test predicted features of nonequilibrium phase transitions in biological systems containing more than two components. John Jeka and I tried—we think successfully—to generalize the theory to more than two components,[7] specifically the arms and legs of humans.[8] Recent studies by Stephan Swinnen and colleagues confirmed our findings.[9]

There are a number of good reasons to study the dynamics of pattern formation and change in this more complex, multicomponent system. The coordination dynamics are potentially very rich because of the greater number of components to be coordinated; anatomical differences among individual components introduce asymmetries into the dynamics; and the multilimb system possesses an obvious capacity to produce a large number of patterns. Moreover, one might think that it's just as easy to switch from any given two-, three-, or four-limb pattern to any other. But this turns out not to be the case: transition pathways or routes between patterns, just like the bimanual case, turn out to be governed by pattern stability. It's easier to switch from a less stable to a more stable pattern than vice versa. Moreover, symmetry of the coordination dynamics is once again the property that structures the transition behavior of the human multilimb system. In general, spontaneous transition behavior in a multicomponent system provides important clues to

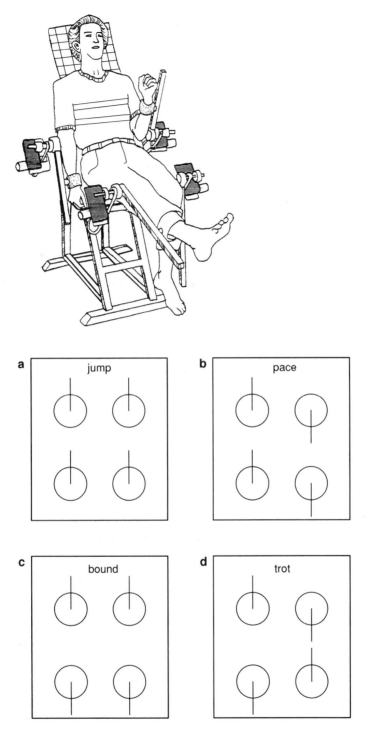

Figure 3.3 (*Top*) The multiarticulator coordination (MAC) device showing a subject in the "trot" pattern. (*Bottom*) Gait pictograms of four of the patterns studied experimentally using the MAC apparatus.[8] Each limb is represented by a circle, with a line indicating the phase of the limb at a moment in time. (a) Jump, (b) pace, (c) bound, (d) trot.

the organizational principles governing an ensemble of oscillators. The opposite is also true, of course. Properties of component oscillators affect how a multioscillator ensemble is coordinated, for example, how fast it entrains. These considerations will prove important when we delve into neural levels of organization.

Before explaining all this, let me describe the kind of experiments Jeka and I did. The apparatus, the multiarticulator coordination device (MAC), otherwise known as the big red machine, is illustrated in figure 3.3 (top) showing a subject performing a trot pattern. Note that the subject is seated, and that when I refer to the trot I am referring to the *pattern of coordination* between the arms and legs, quite independent of locomotory function. Locomotion has all kinds of other requirements (weight bearing, balance, etc.) in addition to the coordination of the forelimbs and hind limbs (arms and legs here). But here we emphasize the pattern formation aspect. You can see that the subject's legs and arms are strapped to aluminum shafts, each of which is attached to a transducer that simultaneously detects the position and velocity of each limb. Cycling movements of all four limbs are limited to flexion and extension about the elbow and knee joints. Apologies for the experimental details, but they are necessary just to understand the kinds of games we play.

In one experiment we started subjects in each of the four patterns shown in figure 3.3 (bottom), which are analogous to common but rather idealized quadrupedal gaits. In the jump (or pronk), all limbs move in phase; in the bound (or gallop) homologous limb pairs are in phase and nonhomologous limb pairs are antiphase; in the pace homologous limbs are antiphase and ipsilateral limbs are in phase; and in the trot both homologous and ipsilateral limbs are antiphase. On each trial run, starting from one of these patterns, cycling frequency was increased by requiring the subjects to synchronize a complete cycle of movement with each beat of a metronome. Subjects were instructed that should they feel the pattern begin to change they should not try to resist, only adopt whatever pattern was most comfortable.

Two kinds of transitions predominated. Subjects always switched spontaneously from trot to pace and from bound to jump. These transitions were always single step: subjects never visited any intermediate patterns (think of all the other phase relations between the limbs that they could adopt, but didn't). Jump was the sole pattern from which no transitions were observed.

Representation of four-limb patterns as a single trajectory on a three-dimensional torus supports the notion that the essential information for coordination is confined or localized to a discrete region in relative phase space. In figure 3.4 we show pace and trot two-torus plots, and an unfolded topologically equivalent two-dimensional version of the same tori. Both representations convert the position data of the four limbs into two angular coordinates $(0-2\pi)$ that are then plotted orthogonally, one as a torus and the other on a rectangular plane. Note that the limbs converge only at certain points in their trajectories, and that convergent points have unique values for each pattern, thereby allowing patterns to be distinguished by using a single graphical

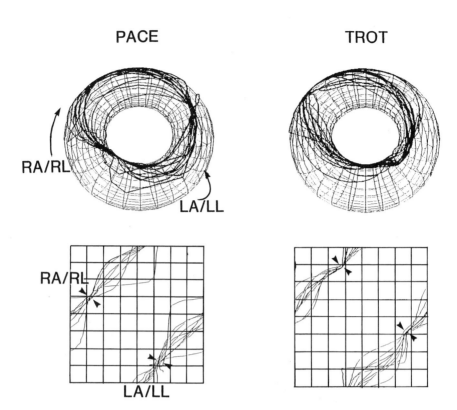

PACE **TROT**

RA/RL

LA/LL

RA/RL

LA/LL

Figure 3.4 Torus plots of two four-limb patterns, pace and trot. Below each torus is an unfolded torus with arrows pointing to places where the limbs are tightly coordinated. RA/RL refers to the right arm-leg combination, LA/LL to the left arm-leg combination.

representation. Thus, not only does relative phase prove to be an adequate collective variable, but the point estimate of relative phase is arguably a more relevant metric for coordination than continuous flow variables.[10]

Three further facts and then we'll move on. First, in the vast majority of runs in which transitions were observed, nonhomologous limb pairs changed their phase relation and homologous limbs maintained their initial phase throughout. Second, four-limb transitions occurred in such a way that limbs of the same side moved in the same direction (e.g., trot to pace, but not vice versa). We'll return to this direction effect later on. Third, guess what? Analysis of relative phase variability among the four legs invariably revealed *fluctuation enhancement* prior to the transition. Instability yet again!

What do these results mean? First, note that all four limb patterns can be classified in terms of symmetry groups.[11] Symmetry, as a colleague Guiseppe Cagliotti is fond of remarking, means no change as a result of change.[12] In fancier terms, all patterns that remain invariant under the same transformations belong to the same symmetry group, *even though they may not be equivalent dynamically* (e.g., they may differ in their stability). A transformation is an operation that maps one pattern into another. For example, in the jump pattern shown in figure 3.3 (bottom), if we interchange left and right phases of

Self-Organization of Behavior: First Steps of Generalization

each oscillatory component, exchange the phases of the arms and legs, and invert all phases (equivalent to time running backward), the jump remains invariant. If the right-left transformation is imposed upon the bound, the pattern is unchanged. We say the bound is left-right symmetric. In short, if the phasings of all limbs are equivalent after a transformation, then the pattern is considered invariant under this operation. In this way, all multilimb patterns characterized in terms of relative phases can be classified. Symmetry thus serves two roles. On the one hand, it identifies the basic patterns to be captured theoretically; on the other, it imposes restriction on the dynamics itself; that is, only certain solutions of the equations of motion are stable.

So why, for instance, does trot change to pace but not vice versa? The fact that at transitions homologous limbs change but nonhomologous ones do not suggests that the coupling between the former is significantly stronger than the coupling between the latter. There is other evidence to this effect too. For example, in earlier work on two limbs, we found that the uncoupled (comfort) frequencies of homologous limb pairs (right and left arms or right and left legs) are approximately the same, whereas the uncoupled frequencies of non-homologous limb pairs (an arm and a leg) differ. Moreover, transitions do not occur in homologous two- and four-limb combinations. Thus, although left-right symmetry appears to be preserved, arm-leg symmetry is broken. Transitions from trot to pace and from bound to jump take the form of *symmetry-breaking* bifurcations. Whereas all four patterns are stationary solutions of the model relative phase dynamics, due to differential coupling among the limbs, not all patterns of coordination are equally stable, *even though they belong to the same symmetry group.*

Here again we see that the same theoretical concepts—order parameters, symmetry-breaking instability, control parameters, multistability, critical fluctuations (and also, by the way, critical slowing down as seen by perturbing one of the limbs)—apply to an entirely different system. The same synergetic principles of self-organization are at work, quite independent of the details. Once again, pattern formation and change are seen to take the form of a nonequilibrium phase transition. Once again, different patterns and pattern complexity are seen to arise from the same nonlinear dynamical structure.

The role of symmetries and broken or reduced symmetry in pattern generation can hardly be overstated. For example, it is not unreasonable to assume that the kinds of spatiotemporal patterns we see here are produced by central pattern generators (CPGs), local or distributed neural circuits located in the central nervous system (see chapter 8). Investigators of CPGs have begun to realize just how complicated even so-called simple networks of relatively few neurons are. Although detailed models of CPGs are an area of active research, the numerous cellular, synaptic, and network properties thought necessary to describe a CPG present a daunting challenge. Even when you know all the anatomical details of a network and which neurons talk to which, this is still not sufficient to understand the CPG. Years ago, the mathematical biologist Leon Glass (and before him W. Ross Ashby) analyzed a system of only four neurons, analogous to our arms and legs, and showed with simple excitatory and inhibitory connections between the neurons that the system could generate cyclic activity through any one out of 256 possible network structures.[13]

In our multilimb experiments we see that general principles are at work that limit the possibilities. The exact realization of these principles depends to some degree on details, but the principles are structure independent. Symmetry, for example, is more an expression of an organizational principle than a physiological property of the neural substrate per se. We can expect symmetries to apply just as much to networks of neural oscillators as they do to the cooperative dynamics among the limbs.[14] Symmetries allow us to address the question of what function-specific constraints the nervous system can impose. A certain symmetry may hold for one biological function, whereas in another the symmetry is broken. Returning to the issue of generalization and Professor Beek, we can now say rather concretely that pattern change in a multilimb system is brought about by an unspecific control parameter, and that such change takes the form of a nonequilibrium phase transition in which loss of stability is the basic mechanism. Is this proof enough to "charm the intellect"? If not, there's more.

Coordination Dynamics Within a Limb

So far, theory and experiment have dealt with coordination between and among components, the hands, the legs, or arms and legs. But the arm itself is a multijointed instrument. Can coordination dynamics be identified for multijoint single-limb behavioral patterns? Are the latter also subject to laws of self-organization? To find out, a paradigm must be invented that allows for the analysis of instabilities, thereby affording a distinction between patterns and the identification of putative collective variables. In other words, we somehow have to implement the synergetic strategy for single-limb coordination.

Taking our inspiration from some early work by Russian scientists Kots and Syrovegin, Steve Wallace, a kinesiologist and cognitive scientist, John Buchanan, and I studied rhythmic multijoint arm movements under four different conditions[15]: flex (extend) the elbow and flex (extend) the wrist together, an in-phase pattern; and flex the elbow while extending the wrist and vice versa, an antiphase pattern (a bit like one of those dancers from Bali or a waiter holding a plate). These two tasks were performed either with the forearm pronated (palm facing down) or supinated (palm facing up). As in previous work, the movements were performed at a slow rhythm that was then systematically sped up. Modern motion analysis technology and computer methods make it possible to implement the experimental strategy of changing a control parameter continuously and studying how coordination patterns evolve and change. (Erich von Holst, the behavioral physiologist working in the 1930s and 1940s, reported that his recordings of fish swimming occupied a total length of about two miles! Our experiments typically involve many more miles, but a lot less paper since the data are digitized and stored inside a computer.)

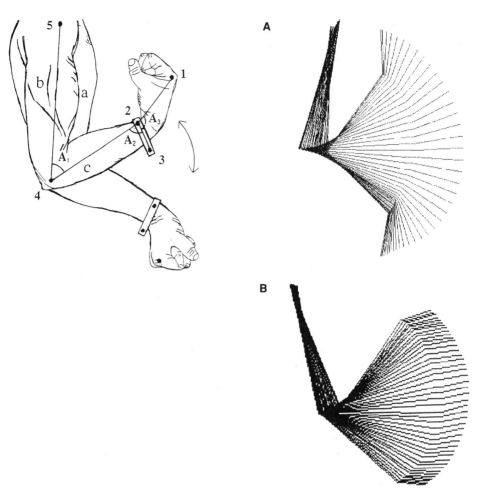

Figure 3.5 (*Left*) Picture showing placement of sensors (1–5) and EMG electrodes (a–c) used to detect movement and neuromuscular activity in Kelso, Buchanan, and Wallace's experiment. Sensors 2 and 3 were mounted on the ends of a small plastic bar attached to the subject's wrist to calculate wrist angle. (*Right*) Geometry of motion for the (A) in-phase and (B) antiphase patterns. Only one-half cycle is shown.

The left part of figure 3.5 provides a graphic depiction of the recording set-up showing where infrared light-emitting diodes (IREDs) were placed on the subject's joints. The right-hand portion nicely illustrates the patterns created by forming vectors among the IRED points placed on the shoulder, elbow, wrist, and hand. One-half cycle is shown for each pattern. As before, subjects were told that should they feel the coordination pattern begin to change, not to resist but rather concentrate on maintaining a one-to-one frequency relation with the auditory metronome.

We discovered, I think for the first time, a phase transition in single-limb multijoint movements. Once again, relative phase proved to be an adequate collective variable, successfully capturing the orderly spatiotemporal patterning between the joints. Both in-phase and antiphase relations between the elbow and wrist were stable before the transition point. As cycling frequency increased, however, an abrupt shift from antiphase to in-phase patterning occurred. No such transitions were seen when the joint movements were begun in phase. Most significantly, the collective variable or order parameter, relative phase, exhibited critical fluctuations indicating that switching was due to loss of stability. There is no better evidence that self-organizing processes in the nervous system underlie coordinative changes than fluctuation enhancement, the hallmark of a nonequilibrium phase transition.

Further Control Parameters?

An intriguing possibility is that HKB coordination dynamics derived for the two hands works also for the single multijoint limb. The two cases differ, of course, in many ways, notably in the brain mechanisms involved. For example, to a coarse approximation, most of the brain structures for controlling a single limb are located in the hemisphere contralateral to the limb, whereas coordination between the hands involves both hemispheres and associated collosal structures. Regardless, the relevant quantities of coordination and the way these quantities relate dynamically are pretty much identical. The same dynamical law appears to work for both.

An interesting novel feature of single, multijoint limb movements that I've not mentioned adds a little bit of spice to the picture. It turns out that not only is the stability of limb coordination dependent on how fast the system is driven, but the *posture* of the limb—a spatial parameter—also plays a role. We know this because when the limb is in the supine position (palm up), antiphase coordination switches to in-phase coordination when the motion is sped up. In the prone position when the hand and forearm are turned over, the opposite is true. What seems to determine the stability of coordination is the *direction* of joint rotation, not some preferred combination of muscles.

Remember, this is also what Jeka and I found in our arm-leg coordination work. Independently, Italian neurophysiologist Baldissera and his colleagues noted the same direction-dependent constraint.[16] Such findings suggest that when the many subsystems are assembled by the central nervous system into coordinated patterns of behavior, a higher priority is placed on direction of movement than particular muscle groupings. The cerebral cortex, it seems, does not plan its actions on the level of muscles. The self-assembly process appears to be spatially determined and hence far more abstract than the language of muscles.

John Buchanan and I decided to manipulate the spatial parameter of forearm rotation directly.[17] Starting with the forearm supine, subjects rotated the arm to a prone position in nine steps of 20 degrees, producing fifteen cycles of wrist and elbow flexion-extension on each step. Similarly, starting with the forearm prone, subjects rotated the forearm through the same step sequence but now in descending order. In the prone condition, they began with the "easy" pattern: wrist extension synchronized with elbow flexion and vice versa. Note that the two initial conditions in this experiment are both stable coordination patterns in the sense that they employ different muscle pairings but generate motion in the same direction. Thus, we encountered a situation very different from previous research that usually varied frequency or speed of motion as a control parameter and observed spontaneous transitions from intrinsically less stable to more stable coordination patterns. Much to our delight, when we manipulated a spatial parameter, forearm posture, from prone to supine and vice versa, we found a spontaneous transition from one stable coordinative pattern to another. This new transition was *bidirectional*; it occurred regardless of whether the motion was initiated with the forearm in either the prone or supine position. Where the patterns switched depended strongly on the direction of forearm angle change, a large hysteresis effect. Highly significant fluctuation enhancement was observed prior to the transition, regardless of whether forearm angle varied from supine to prone or from prone to supine.

Two further lines of evidence allowed us to converge on an instability mechanism for coordinative change in single-limb trajectories. Because every shift in forearm angle *perturbed* the continuing coordinative pattern, it was possible to introduce a measure that we called *settling time*, the time it takes to return to a stable pattern after a parameter change. If spontaneous switching is due to instability of motion, we reasoned that settling time should increase as the system nears the transition point. Similarly, near an instability any perturbation should kick the system from one attractor into another. In other words, the likelihood of perturbation-induced transitions increases as the system approaches criticality. (All this is terribly intuitive: a stable system recovers quickly when it is perturbed, but the same perturbation applied to a system near the brink takes a lot longer to recover from and may produce dramatic qualitative effects. See chapter 2.) Both measures—settling times and perturbation-induced transitions—test the *critical slowing down* prediction of theories of self-organization in nonequilibrium systems. It is remarkable that both measures behaved as predicted.

How do these results affect current work on trajectory formation in human movement and the design of flexible robots? What neural processes are responsible for sustaining coordination between the joints? What physiological mechanisms underlie direction-dependent transitions? Which parts of the central nervous system are involved? These are all questions that we've addressed in the technical literature and, of course, the answers are important in their own right. But that's not the main goal here. Rather it is to bolster the emerging paradigm of self-organization for understanding biology and behavior by showing, in a variety of contexts, how complex, multivariable systems exhibit low-dimensional dynamics, especially near instabilities. Here a single collective variable and a couple of control parameters are sufficient to capture quite a variety of multijoint limb coordination patterns produced by the nervous system. It's hard to resist hypothesizing that this is nature's way of solving the degrees of freedom problem in complex biological systems. We see, albeit dimly, how lawful constraints on biologically relevant degrees of freedom allow diversity of behavioral pattern. To quote the composer Igor Stravinsky:

The more constraints one imposes, the more one frees one's self of the chains that shackle the spirit.[18]

Of course, the constraints that I am talking about are not of the rigid kind. Rather they refer to task-dependent dynamical laws, equations of motion that are intrinsically nonlinear, and thus capable of generating considerable co-ordinative complexity.

Spontaneous Recruitment/Annihilation of Task Components

All the examples I've treated thus far deal with bifurcations within systems whose components are already active. Organization among these compo-nents, as in gait transitions, turns out to be a critical phenomenon. Transitions are always of the order-to-order type: the same biomechanical components are spontaneously reordered at a critical value of a control parameter. But how are previously quiescent biomechanical degrees of freedom (df) spontaneously activated according to task demands? An essential feature of complex biologi-cal systems is their ability to engage and disengage appropriate components according to functional requirements. For example, when I reach for a cup of coffee that is close by it may only be necessary to extend my elbow and shoulder. A cup located farther away, however, may require forward lean of my trunk or even, if I am seated, rising from the chair (a transition from a three- to two-point stance, the word "point" being loosely defined; maybe a better description is from a one-bottom to a two-foot stance!). Such flexible and spontaneous recruitment and annihilation of df is accomplished effort-lessly by human beings and animals, but not by robotic devices. And, to be quite frank, scientists studying how movements are controlled have hardly tackled the issue, possibly because it seems too complicated. The problem strikes me, however, as reminiscent of one of the most fundamental processes in dissipative dynamical systems described in chapter 1, namely, the simulta-neous creation and destruction of states as a parameter is varied.

To pin this idea down, let's build on a serendipitous phenomenon that John Scholz and I first observed ten years ago.[19] For reasons more mundane than the present topic (a desire to obtain better measures of EMG activity from finger muscles) we studied repetitive bimanual motion in the (x-y) plane. In other words, instead of confining motion to the horizontal plane alone, as I had done in my original work, we allowed the fingers to move in both the horizontal and vertical planes of motion. The serendipity arises from the fact that we observed new bifurcations. Specifically, after the typical antiphase to in-phase transition in the horizontal (x) plane, we found that when cycling frequency was increased still further, yet another transition emerged, this time from the horizontal to the vertical (y) plane! Occasionally, the motion of the fingers became rotary before the transition to the vertical plane.

Self-Organization of Behavior: First Steps of Generalization

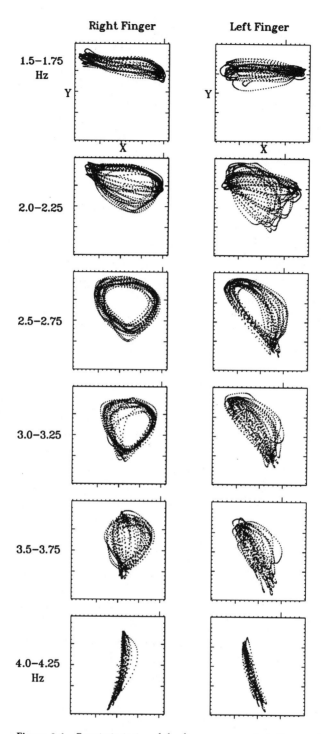

Figure 3.6 Raw trajectories of the finger movements in the x-y plane. In this example, the initial condition is horizontal in-phase coordination, and the frequency of motion increases from top to bottom. Note the transition from horizontal to rotary to vertical motion as frequency (in Hz) is increased.

More recently, my colleagues and I designed and conducted more detailed experiments.[20] The raw trajectories of the two fingers recorded in this newer research are shown in figure 3.6 and illustrate the variety of transitions within and across planes of motion as cycling frequency is increased. What's going on here? Qualitatively speaking, when motion is restricted to the horizontal plane, only the symmetric and antisymmetric modes of coordination are stably performed over a range of frequencies. That much is known and well understood. For frequencies above this range, however, no comparably stable pattern is available on the horizontal plane of motion. The system consisting of an already active set of components that have been self-organized for particular movement patterns is now no longer able to support such behavior. To achieve stability, previously quiescent df (i.e., in the vertical dimension) are spontaneously recruited. With the availability of these added df the system can perform stably again. Note the new movement pattern may still be topologically equivalent to the previous one—both may be limit cycles, for example—but additional df are required to perform the task.

What generic mechanism underlies this kind of self-organization? We propose that these *spatial transitions*, first the recruitment of previously quiescent df in the vertical dimension and then the annihilation of horizontal oscillation, are the result of two consecutive Hopf bifurcations. A similar mechanism was proposed by Ian Stewart and Jim Collins to model the transition, for example, from standing to walking in quadrupedal gait.[21] The essential difference between their theory and ours is that we use bifurcations as a vehicle to maximize the stability (i.e., reduce fluctuations) of the performed pattern while fulfilling task requirements stipulated by the environment. In the case of gait changes, bifurcations provide a mechanism that converts one functional state to another. Bifurcations in coupled oscillator models of gaits pertain to an *already active* set of dynamical variables, but don't speak to the self-organized, task-related recruitment (and annihilation) of biomechanical df.

What's a Hopf bifurcation? Thus far I've only discussed the pitchfork bifurcation in the context of coordinative behavior. Although I didn't mention it, the results discussed earlier of single, multijoint limb movement transitions, in which one *stable* coordination pattern switches to another *stable* pattern at a critical value of spatial orientation, corresponds to a *transcritical* bifurcation. In transcritical bifurcations, two stable solutions coexist, and an exchange of stability occurs at the bifurcation point.

The Hopf bifurcation, named after German mathematical physicist Eberhard Hopf, one of the pioneers in the field of dynamical systems, is the complex equivalent of the pitchfork and comes in two flavors.[22] The *supercritical* Hopf bifurcation, or *soft excitation*, refers to the fact that a limit cycle attractor (a circle on the x-y plane) emerges gradually from a stable fixed point that loses stability as a parameter is gradually varied. A diagram of the Hopf bifurcation and the resulting trajectories are shown in figure 3.7. For initial conditions inside and outside the solid circle the system generates rhythmic oscillations. The other flavor of Hopf is called a *subcritical* bifurcation, or *hard excitation*

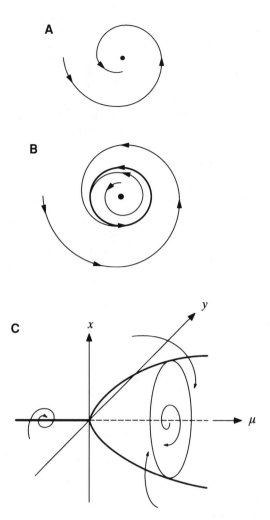

Figure 3.7 The Hopf bifurcation. (A) Fixed point attractor at the origin exhibits a limit cycle (B) at a critical value of the control parameter, μ. (C) Diagram of the Hopf bifurcation. Arrows show lines of force of the vector field.

transition to oscillation. That term refers to the fact that the limit cycle attractor appears suddenly with finite amplitude at a critical parameter value.

How do Hopf bifurcations help explain our results? Sparing the mathematical details, consider the motion of just a single finger on the x-y plane. The dynamics in both x and y directions can again be modeled by nonlinear oscillators controlled by parameters such as the movement rate. For a slow rate, the trajectory is confined to the x direction in the x-y plane. Thus, x motion is oscillatory, and y motion barely fluctuates around a stable fixed point. As rate increases, the stable fixed point at y becomes unstable, and a limit cycle appears giving rise to oscillatory y motion. This is the first Hopf bifurcation. The trajectory forms an ellipse whose orientation in the x-y plane is determined by the relative phase between x and y oscillations. As speed

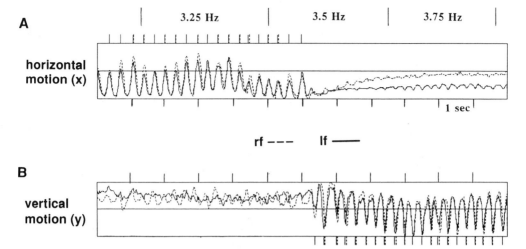

Figure 3.8 Experimental records of (*A*) the *x* and (*B*) *y* components of the left and right index fingers showing the damping of horizontal motion and subsequent recruitment of vertical motion as the cycling frequency is increased. In this case, the transition is from horizontal in-phase to vertical in-phase coordination.

increases further, the oscillation in the *x* direction disappears. This is the second *inverted* Hopf bifurcation: motion is now confined to the *y* direction.

About half of the time oscillations are created or annihilated through supercritical Hopf bifurcations, their amplitude increasing gradually from zero or decreasing gradually to zero. Thus, rotary motion is observed. But in the other 50% of the trials, abrupt transitions from horizontal to vertical motion occur, typical of subcritical Hopf bifurcations. Figure 3.8 shows an experimental time series of the *x* and *y* components of both index fingers. It is quite clear that damping of horizontal motion and growth of vertical motion occur simultaneously, exactly as the Hopf mechanism predicts.

The Hopf mechanism offers an intriguing way to spontaneously create and dissolve dynamic patterns of behavior. It is potentially important because this recruitment-annihilation process appears to be one way the nervous system achieves stability while flexibly adapting to environmental requirements.

Some Applications: Robotics and Machine Pattern Recognition

A frequently expressed view is that the fields of robotics and pattern recognition by machines can benefit from studies of biological function. As a short digression, let's consider the implications of our findings thus far for these more applied fields. Robots, of course, are constructed by engineers, to perform specific tasks and functions. They are *organized* rather than self-organized devices. Excess or redundant degrees of freedom induce enormous computational complexity for the roboticist. In other words they are a big headache. Yet it's clear that this very redundancy provides an essential source of versatility. The irony is that whereas the designer is faced with computing, say, the joint angles and forces (torques) to generate all desired trajectories of the robot arm, biological systems handle the problem effortlessly by typically choosing just one.

One thing that might be helpful to reduce the redundancy problem in robots is to incorporate the kinds of dynamical constraints (e.g., phase relations) that biological systems employ into the equations of motion that are used to control robot arms. When this is done, the mapping between a given hand trajectory and the joint angles of the arm may no longer be redundant and "ill posed." The so-called inverse problem of robotics—given the x, y, and z coordinates of the end effector over time, determine the time course of the joint angles—effectively disappears.

Robotics research has long recognized that additional constraints must be formulated to resolve redundancy problems.[23] Nevertheless, the constraints employed so far are not of the coordination kind I am talking about. Similarly, every time a robot must switch from one coordination pattern (time course of joint angles) to another, the entire problem has to be recomputed. This is computationally expensive in terms of real time coordination needs. An alternative strategy, as we've seen, is to build nonlinearity and multistability into the coordination dynamics. Robots must necessarily be designed on the component level. By nonlinearly coupling these components, desired spatio-temporal patterns can be stabilized. Moreover, spontaneous, task-related switching and multifunctionality are natural outcomes of such coupling without the need for any pre-computation or recomputation whatsoever. Maybe.

In the previous section we saw how it is possible to construct a coordination dynamics that spontaneously recruits and/or eliminates biomechanical components according to task demands. This would be a wonderful feature to try to incorporate into robots, and may hold the key to genuine flexibility in machines. These short remarks are intended only as hints of how principles of self-organization for biological control and coordination might affect the design and control of robot manipulators.

When it comes to pattern recognition by computer, we're on firmer ground, and it's really quite exciting. I'll talk about *perception* a lot more in chapter 7, but for now I want to make a fairly simple point about the mutuality between pattern generation and pattern recognition. Figure 3.9 shows a stimulus called a point light walker. It consists of small patches of light mounted on a person's body as they perform certain actions. When observed as a static array of points, you don't see a coherent pattern; only the points are viewed, not the person. Once in motion, however, the eminent Swedish psychologist Gunnar Johansson showed that naive observers rapidly perceive the human form and what it is doing.[24] This is true even when the stimulus is masked by randomly moving dots superimposed on the target. How is this accomplished?

When you realize that a 2-D retinal (or camera) image may consist of an enormously high-dimensional array of image intensities, wavelengths, and the like, varying in time, pattern recognition by the human visual system is a kind of miracle. The intuition, of course, is that even though the retinal image

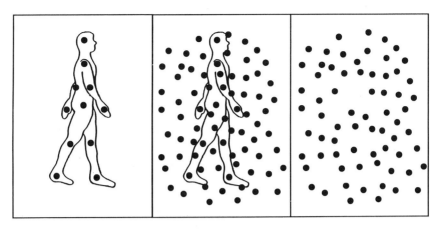

Figure 3.9 Point light walker stimulus displays created by computer (Courtesy of Bennett Bertenthal). Left panel shows the canonical target of eleven point lights moving as if attached to the major joints and head of a human being. Once in motion, observers can quickly detect that the point lights specify biological motion even when masked by randomly generated dots. Note that the observer does not see the outline of the walker shown in the left and middle panels.

seems vastly underdetermined,[25] the visual system has ways to organize elements into unitary percepts. One hypothesis, put forward in 1979 by Haken, is that this organization itself is a self-organized pattern-formation process, hence the duality between principles of pattern *formation* and pattern *recognition*.[26] If Haken is correct, we are in a unique position: because we know the relevant collective variables for coordinated motion, this gives us a low-dimensional description on which to design computer algorithms for pattern recognition of all kinds of biological motions. Some examples are shown in figure 3.10.

The very fact that we can get a computer to recognize such dynamic patterns, even when embedded in noise, suggests that we've latched on to the right order parameters. In the synergetic computer, a set of prototype vectors composed of magnitudes and relative phases for the different patterns constitutes attractors of a (multistable) dynamical system. According to the dynamics, any test pattern will be attracted to the one and only prototype to which it is most similar. Thus, when a motion pattern such as a multijoint limb trajectory is offered to the computer, it is pulled into its own attractor. Once that attractor is reached the pattern is recognized.[27] In other words, meaningful information for behavioral action and pattern recognition by people and machines may be said to lie in attractors of the order parameter dynamics. In the case of human perception, of course, the organism has to be sensitive to relative phase. From visual input, it has to self-organize the vector of relative phases as a pattern-formation process. Remarkable evidence garnered by Bennett Bertenthal and colleagues indicates such processes may be going on in humans as early as age three months.[28]

Self-Organization of Behavior: First Steps of Generalization

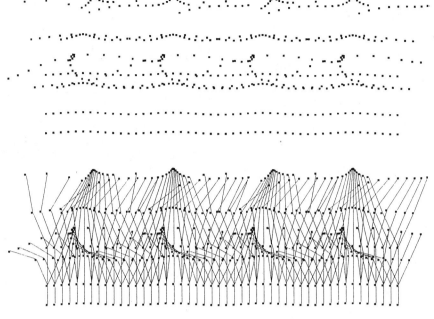

Figure 3.10 (*Top*) Point light displays presented to the synergetic computer. (*Bottom*) The computer "connects the dots" and recognizes the pattern correctly as walking.

It's a far flung connection from the structure of biological motion to the field of city planning. But, in *The Death and Life of Great American Cities*, written thirty years ago, Jane Jacobs wrote:

The leaves dropping from the trees in the autumn, the interior of an airplane engine, the entrails of a dissected rabbit, the city desk of a newspaper, all appear to be chaos without comprehension. To see complex systems of functional order as order, and not chaos, takes understanding.[29]

That's the problem of pattern recognition in a nutshell, expressed much better than I am able to. The image impinging on the retina is always undergoing transformation and change. Without an appreciation of the relevant variables characterizing the coherence of different dynamic patterns we're lost, and not just in terms of scientific understanding. That's the message I've tried to convey with these examples of biological motion. Here, unlike the city desk of a newspaper, we have some understanding about how the patterns are organized.

COORDINATION BETWEEN ORGANISMS

Speech Production and Perception

Let's change gears a bit and consider an entirely different system, the many different articulators (tongue, lips, jaw, etc.) whose motion changes the shape of the vocal tract, which in turn creates sounds that allow us to talk and listeners to understand. It takes the coordinated action of about thirty odd muscles in the tongue, the larynx, the respirators, and so forth for a baby to say "ba." As all parents know, this is a major accomplishment of nature that only pales when you compare the 40,000-odd muscles that the baby elephant must coordinate to move his trunk. Talk about the need to identify relevant collective variables in complex systems! To measure every single biomechanical degree of freedom in the human mouth and the elephant's trunk would be impossible and irrelevant.

In the case of a human speaker, somehow from a space of huge dimensionality, sound emerges as a distinctive and well-formed pattern. Of course, how this happens is a complicated business and too long a story to tell here. But a key part concerns how the articulators are coordinated. Let's take the problem a step further. When a speaker produces a word, the variation in the resulting sound is subject to many factors. Among these are the rate at which the speaker talks, the stress pattern, the accent (if he's Irish, like me, or from Brooklyn, like my wife), the influence of surrounding words and so forth. Make a long distance phone call over the usually noisy telephone system and the problem looms even larger. How do we identify the word in the face of all these acoustic variations? What holds it together? At this point, the reader is entitled to chime in: synergetic concepts of collective variables, control parameters, phase transitions, and all don't apply here, too. Or do they? If so, how do they help us understand communicative acts?

The fact that speech perception is so robust is lucky for humans, who seem to extract the message no matter what, and has sparked a long and continuing search for the source of such invariance. To be blunt, no one has found this particular holy grail. Betty Tuller and I think that part if not all of the problem rests with the entire idea of invariance in language and speech.[30]

Typically, in speech research if a variable does not change much over a range of parameter values it is assumed to be invariant. But, one may ask, invariant with respect to what? Or, how much can the measured value of a variable change before one decides that a particular candidate is or is not invariant? The whole issue of invariance in speech is plagued with measurement problems and false starts. We argue, based on empirical and theoretical grounds, that a far more appropriate concept in this case is stability. I personally think that the powerful concept of invariance should be reserved for symmetry operations (as in our earlier analysis of locomotory gaits). The reason is that fluctuations are a conceptually crucial part of the dynamics and an absolutely necessary part of the measurement process. Speech, as a dynamic process, is no exception.

If stability is a key concept for speech, what's the stability of? Consider an example discussed by the brilliant American phonetician R. H. Stetson[31] in which a subject produces the syllable "at" repetitively. As speaking rate is gradually increased, the syllable switches to "ta." Stetson thought that the need to simplify coordination caused the syllable-final consonant to become syllable-initial because the final consonant was "off-phase, out of step with the syllable movement." Tuller and I showed that the relative phasing among the movements of the articulators is the collective variable that characterizes stability and change in syllable structure. Just like the previous examples, we demonstrated that the cause of switching was due to an instability, yet another phase transition.

I won't go into all the details of our experiments, which are a bit gory. We put a fiberoptic bundle up our subjects' noses and down the back of the throat to aim a light toward the vocal cords. The light transmitted between the cords is measured by a photoresistor placed against the neck that generates a time-varying signal indicating glottal width. We monitored the air pressure inside the mouth by inserting a plastic tube into the oral cavity and connecting it to a pressure transducer. Lip movements were measured using IREDs attached to subjects' lips. The key collective variable, as I mentioned, turned out to be the relative phasing between the opening of the glottis and the closing of the lips. This was pretty stable (less variable) and took on a different value for the syllable "eep" versus "pee." At a critical speaking rate, the phase relation underlying "eep" jumped to that of "pee," which otherwise remained stable for all speaking rates.

The parallels with previous work are extraordinary. The collective variable dynamics are bistable in one parameter regime and monostable in another. The shift in pattern is a result of loss of stability and the system is again hysteretic, signatures of a pattern-forming, nonlinear dynamical system. These results and analyses put the entire invariance notion in a new light. But there's more. Tuller and I reasoned that if the relative phasing among articulators is a collective variable relevant to speech communication, then there should also be *perceptual* consequences of a change in articulator phasing. Is a shift in relative phase from "eep" to "pee" actually perceived as such? We prepared a stimulus tape and presented it to listeners through headphones. After hearing each sound, listeners were required to indicate whether they heard "eep," "pee," "eeb," "bee," or some other sound. The results revealed an intimate production-perception linkage. Most tellingly, perception switched from "eep" to "pee" at *exactly the same point* at which a jump in articulatory relative phase occurred. Changes in the *articulator* phasing structure result in sound (and meaning) changes.

So why does the vowel-consonant (VC) form shift with speaking rate but the consonant-vowel (CV) form does not? We can only speculate, but I think the possible answers are intriguing. There appears to be something special about the CV syllable in language. It is the only syllable type present in all the known languages of the world. Other forms are more sparsely distributed. Also, babies babble in CVs (ba, ma, da, etc.) not VCs, causing some linguists to refer to CVs as favorite sounds or preferred movement patterns. Whatever the reasons, there are strong hints from these results that language evolution and development take advantage of the kinds of articulatory-perceptual stabilities we identified. But there might also be a negative side to all this.

Syllabic forms that are too stable aren't easy to modify, making learning a second language a tough job, especially the older one becomes.

As a final point, the results, limited though they are and certainly a challenge for future research, nevertheless suggest that the relevant information for perception of dynamic speech patterns lies in the collective variables that characterize the pattern's generation. What inferences might we draw from this hypothesis? Harking back to our earlier discussion of (visual) pattern recognition by computer, there may also be a message for (biologically motivated) speech recognition by computer, namely to focus on those variables in the acoustic signal that are coupled specifically to relevant aspects of articulation. Articulatory gestures place constraints on what can be acoustically realized and perceived. The more we learn about these constraints the better we will understand the entire communicative act. Similarly, the sooner we can show people with communicative disorders what they are doing inside their mouths—how to coordinate their vocal tract in space and time—the better off they'll be. In particular, relative timing information appears to be crucial. New imaging techniques[32] can play a big role in helping overcome communication disorders, especially when we know the relevant variables. As we've seen, using our phase transition methodology, such collective variables encompass production, acoustics, and perception, and offer new light on the time-honored invariance issue.

Social Coordination

Not very long after discovering transitions in bimanual coordination, I used to play a game with my colleagues. Like many of the games scientists play, it eventually led to formal experiments. From these experiments, a rather profound message emerged that will allow me to bring this chapter to a close. What was this game? With a colleague in one booth and me in another, sight unseen but not sound unheard, I tapped out a rhythm with my hand. The colleague's task was to syncopate with me. "Try to produce a beat in between my taps," I requested. As in the hand movement experiments, I gradually increased the tapping rate. By now the reader will have guessed what happens. (My colleague says, "Go to hell, I've got work to do," yes? Well, actually, no.) At a certain critical rate my colleague spontaneously starts to synchronize with me. He can't help himself. It's the old antiphase to in-phase switch again.

But look what's going on: the only connection between my colleague and me is through sound, yet the dynamics appear to be the same as or similar to when a *single* individual coordinates both hands. Moreover, our game involves an admittedly primitive social interaction between two people (possessing, however, very different life histories, expectations, minds, brains, behavioral repertoires, cognitive representations, etc.). Could this game provide a framework for understanding social interactions, say between speakers and listeners, mothers and infants? Whether it does or not is a challenging

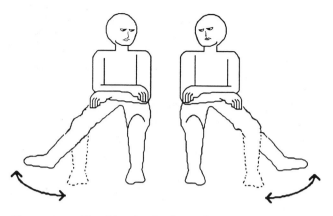

Figure 3.11 The Schmidt, Carello, and Turvey (1990) social coordination paradigm. (Adapted from reference 33).

issue that doesn't really concern us here. Rather, the game is significant because it returns us to the central problem of coordinated behavior, and once again suggests a solution in terms of collective variables and their self-organized dynamics. It begins to look as if the dynamics that describe coordination *within* an organism is sufficiently general to also cover basic interactions *between* organisms. Put another way, the solution to the degrees of freedom problem seems to be similar in kind.

Richard C. Schmidt has tackled the problem of interpersonal coordination systematically.[33] The set-up he devised is shown in figure 3.11. The idea is probably quite clear from the picture. Instead of one person oscillating, say, two fingers, two people each oscillated a leg with the common goal of coordinating these body parts antiphase or in phase. Frequency was increased by a metronome. Of course, to do the task the pair had to watch each other carefully. When they didn't, there was no coordination and no transitions. When they did, all the typical phenomena predicted by the HKB model and the theory of nonequilibrium phase transitions in general were observed. The lone exception, critical slowing down, required perturbation methods and wasn't tested.

Is this just another generalization, a little mud in the eye of the skeptic? Yes and no. Generalization is, of course, a central goal of any scientific theory. Here, we've shown that the same dynamic pattern principles apply to very different kinds of things. What matters is that the same order parameter, relative phase, captures the cooperative behavior between the parts, regardless of the parts themselves (parts of the body, of the nervous system, of society). The order parameter dynamics do not have to be exactly the same, but the common features they share are nevertheless pervasive. As elaborated so far, synergetic principles of self-organization have been shown to apply not only at different levels and among different components but for very different functions, such as speech and locomotion. They appear scale independent. Such a statement does not preclude the emergence of new phenomena at scales yet to be examined or for functions yet to be considered.

These are early days for the theory of behavioral self-organization. A limitation of the treatment thus far is that all the phenomena of self-organization demonstrated here arise due to nonspecific changes in control parameters. We have yet to incorporate specific parametric influences on the dynamics, due, for example, to intentionality, the environment, learning. This task is in front of us.

ON COUPLING

It might be said that the concepts, tools, and methodology used here are adequate for treating the coordination of the body, a new kind of behaviorism, as it were, but don't say much about the mind. This really misses the point, which is that level-independent principles of coordination in complex systems apply to bodies, brains, and minds. Aside from putting to rest the concerns of Professor Beek, stated earlier in this chapter, my goal was to elucidate the elementary form that these principles of self-organized coordination take and show how they are expressed in a variety of experimental model systems. The remarkable connection among these different systems is that they adhere to the same coordination dynamics. The key lies in what may be a fundamental coupling. That is, the original HKB coupling provides the simplest way to guarantee properties that are critical to living things: multistability, flexibility, and transitions among collective states. This is true even though the components themselves may be anatomically dissimilar and play different functional roles.

There is one further remarkable fact. In contrast to classical dynamics that deals with fundamental quantities such as mass, length, and time and their relation, coordination dynamics deals with *informational* quantities of a *relational* kind. Of significance, relative phase specifies the relation among the parts of the system, *independent* of the nature of the interactions themselves (e.g., visual in the between-person case, proprioceptive in the within-limb case, etc.). Perhaps this independence is only superficially true. Every example described in this chapter involves the nervous system: these are not entirely mindless acts. Could it be, then, that the essentially nonlinear coordination dynamics, written in the language of the order parameter, relative phase (a purely mathematical quantity that lives on the interval 0 to 2π), is one of the "new laws to be expected in the organism" that Schrödinger, Bohr, and others in the 1940s and 1950s hoped might emerge? I raise this possibility only to tantalize. Biologists are used to looking for switches—operators and regulators—that can turn genes on and off. Coordination dynamics is strange: it explains switching in open, nonequilibrium systems without any switches at all.

4 Extending the Basic Picture: Breaking Away

I play with arrangements that lie on the brink of instability to achieve lyrical motions.
—George Rickey, sculptor

Coordination and other similar function words such as integration and orchestration do enormous scientific duty in the behavioral and brain sciences. The word "coordination" seems to hold the key to everything that is going on at nearly every level of description. Although coordination is a word that often carries the onus of explanation, in my opinion it is a word that demands explanation. It trips off the tongue so easily as a way to connect things that, because of the natural tendency to focus on the things themselves, the significance of the *coordinative relations* between things is lost. One is reminded again of the words of the French mathematician Henri Poincaré, the discoverer of what we now call chaos, that "the aim of science is not things themselves, as the dogmatists in their simplicity assume, but the relations among things; outside these relations there is no reality knowable."[1] As a nonmathematician, I'd say we need both, mindful of the possibility that what is a thing at one level may be relations among (different) things at another.

The aim of this chapter is to build on and elaborate the basic picture of coordination that emerged from our hand movement experiments in chapter 2 and the various generalizations described in chapter 3. There we saw in an enormously complex living system (the human being) contraction of the dynamics to an evolution equation involving the relative phase alone. I have to admit, although I find it quite beautiful, the picture drawn in chapters 2 and 3 is a bit idealized. This may be a natural, even mandatory, step on the road to understanding.

To a large degree all scientific progress rests on idealizations. Physics uses nature's simplest atom, hydrogen, to illustrate the power of quantum mechanics. Chemistry and biology have their idealizations, too. The unit of inheritance is a very large molecule, DNA, which takes the geometrical form of the famous double helix deduced by Crick and Watson. The genome itself is far more complicated and dynamic than the static geometrical structure of DNA might lead us to believe. But none of this minimizes the significance of the hydrogen atom and DNA as means of revealing important physical and biological insights at their respective levels of description.

So too, when we come to the problem of understanding coordination in complex living things, our torus with a coordination dynamics running around inside is a very simple idealized structure. But not so simple that it lacks essential properties. Already we've seen the importance of concepts such as symmetry (in determining basic modes of coordination), multistability (the coexistence of several coordination modes for the same parameter value), hysteresis, switching between modes at critical points, transition pathways, and all those predicted phenomena associated with fluctuations and dynamic instabilities. They allowed us to arrive at a primitive nonlinear structure for coordinated behavior. Now it is time to expand it.

To take the next step, imagine the following situation. An adult is walking along the beach with a small child. The two are not physically coupled (though they might be if they held hands), and they are not necessarily biologically coupled (though they probably are). Let's say, rather, that they are informationally coupled. Perhaps they are talking to each other, or one is telling the other a story. To remain together (coordinated), either one or both must adjust their step frequency and/or stride length. Unlike experienced lovers and dancing couples, synchronization is difficult unless one or both spontaneously adjust to the circumstances. For example, to keep up, the child may sometimes skip a step or the adult slow down just so the two can remain together. This form of coordination is far more variable, plastic, and fluid than pure phase locking. Certainly, *tendencies* toward phase and frequency synchronization are still present, but sometimes the phase slips before it is reset again to some regular rhythm.

How can we go about understanding this less rigid, more flexible form of coordination? The brilliant German physiologist Erich von Holst (figure 4.1) coined the term *relative coordination* for this kind of behavior:

Relative coordination is a kind of neural cooperation that renders visible the operative forces of the central nervous system that would otherwise remain invisible.[2]

In relative coordination, an *attraction* to certain phase relations among coordinating components may exist (which von Holst called the magnet or M effect) but it is offset by differences between the components themselves. In other words, individual biological components possess intrinsic properties that tend to persist even when the components are coordinating with each other (von Holst called this the maintenance tendency). Thus, relative coordination was the outcome of a latent and never-ending struggle between maintenance and magnet effects.

Given this beautiful but somewhat qualitative description, I believe that the phenomena von Holst discovered are enormously important in at least two senses. First, in modern-day jargon, relative coordination falls under the heading of correlated neuronal activity, which is suspected of playing a highly

Figure 4.1 Eric von Holst (1908–1962) in a picture taken in 1958. Von Holst was one of the most original systems physiologists and was a friend of K. Lorenz and the physicist W. Heisenberg. (Reprinted with permission from MIT Press)

influential organizing role in functions such as perception, memory, and learning, as well as the development and plasticity of structural-functional linkages in the nervous system.[3] Striking coordinative relations also exist within and among cardiovascular, respiratory, and vegetative functions.[4] Second, and more important at present, is that relative coordination suggests that biological systems have access to other dynamical mechanisms (beyond bifurcations or phase transitions) for going in and out of coherent states.

I connect the phenomenon of relative coordination to a generic feature of dynamical systems called *intermittency*. The difference between intermittency and phase transitions is that the phase transition mechanism uses an active process (a parameter change or fluctuation) to switch the system from one stable state to another. The intermittency mechanism does not. Rather, the system is poised *near* critical points where it can spontaneously switch in and out. Strictly speaking, in the intermittent regime it no longer possesses any stable states at all.

In this chapter I will show how intermittency may be built into the elementary laws of coordination. A great advantage of living near but not in phase-locked states is that the system is (meta)stable and flexible *at the same time*. I'll explain what this means shortly. However, before we get too far ahead of ourselves, let me explain what relative coordination is in a bit more detail. Then I'll show, by way of a specific experimental example, how to model it.

There won't be much fancy mathematics here; only a way of looking at old phenomena from a new point of view.

RELATIVE COORDINATION

The time series displayed in figure 4.2 show the essential difference between absolutely and relatively coordinated behavior. The former, rigid and machine-like, involves phase relationships that are constant. The corresponding histogram has a single large peak concentrated at a single relative phase. Relative coordination, in contrast, displays *all* possible relative phase values in the S-S interval spanning 360 degrees ($0-2\pi$ radians) even though a common phase characteristic of absolute coordination is still present. I have drawn a box around the relative coordination time series to show how easy it is to make a mistake and confuse the two classes of coordination. (One always has to ward against the unconscious tendency to display the data that look the most orderly.) It would be quite easy, for example, to cut up the time series, as I have, and claim that mode locking or absolute coordination (phase and frequency synchronization) is observed!

But relative coordination is not mode locking. In particular, inside the box we see a *progressive* and *systematic* slippage in the phase relation between the components, and then when we increase the time scale of observation, the insertion of an extra step to keep the components together. It's just like the adult and child walking together. Rigid entrainment (absolute coordination), which reflects asymptotic convergence to a mode-locked attractor, may be more a feature of biological oscillator *models* than of reality.

(Broken) Symmetry Again

How ubiquitous is relative coordination in natural systems? The list is likely to be endless, in large part because the phenomenon is possible in any system containing two or more components whose frequencies couple nonlinearly. Yet by far the greatest attention has been given to absolute coordination or frequency locking, which was discovered over 300 years ago by Dutch physicist Christian Huygens when he noted that two pendulum clocks placed near each other tended to synchronize due to tiny coupling forces transmitted by vibrations in the wall on which they hung. Physicists James Glazier and Albert Libchaber have provided a representative list of the "almost bewilderingly common" occurrences of frequency locking in the natural world.[5] It includes in mechanics, the damped driven pendulum; in hydrodynamics, the vortices behind an obstacle in a wind tunnel or an airplane wing, the dripping of a faucet, our familiar convective rolls, and the oscillations of acoustically driven helium; in solid-state physics, Josephson junctions and oscillations in other materials; in chemistry, the Belousov-Zhabotinsky and many other reactions; and in biology, cardiac rhythms, brain rhythms, slime molds, and menstrual cycles.[6] To this one could add electrically stimulated giant squid axons, neural

a   **Relative Coordination**

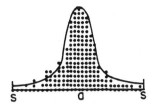

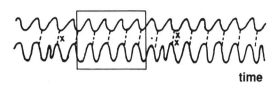

time

b   **Absolute Coordination**

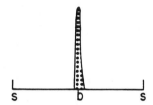

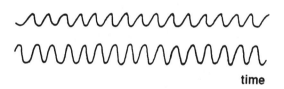

time

Figure 4.2 (a) Relative coordination. (*Top*) The distribution of possible phase relations between two signals. (*Bottom*) The corresponding time series from which the phase relation is extracted. The distance from S to S on the abscissa spans the phase interval 0 to 2π rad. The signals come from pectoral (upper time series) and dorsal (lower time series) fin movements of a fish. (b) Absolute coordination. Distribution of phase relations and corresponding time series as described in (a). (Adapted from reference 2.)

and muscle membrane oscillations, locomotor-respiratory rhythms, speech-hand movement patterns, and cell populations in primate and cat visual, auditory, and sensorimotor cortex, to name just a few examples in neuroscience and psychology.[7]

But how is this more plastic and fluid form of organization called relative coordination to be understood? The basic reason for relative (rather than absolute) coordination is that the component parts of complex biological

Extending the Basic Picture: Breaking Away

systems are seldom identical, thereby introducing *broken symmetry* into coordination dynamics. Nature thrives on broken symmetry for its diversity, and coordination it turns out, is no exception. As we'll see, any influence that causes the components of the system to differ is a potential source of symmetry breaking. Handedness and hemispheric differences in the brain are obvious examples.

Coordination often occurs between different structural components, as in the case of speech, for example, or between the same components put together for different functions (e.g., playing the piano versus playing the flute). In general, symmetry breaking occurs when different (neuro)anatomical structures, each possessing a different intrinsic frequency, must to be coordinated. Alternatively, task requirements may dictate that some response must be coordinated in a particular fashion with an environmental event. Just this situation provided an experimental insight that enabled my colleagues and me to understand relative coordination as a consequence of broken symmetry in the coordination dynamics, and to formulate the corresponding law.[8]

Action-Perception Patterns: An Example

I once described this experiment to a professor of music who told me that I had (re)discovered a test used at the Juilliard School in New York City to evaluate musical talent. I have no idea whether this is true or not. According to my source, sometimes it is difficult to decide between two prospective pupils so the following test is conducted. "Clap *between* the metronome beats," the student is instructed, as the teacher turns the knob on the metronome, making it go faster. The one who keeps out of time the longest wins!

In our experiment the task for the subject was to synchronize peak flexion of the index finger with a metronome in two modes of coordination: on the beat and off the beat. For each mode, the pacing frequency was monotonically increased (or, in another condition, decreased). Figure 4.3 shows representative plots of the relative phase between the metronome and the hand for the off the beat, syncopated condition. Several different kinds of patterns are observed.

The subject shown in figure 4.3A would win no prizes at Juilliard. Beginning in syncopation (strict syncopation would fall on the dotted line, i.e., relative phase $= \pm \pi$), spontaneous switching to synchronization (relative phase $= 0$) occurs near the end of the first plateau. He can't keep time for very long, at least, off the beat. After that, the subject sustains synchronization (on the beat, in phase with the metronome) for the remainder of the run. The time series shown in figure 4.3B is similar: here again there is a transition from one mode to the other, but at higher frequencies a new phenomenon occurs. Synchronization is lost and the relative phase "wraps" continuously in the interval between 0 and 2π radians. This means that the subject can no longer maintain a one-to-one relation with the metronome. In figure 4.3C the subject keeps the syncopation mode very stably until at higher frequencies the relative phase

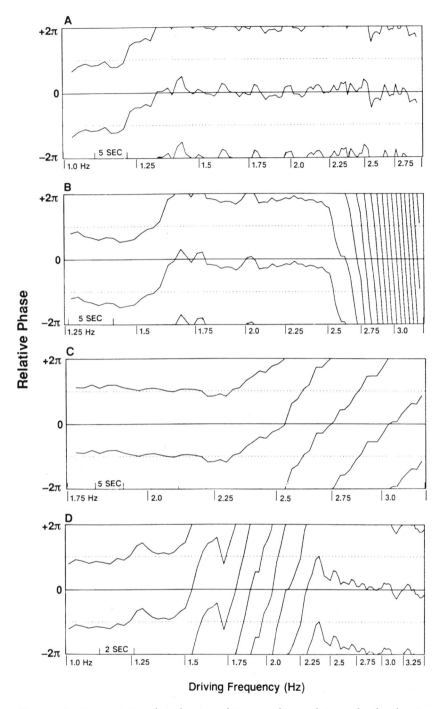

Figure 4.3 Representative plots showing relative coordination between hand and metronome in the Juilliard experiment of Kelso, Del Colle, and Schöner.[8] Relative phase is normalized to the unit interval $[0, 2\pi]$ and the plot is duplicated in the interval $[-2\pi, 0]$. The tendency for phase attraction persists even though the oscillatory components exhibit different periodicities. The characteristic features are interspersed intervals of strict mode lockings, occasional phase wandering, and longer epochs of phase wandering as metronome frequency is increased.

Extending the Basic Picture: Breaking Away

starts to slip. Notice that this is a *directed*, not random, drift with brief pockets (the little flat parts) of *nearly* mode-locked behavior. Finally, in figure 4.3D we observe syncopation followed by slow drift and eventual synchronization.

The phenomena shown in figure 4.3B through D are typical of relative coordination or what a dynamicist would call loss of entrainment (phase and frequency desynchronization). The tendency is for phase attraction to persist even though the oscillatory components exhibit different periodicities. Typical features associated with desynchronization are interspersed intervals of nearly mode-locked behavior and occasional phase wandering, with longer epochs of phase wandering as the system is forced to coordinate itself at higher movement frequencies.

One's belief in the scientific method is enhanced when a given set of findings is confirmed or replicated by others, especially if it happens to be in a slightly different paradigm. In this case, Wimmers, Beek, and Van Wieringen used a visual tracking task rather than an auditory-motor task, but found the same basic transitions as reported here, including critical fluctuations. They didn't, however, look for any relative coordination effects.[9]

RELATIVE COORDINATION EXPLAINED

By now the reader will have noted the similarity among coordination of hand movements, coordination between people, and coordinating one's action with an environmental event. Comparing these cases one can appreciate that biological coordination is essentially a synthetic process; it deals with how the components are coupled together independent of the material structure of the components themselves and the physical nature of the coupling. This is not to say that the specific components do not constrain or shape coordinative patterns and the pattern dynamics. In the action-perception case, metronome and limb are obviously different components. As a consequence, we can no longer assume symmetry of the dynamics under the operation $\phi \rightarrow -\phi$. The fact is that any situation that creates differences between the interacting elements is a potential source of symmetry breaking.[10] Moreover, as we will see in later chapters different functional requirements arising, for example, due to environmental demands, intentionality and learning (chapters 5 and 6) break the symmetry of the relative phase dynamics.

To accommodate relative coordination effects in the action-perception example requires only a single, apparently trivial change in the HKB symmetric coordination law described in chapter 2. The manifold consequences of this symmetry-breaking step, however, turn out to be subtle and nontrivial. The sole change we have to make is to include a term, $\delta\omega$, that takes into account intrinsic differences between the frequency that the moving limb generates spontaneously (call it ω_0) and the metronome frequency (call it ω).[11] This term is derived, of course, from a detailed (coupled) oscillator analysis, which the reader is spared here. When uncoupled, the individual components behave according to their respective (natural) frequencies. If $\delta\omega$ is the frequency difference, the rate of phase change between the components is just this difference. When coupled, the modified, symmetry-breaking version of the phase dynamics reads as[12]

$$\dot{\phi} = \delta\omega - a\sin\phi - 2b\sin 2\phi.$$

It is easy to see that if $\delta\omega = 0$, this is just the original symmetric-coordination law. For small values of $\delta\omega$ one can obtain phase locking as before, but the fixed points are slightly shifted away from the pure in-phase and antiphase patterns of coordination.

Our new equation corresponds to motion in a potential

$$V(\phi) = -\phi\delta\omega - a\cos\phi - b\cos 2\phi,$$

which is plotted in figure 4.4. (The minus signs allow us to interpret the potential as a landscape with attractor states at the minima for positive $\delta\omega$, a and b.) This picture and others that follow tell the whole story.

The effect of the first (linear) term is to tilt the whole curve along the line $V = -\phi\delta\omega$. For a given movement frequency, as $\delta\omega$ is increased, there is a point at which the curve loses its stable fixed points (minima of the potential), the system is no longer phase locked—synchronization is lost, and running or wrapping solutions predominate. The exact point at which the detuning

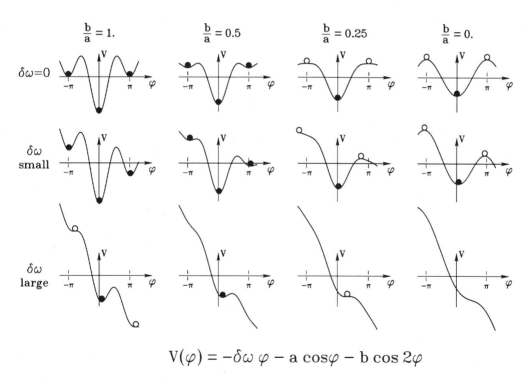

$$V(\varphi) = -\delta\omega\,\varphi - a\cos\varphi - b\cos 2\varphi$$

Figure 4.4   The potential, $V(\phi)$, of the coordination dynamics with broken symmetry. The region around each local minimum acts like a well that weakly traps the system into a coordinated state (see text). Black balls correspond to stable minima of the potential, white balls symbolize unstable states.

Extending the Basic Picture: Breaking Away

parameter, $\delta\omega$, causes the running solution to appear, and whether it appears at all, depends on the other parameters, a and b. This simply reflects the fact that *both* the frequency difference, $\delta\omega$, and the basic frequency of coordination affect the onset of the running solution.

The reader will remember from our previous analysis that the ratio b/a expresses the relative importance of the intrinsic phase attractive states at $\phi = 0$ and $\phi = \pi$, or 180 degrees. Note that in figure 4.4 the local minima serve to trap the system into one of the intrinsic phase states, depending on the initial condition. Then as parameters are changed, only the coordination mode near $\phi = 0$ remains, until eventually, when even that localized well becomes shallow (figure 4.4, bottom), the system escapes and runs. This behavior is especially interesting because even though all the stable fixed points are gone, *remnants* of the minima remain. One can imagine the system residing in these valleys for variable times before continuing to run depending on the curvature of this washboard potential.

Figure 4.5 shows the behavior of the relative phase in time corresponding to the potential pictures of figure 4.4. The running solutions have a fine

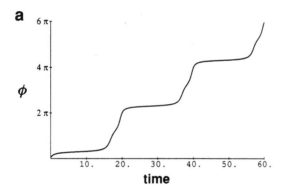

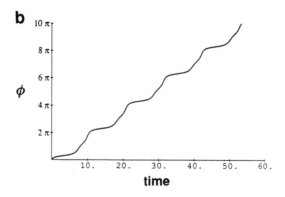

Figure 4.5 Running solutions in which there is still a tendency to maintain preferred phase relations. (a) The system is closer to a stable fixed point (mode-locked state) and therefore stays longer than in (b), where the plateaus are of shorter duration.

structure, spending more time at those relative phase values where the force, the derivative of ϕ with respect to time ($\dot{\phi}$) is minimal. This reproduces exactly the phenomenon of relative coordination. Although there is no strict mode locking and the system is nonstationary, it displays a partial form of coordination *in between* rigid mode locking and completely uncoordinated behavior. The reason there is a longer flat portion in figure 4.5a than figure 4.5b is that the system is closer to the stable fixed point near $\phi = 0$, 2π, 4π, and so on in the former than the latter. The closer the system is to the fixed point, the longer it hangs around. And the reason, of course, that the system runs (ϕ ever increasing) is that the component frequencies are no longer the same. Eventually, one component takes an extra step, and then both are nearly, but never totally, locked again.

The Bifurcation Structure

The phenomena of relative coordination can readily be understood in dynamic language. Consider the plot of $\dot{\phi}(d\phi/dt)$ versus ϕ (figure 4.6) for a fixed frequency difference, $\delta\omega$, as the coupling ratio b/a is varied. (It might be useful to compare the same plot for the symmetric case; see figure 2.8). Once again,

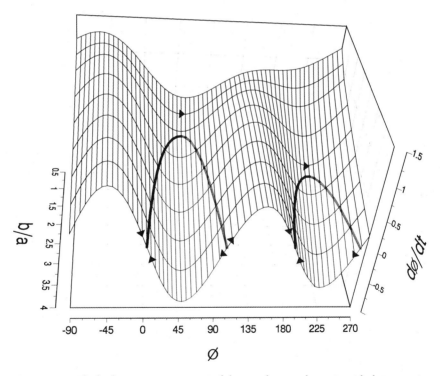

Figure 4.6 The broken symmetry version of the coordination dynamics with the parameter, $\delta\omega \neq 0$. When the symmetry of the coordination dynamics is broken, inverse saddle node bifurcations are seen. Eventually no stationary solutions exist (see text for definitions and discussion).

the system contains stationary patterns or fixed points of ϕ where the time derivative $\dot{\phi}$ is zero and crosses the ϕ axis. When the slope of $\dot{\phi}$ is negative, the fixed points are stable and attracting; when the slope is positive, the fixed points are unstable and repelling. Arrows in figure 4.6 indicate the direction of flow. Thick solid and dashed lines signify stable and unstable fixed points that establish the boundaries of the system's basin of attraction.

Now the multiple manifestations of broken symmetry in the coordination dynamics emerge. The bifurcations in this asymmetric system are called *saddle node* bifurcations ("saddle" referring to the repelling direction, "node" to attracting direction). On the right side of the figure around $\phi \approx 180$ degrees, stable and unstable fixed points coalesce onto a tangent, causing a transition to the only other stable fixed point available near $\phi = 0$ and then, *by exactly the same mechanism*, the last remaining stable fixed point disappears.

The broken symmetry law generates rich dynamics. Whereas the stable fixed points in the symmetric case do not change their value as the control parameter is varied (cf. figure 2.8) *systematic drift* is evident in the asymmetric case as parameters change. Moreover, due to the greater slope of the function surrounding the fixed point near $\phi = 180$ degrees, it is theoretically easier for the system to transit to the other stable fixed point in one direction (ϕ increasing from ~ 180 deg) than the other. Experiments by Jeka confirmed these drift and *transition path* predictions by directly manipulating the asymmetry between arms and legs using different loads applied to the limbs. Dagmar Sternad and colleagues used the hand-held pendulum paradigm to establish the systematic effect on relative phase due to frequency differences between the limbs.[13]

Loss of Entrainment

In figure 4.6 it is easy to see that as the function is flattened by decreasing a control parameter, stationary solutions eventually disappear. In this regime (exemplified by the solitary arrow at the top of the figure) there is no longer any phase or frequency locking, a condition called *loss of entrainment* or *desynchronization*. Such desynchronization is not present in the symmetric version of the coordination dynamics, but is again a consequence of symmetry breaking. Desynchronization does not always mean irregular behavior; its magnitude depends on how close the system is to its critical points, which depends on all three parameters, $\delta\omega$, a, and b (see figure 4.4). This simply reflects the fact that in broken symmetry coordination dynamics, both the frequency difference between components and the basic frequency of coordination are control parameters affecting the onset of the running solution.

Intermittency

When the saddle nodes vanish, indicating loss of entrainment, the coordination system tends to stay near the previously stable fixed point. It's as though the fixed point leaves behind a remnant or a phantom of itself that still affects the overall dynamical behavior. Thus, in figures 4.4, 4.5, and 4.6 there is still attraction to certain phase relations even though the relative phase itself is unstable ($\dot{\phi} > 0$). Such behavior is especially significant because it shows that although there is no longer any strict mode locking, a kind of partial coordination exists in which the order parameter is temporarily trapped. Motion hovers around the ghost of the previously stable fixed point most of the time, but occasionally escapes along the repelling direction (phase wandering). A histogram of the phase relation in this intermittent regime of the coordination dynamics contains all possible phase values, but concentrates around preferred phase relations (the previously stable fixed points) exactly as shown in figure 4.2 and 4.3 for relative coordination![14]

As a scientist, there are occasions—usually few and far between—when one gets an insight or reaches a level of understanding about a problem that gives one a feeling of genuine delight and satisfaction. One of them was when I realized the theoretical connection between relative coordination, which I'd studied for many years, and the dynamical mechanism of intermittency, one of the generic processes found in low-dimensional dynamical systems near tangent or saddle node bifurcations.[15] What's the excitement about? Well, there are a few reasons.

One is that relative coordination has been described for over sixty years without a satisfactory explanation. So an intermittency mechanism for relative coordination seems like an idea whose time has come, and is certainly worth exploring. Another reason is that the connection between relative coordination and intermittent dynamics suggests that biological systems tend to live near the *boundaries* separating regular and irregular behavior. They survive best, as it were, in the margins of instability. Several authors, notably the theoretical biologist Stuart Kauffman, have proposed a similar idea independently in an entirely different context, namely "rugged fitness" models of adaptive evolution. I'm coming from a direction that Kauffman admittedly eschews—whole organisms acting in their environment—but, as others have pointed out, the similarities are quite compelling.[16]

Why should a biological system occupy the strategic position near boundaries of mode-locked states rather than residing inside them? The answer given by the intermittency theory of relative coordination is that by residing near the edge, the system possesses both flexibility and metastability. There is *attraction* (the ghost of the fixed point), but no longer any attractor.

ABSOLUTE AND RELATIVE COORDINATION UNIFIED

Both relative and absolute coordination fall out of the broken symmetry version of the coordination dynamics. Obviously, the solid lines in figure 4.6 depict absolutely coordinated phase- and frequency-locked states. Relative coordination exists in the intermittent regime, just beyond mode-locking regions, where stable (attracting) and unstable (repelling) fixed points collide.

The symmetric version of the coordination law sires only absolute coordination; the asymmetric version provides a single coherent theory of both absolute and relative coordination. Which version holds depends on the system's symmetry. These basic coordinative forms emerge as two sides of the same coin, inhabiting different regimes of the same underlying law.

The broken symmetry dynamics encompasses all the effects we encountered in chapter 3 where the coordinative interactions were mostly among dissimilar components, including multistability, the coexistence of several coordination modes for the same parameter value; switching among modes at a critical parameter value due to dynamic instability; and hysteresis, where the coordination mode observed depends on which direction parameters are changing and so forth. But in addition to these effects—in some sense as vestiges of them—there is a further transition from absolute to relative coordination.

The individual components now express themselves freely, and/or the coupling between them is not strong enough to suppress their individuality completely. Only *tendencies* to coordinate in a strict fashion remain. When coordination slips, extra steps have to be inserted to keep the pattern cohesive. And then, if it's pushed further, the whole system breaks up and becomes desynchronized. Only a shadow of its previous coordination remains. Like in some marriages attraction fades, here due to the weakened curvature of a mathematical function. The mind boggles.

RELATED MODELS: FIREFLIES, LAMPREYS, AND LASERS

Nancy Kopell and Bard Ermentrout are two applied mathematicians who have worked extensively on coupled nonlinear oscillator models of biological systems. Their goal is to develop a body of mathematics that can help biologists decide whether differences they observe are crucial or not. Their approach is complementary to that taken here: these investigators work with so-called robust classes of equations and, within that framework, sort out which features are essentially universal and which depend on further structure. "Further structure" means adding more terms to the equations. Coming more from an experimental background that is inspired by the physical concepts of synergetics, I look for robust phenomena and seek a minimal, bare-bones set of principles (mathematically instantiated, of course) that embraces as many of the phenomena as possible. Although simplicity is an admirable sought-after property of mathematical models, a fine line exists between the goal of simplicity and omitting important details. Nevertheless, it is generally agreed that the best theories are those that explain the known facts and predict new facts of the same kind. Insight isn't necessarily gained by making mathematical models more complicated. Again, it comes down to determining variables that are essential and their dynamics. Here I'll describe two situations in which an identical model successfully captures certain key phenomena, although certainly not all the particulars. Then I'll make some remarks about the relation of these models to our coordination dynamics, although by the end it should be pretty obvious.

The astute reader may have connected many of the coordinative phenomena discussed thus far with one of the most spectacular sights in all of nature: the synchronous flashing of huge swarms of fireflies that occurs in places such as Malaysia and Thailand. These insects have the ability to synchronize their flashing with either an outside signal or with other fireflies of the same species. A propensity for rhythmic communication is evidently shared by humans and fireflies. Of course, on a cellular level, as we'll see, such behavior is not only ubiquitous but important, as in the case of pacemaker cells that coordinate their electrical activity to maintain the heartbeat.

Ermentrout and John Rinzel consider firefly entrainment as a problem of entraining the free-running firefly oscillator (of a certain periodicity) with a stimulus (a *Zeitgeber*) of a different period.[17] Fireflies differ considerably in terms of their flashing periodicity, but the typical interval between flashes is about a second. As long as the periods of the firefly and the *Zeitgeber* are close enough, the creature will entrain and flash at a distinct phase of the stimulus cycle (exactly like my Juilliard experiment). But if the periods (the inverse of frequency) are too different, the firefly reaches the limit of its entrainment ability, and desynchronization occurs. In many cases, this loss of entrainment is only transient (intermittency?), but in other cases, the firefly-stimulus phase difference may cycle repeatedly through all values, a phenomenon known as phase walk-through in the firefly literature (see figures 4.3 and 4.5). Phase walk-through, of course, occurs because the discrepancy between the natural period of firefly flashing and the *Zeitgeber* period is too large.

Ermentrout and Rinzel model firefly entrainment, its transient loss, and phase walk-through with the following one-variable model for the phase dynamics:

$$\dot{\phi} = \delta\omega - a\sin(\phi),$$

where $\delta\omega$ is the frequency difference, and a measures the relative influence of the *Zeitgeber* on the firefly. Figure 4.7 shows the phase dynamics when entrainment occurs (top) and when the entrainment limit is reached, giving rise to phase walk-through (bottom). In the former case, ϕ has a stable fixed point near zero (where $\dot{\phi}$ crosses the x-axis); in the latter, due to increasing $\delta\omega$, the function is lifted off the line and entrainment (the stable fixed point) is lost. Note, however, that the relative phase changes very slowly close to the horizontal axis, increasing beyond the shaded interval. Note also that the bifurcation is a saddle node, stable and unstable fixed points moving toward each other, colliding, and then disappearing, exactly as in the analysis of my Juilliard experiment.

Ermentrout and Rinzel's model is even simpler than our coordination dynamics, because, at least as described, it does not exhibit multistability, bistable to monostable phase transitions, or hysteresis. Nor does it consider the role of stochastic fluctuations. Nevertheless, it is a beautiful little model

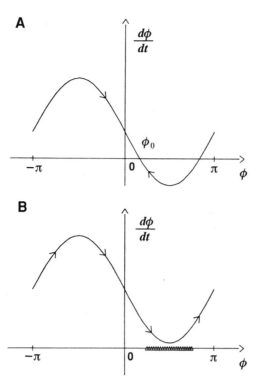

Figure 4.7 Dynamics of phase relation in Ermentrout and Rinzel's model of firefly entrainment. (A) Entrainment occurs at ϕ_0 where stimulus and pacemaker periods are close. When the function lifts off the horizontal axis (B) there is no longer any stable phase relation. Phase walk-through occurs, and ϕ continually increases, although slowly in the shaded interval. (Reprinted by permission from the American Physiological Society.)

that occupies its own niche in the spectrum of possible mathematical models. I like it because it prunes away all the details, including all the physiology and anatomy of these tiny creatures, but captures the essence of their relation to each other and to the environment.

Of course, the model leaves out many subtle aspects of firefly synchronization. Recent work addressed some of these, including how synchronization occurs in a very large population of fireflies regardless of when they start flashing.[18] This research shows that synchronization is the rule in a system of numerous identical oscillators regardless of initial conditions. But many questions remain. For example, what happens when the population of oscillators is not identical? Do traveling waves occur? What is the nature of the coupling mechanism? and so forth.

The lamprey is a strange beast: it's a bona fide vertebrate like humans; and it swims like a fish, but lacks fins and other complicated appendages. As far as we know, the lamprey swims with just one basic pattern of alternation between local opposing muscles. It turns out that neurobiologists, like Avis Cohen and Sten Grillner, can isolate the animal's spinal cord and keep it alive in a bath containing various compounds that acts as neurotransmitters.[19] This makes the lamprey a great preparation for studying how the vertebrate nervous system generates patterns of neural activity (see chapter 8). The motor patterns of the intact lamprey and the isolated spinal cord are basically the same, raising hopes that the study of this creature might help us better understand spinal cord injuries in humans.

The overall organization of lamprey pattern generation is believed to be that of a chain of segmental neural oscillators coupled across the creature's spinal cord. This gives the strict alternation pattern that produces a traveling wave up and down the cord. The lamprey needs only four segments to generate the traveling wave (the adult has about 100) and two to produce alternating activity. So, as far as theoretical modeling is concerned, we're back in the ballpark of nonlinearly coupled oscillators. Cohen, Kopell, and colleagues are world-famous for their work on modeling central pattern generation using systems of coupled oscillators.[20] There's more to this than meets the eye, but what I like is that they have come up with a remarkably simple model for the lamprey that is formally similar to that of firefly entrainment. Considering each pair of segmental oscillators, they show that a phase-locked, alternating pattern will occur if the difference between the oscillator frequencies is small relative to the coupling between them. They also show transitions from drifting to phase-locked motion between two oscillators of different frequencies, ω_1 and ω_2, when the coupling is increased. Their model reads as follows:

$$\dot{\phi} = (\omega_1 - \omega_2) - (a_{12} + a_{21}) \sin \phi,$$

where ϕ represents the phase lag between the oscillators, and a_{ij} is the coupling between them.

Notice that the lamprey model is of exactly the same form as the firefly entrainment model, but the coupling is different. Whereas the *Zeitgeber* to firefly coupling is one-way (the *Zeitgeber* drives the firefly but not the other way around), the segmental oscillators of the lamprey are *mutually coupled*, each affecting the other's behavior. Thus, in the Rand et al. model, if the net coupling $(a_{12} + a_{21})$ is positive or excitatory, oscillator 1 (the faster one) leads. If the coupling is negative or inhibitory, the slower oscillator leads. This model provides a mechanism for reversing direction of motion from forward to backward by simply changing the relative frequencies of the oscillators.

I don't know if lampreys swim backward, but I do know why I find these models of firefly entrainment and lamprey coordination interesting. First, they are simple and aesthetically pleasing. Second, they show that the same basic principles of coordination are in operation for very different kinds of things. Third, they exhibit the same behavioral patterns—phase locking and entrainment, relative coordination, and desynchronization—but by very different physiological mechanisms. All the patterns of firefly entrainment and lamprey pattern generation boil down to mathematical patterns of symmetry breaking in coupled oscillators. Was it Einstein or Wigner who asked why mathematics works so well to describe nature?

Excerpt from—Haken, H. et al. (1967).
Theory of laser noise in the phase-locking region.
Zeitschrift für Physik, 206, 369–393.
The essential noise source in lasers is so-called spontaneous emission noise which is of quantum mechanical nature. We have shown in previous papers (Haken, 1965) that spontaneous emission noise may be introduced from first principles by using quantum mechanical Langevin forces. In the present paper, we derive for the most complicated example, namely the self-mode locking, the basic equation for the relative phase including noise. As we will show, this equation has the form

$$\dot{\psi} = \delta - \beta \sin \psi + f(t),$$

where the fluctuating Langevin force $f(t)$ represents white noise and is gaussian. *The treatment of one example makes clear that this force is quite universally determined for all kinds of frequency locking phenomena* (italics mine) (p. 371).

Chronology aside, why didn't Haken, Bunz, and I use the simpler coordination law characteristic of lampreys and fireflies to model the basic forms of observed coordination within persons, between persons, and between organisms and environments? The equation is well known to Haken. Nearly thirty years ago he and his colleagues derived it from quantum mechanics (see box). But the simpler equation does not exhibit multistability, the fact that the two basic modes of coordination, in-phase and antiphase, may be produced for the same parameter value. Nor does it produce order-order transitions. For this the sin 2ϕ term is necessary. Multistability, as I've stressed, is the dynamical equivalent of multifunctionality in biology: organisms meet the same functional or environmental requirements in different ways. How else might one imagine writing one's name with one's hand, one's big toe, or one's nose?

INSTABILITY AND THE NATURE OF LIFE: THE INTERMITTENCY MECHANISM EXPOSED

The title of this section, at least the part before the colon, is stolen from an old paper by Arthur Iberall and Warren McCulloch, each one a seminal thinker.[21] Their article, which reads like a primitive poem, resonates with the intermittency theme that I want to expand in this section. I will quote only a few sentences to convey the essence:

An essential characteristic of a living system is its marginal instability.... As a result, the motor systems of the organism are plunged into intermittent search modes to satisfy all of its hungers.

The living system unfolds its states, posture by moment. In each posture (the action of the body on the body) the system is temporarily locked into an orbital constellation of all its oscillators. The psychological-physiological "moment" then changes from instant to instant.

The function of the central nervous system with its memory, communications, computational and learning capabilities is to ... modulate the system into behavioral modes.

This language must seem terribly outdated and vague to the sophisticated reader, but to me it has a solid ring of truth, nay, inspiration. McCulloch and Iberall's admittedly qualitative emphasis on instability, nonlinearity, information compression, and collective behavioral modes was ahead of its time. Of course, my primary reason for including them here is that I like the idea that organisms are "plunged into intermittent search modes." One of the goals now is to put more clothes on the frame of intermittency because it will be crucial later when I discuss perception (chapter 7) and the brain (chapter 9). The other is to deal with "orbital constellations" of oscillators whose frequencies are not locked one to one (1 : 1).

So far all the situations I've described deal with one-to-one coordination between the interacting components. Yet in many cases in living systems the components are not so coordinated. Staying within the context of rhythmic activity, it's possible for the components to be coordinated stably with other frequency ratios, such as 2 : 1 or 3 : 2, or the frequency relation may even be irrational, giving rise to quasi-periodicity. How do we understand these multifrequency situations, and where does intermittency come in? Our coordination law as it stands doesn't handle such situations, which are actually quite widespread.

From Continuous Flows to Discrete Maps

Our coordination dynamics is written as an ordinary differential equation (ODE), the conventional way in which many physical laws are stated. Of

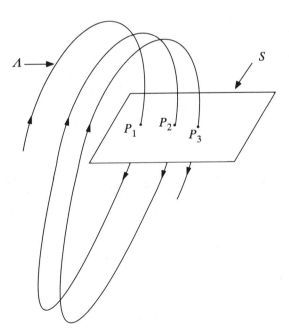

Figure 4.8   Poincaré surface of section for an ODE (see text).

course, it's possible to extend this law to accommodate oscillations of different frequencies, but this is quite complicated, technically speaking. Earlier, I mentioned an important class of dynamical system called *difference equations* or *maps*. It is this tool that I will use here as a simpler way to gain understanding of multifrequency processes and intermittent behavior in neurobiology and behavior. Instead of flowing continuously, as in ODEs, time is a discrete integer in maps. Maps and ODEs are intrinsically related to each other. In particular, Poincaré comes up again because the way they are related is named after him (Poincaré return maps and Poincaré cross-sections).

Just to get the idea, figure 4.8 shows a smooth trajectory Λ of an ODE. The plane S is designated as a surface of section and is pierced every time the trajectory cuts it, shown as the points P_1, P_2, and P_3 in the figure. Obviously, the point P_1 is related to the point P_2 which is related to P_3 and so on. Thus, a map G exists such that $P_{n+1} = G(P_n)$. We can learn a lot from such maps—physicists and mathematicians already have—because they can capture many important aspects of the real situation. Of course, the map that we will use is for the relative phase, ϕ, the collective variable that captures the coordination between nonlinearly coupled components. We call this the HKB or phase-attractive map because it's a reduction of the nonlinear oscillator model formulated by Haken, Bunz, and me (see box).

Instead of studying the entire trajectory on a 2-frequency torus, we simply encode the entire dynamics in the form of a map, $\phi_{n+1} = F(\phi_n)$. This return map is a map of the circle onto itself. It is easy to study by plotting ϕ_{n+1} versus ϕ_n as parameters are varied. The best way to do that is on a computer. Some people even believe that because we have computers, differential equations, especially in areas such as biology and the social sciences, might be supplanted by discrete maps, which are better suited to the digital computer. Whether this happens or not, of course, hinges on how well such maps model the phenomena of interest.

The Discrete Nature of Coordination

In our case, we don't need the digital computer to motivate introducing the phase-attractive map. The reason is that coordination itself is often of a discrete nature. For example, when we look at bursts of activity in muscle or brain recordings (and, incidentally, many other kinds of records), event *onsets* in different components are often related to each other. In other situations, it may be a well-defined peak of activity that enables analysis of coordination. The way we calculate relative phase is usually but not always discrete and local (see figure 2.5). In such cases, detailed trajectory information is disregarded because the essential information for coordination is localized around discrete regions. In our studies of multilimb coordination in humans, Jeka and I found that the relative phase converges only at certain points for a trot or a pace. Peter Beek noticed the same feature in skilled jugglers.

FROM FLOWS TO MAPS

How may a pair (or a system) of N-coupled nonlinear oscillators be reduced to a map for the relative phase only? The phase of each oscillator, θ_i, is viewed as living on a circle, $0 \le \theta_i < 2\pi$, rotating at a frequency, ω_i, and a period, $2\pi/\omega_i$ (figure 4.9). Since motion is confined to a circle, there is no distinction between one oscillation and the next. The shape of the waveform and its content, which may be important at the oscillator level, is ignored. That is one price that has to be paid in this kind of analysis. A system of N oscillators will inhabit N circles, but here I just show two (step 1). Mathematically, the product of N circles is an N-dimensional torus T^N. For the case of $N = 2$ the state space is a 2-torus, T^2 (step 3). The bridge between steps 1 and 3 is a square (step 2) in which the phase of each oscillator is displayed on the horizontal and vertical axes in the interval $[0, 2\pi]$. A constant relative phase, ϕ, between the oscillators, ($\phi = \theta_2 - \theta_1$), is reflected by a straight line: this is called phase locking. It is quite possible, even usual, for each oscillator to complete a cycle in the same time, but the phase difference between them may not be constant. This is called phase entrainment. The vertical plane in step 4 cuts the torus at some reference time, "freezing," as it were, the motion of one oscillator, but allowing us to see where its partner strikes it every time it traverses the torus. Each strike makes a point on the circle, allowing the entire dynamics to be encoded in the form of a return map in which each iteration takes a point, ϕ_n, on the circumference of the circle to the next point, ϕ_{n+1}. The problem of coupled oscillators is thus reduced from a flow on the torus to the study of a map from the circle to itself, $\phi_{n+1} = F(\phi_n)$.

We might expect coordination with the external environment to be of a discrete nature in situations where there are heavy constraints on timing; for example, when getting on an escalator one has to step at the right moment or one can fall on one's nose. Even on neurobiological grounds, a discreteness assumption seems valid. Neurons either fire or they don't. And they communicate with each other by releasing discrete packets of neurotransmitter in a quantal fashion.

Treating the dynamics in terms of a return map is tantamount to saying the system is highly *dissipative*. All other variables are subservient, as it were, to the collective variable. Discrete phase dynamics therefore represents the contraction of a higher- to a lower-dimensional space. Even our bagel—the two-oscillator torus—is reduced from a state space of four variables (x and $\dot{x}$ for each oscillator) to just one, the relative phase.

With these preliminaries over, we can get down to business. Our phase-attractive circle map

$$\phi_{n+1} = f(\phi_n) = \phi_n + \Omega - K/2\pi\{1 + A\cos(2\pi\phi_n)\}\sin(2\pi\phi_n) \qquad \text{(mod 1)}$$

turns out to be a function of three parameters, the meaning of which is inferred from general properties of circle maps as well as corresponding parameters in the continuous differential form of the coordination dynamics proposed by Haken et al. Thus, Ω is the frequency ratio of the components, K is the strength of coupling between components, and the parameter A is a measure of the relative stability of the intrinsic phase states, $\phi = 0$ and $\phi = \pi$.[22] The way the map works is illustrated in figure 4.10. One feeds some

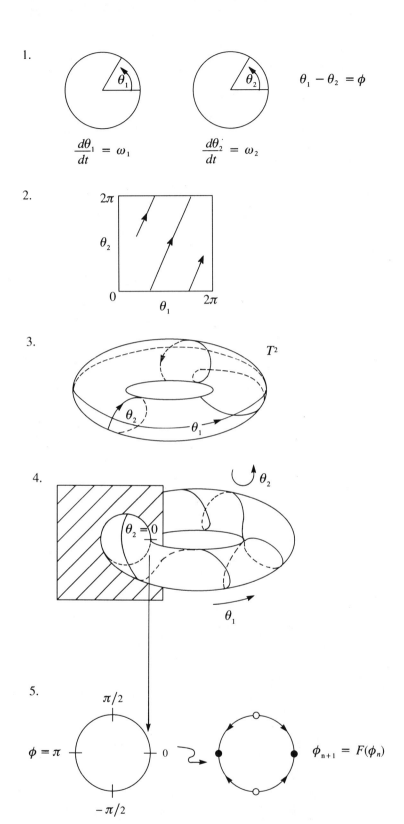

Figure 4.9 From flows to maps (see box for steps 1 to 5).

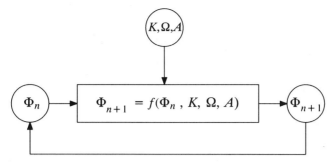

Figure 4.10 How iteration of the phase-attractive circle map works. See text for explanation.

initial phase, ϕ_n, as input into the function, then computes the output, ϕ_{n+1}, which is then fed back into the function in an iterative fashion. In short, by studying the properties of our phase map we can more readily understand how coupled oscillators behave, especially when the component frequencies are different and the coupling is subject to change.

Multifrequency Stability

Everything that we've observed for $1:1$ coordination (e.g., bistability to monostability phase transitions) can be reproduced by the map. But there's a wealth of other phenomena besides, due to the presence of the Ω and K parameters. Consider, for instance, the ubiquity of low-order frequency ratios in physiology, neurobiology, and psychology.[23] Just to get the idea across, try tapping $2:1$ with your two hands. Some people find it easier than others, and it depends on which hand beats the faster rhythm. (By now the reader has become an experimental psychologist and has tried these tasks with left- versus right-handers; musicians versus nonmusicians; brain-damaged versus normals; under the influence of alcohol … ad infinitum.) Many studies of temporal organization in humans, especially by Michael Peters in Canada, Jeff Summers in Australia, Peter Beek in Holland and Diana Deutsch and others in the United States, have shown that the low-order ratios ($1:1$, $2:1$, $3:2$, etc.) are easier to perform than higher-order ones ($4:3$, $5:3$ …). I confess to feeling a secret pleasure with such results. Here is the most complex system of all, the human brain, yet a typical person is only able to perform, stably at least, low-integer frequency ratios between the hands!

Ignoring questions about skilled musicians and learning for the moment, the fundamental reason for this restriction has to do with the *structural stability of the rational frequency ratios*.[24] Structural stability means that slight modifica- tion of the system does not alter the stability of ratios. It accounts for similar behavior across very different systems, such as clock mechanisms, moon-earth phase locking, walking and breathing in humans, electrically stimulated nerve membranes, frequency locking in mammalian visual cortex, and so on.

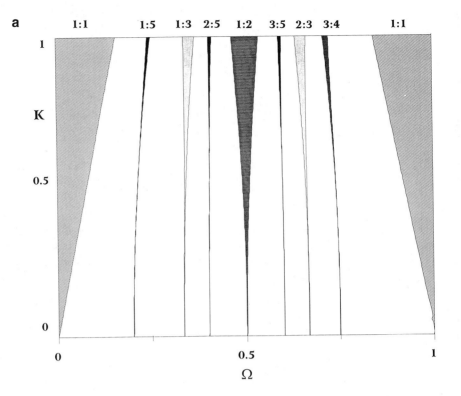

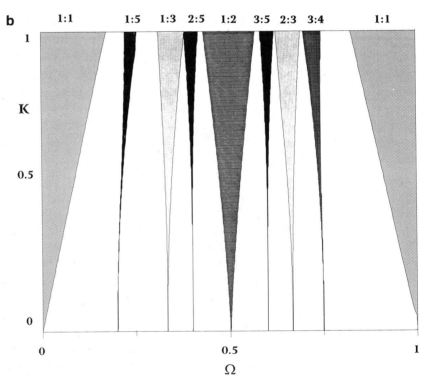

Numerical studies of coupled oscillators or circle maps allow us to calculate the size of mode-locked regions. Technically, the asymptotic value of the frequency ratio is called the *winding number, W*, and is, in fact, the measured frequency ratio between the oscillators. When *W* is a rational number, the underlying oscillations are mode-locked or frequency synchronized. When *W* is irrational, which means it cannot be expressed as the ratio of two integers, the oscillations are desynchronized. The outcome is either quasi-periodicity or chaos.

Figure 4.11 (see also plate 3) shows why low-integer frequency ratios are ubiquitous in nature and why they are the easiest for humans to perform. The wedges emanating from the horizontal axis, called *Arnol'd tongues* after the Russian mathematician who first described them, correspond to pure mode locking. Tongues represent regions of the (K, Ω) parameter space that have asymptotic solutions to our phase map equation. In between these tongues are quasi-periodic regions where the frequency ratios are irrational. As *K* increases above zero, the width of each locking increases, culminating in a situation where locked states fill the entire interval (see below). The widest mode-locked regions correspond to low-frequency ratios and are the most (structurally) stable and attractive. Because of their width, any parameter variation inside these larger Arnol'd tongues will not kick the system somewhere else. As a consequence, fat Arnol'd tongues containing the $2:1$, $3:1$, $3:2$ ... coordination modes are relatively easier to perform than slim ones $(4:3, 5:2 ...)$.[25]

The Farey Tree: Hierarchical Complexity

Notice in plate 3 and to a lesser extent in figure 4.11 how the widths of the mode-locked regions are ordered. Scanning from left to right, the biggest tongues order as

0/1, 1/4, 1/3, 1/2, 2/3, 3/4, 1/1.

This *Farey sequence* represents an increasing succession of rational numbers p/q such that $q \leq n$ where $n = 4$ in this case. It is possible to order the mode-locked regions (the rational numbers between 0 and 1) into a hierarchy called the *Farey tree*. The power of mathematics shows up again: the branching structure of the tree encapsulates all the possible mode-lockings in this entire class of dynamical systems. To grow the tree, start with the parents 0/1 and 1/1. Add the numerators and denominators together according to the formula $p/q \oplus p'/q' \equiv p + p'/q + q'$, forming successive levels as follows:

Figure 4.11 Arnol'd tongues for the phase-attractive circle map showing mode-locked regions for some of the lower-order (non-1:1) mode-locked regions. (a) A = 0 yields the familiar result for the sine circle map. (b) A = 0.5 produces a widening effect on the tongues. In both cases, the relative widths of the tongues provide a basis for the differential stability and complexity of multifrequency coordination. Quasi-periodic dynamics exists between the tongues.

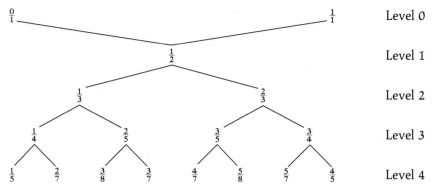

$$\frac{0}{1} \qquad \frac{1}{1} \qquad \text{Level 0}$$

Level 1 is created by the Farey sum $0/1 \oplus 1/1 = 1/2$. The branches of level 2 are given by $0/1 \oplus 1/2 = 1/3$ and $1/1 \oplus 1/2 = 2/3$ ... and so on. Notice in figure 4.11 and plate 3 the lowest denominator tongues are the widest. No wonder they are the easiest to get into and stay in. In our theory, pattern stability and complexity are related to the level in the Farey tree hierarchy and are inversely proportional to tongue width.

Multifrequency Transitions

How does the Arnol'd tongue structure envisage *transitions* in coordination between multifrequency states? How does the rhythm change? In principle, all the system has to do is cross from one tongue to the other. This might sound easier than it looks. It is clear, for example, that a system buried deep inside one of the tongues is pretty stable despite parameter changes or the influence of noise. As I said before, this fact is significant for rationalizing the dominance of the low-order frequency ratios. But it is also obvious that the width (stability) of the mode-locked regions plays a role in determining which patterns are easiest to switch into and out of.

In our experiments, spontaneous transitions are far more likely from the less stable frequency ratios to the more stable frequency ratios than vice versa, just as the Farey tree would predict.[26] This may be one of the reasons why people (even the best musicians) have to practice so hard to produce intricate rhythms: any little noise will kick the system into one of the more stable ("easier") Arnol'd tongues. Learning complex rhythms in this context involves stabilizing higher-order frequency ratio states (there's much more to it than that, of course, as we'll see in chapter 6). Just getting into these thin Arnol'd tongues is tricky. The fat ones are far stickier.

Between the tongues sprouting from $K = 0$ in figure 4.11, the dynamics are quasi-periodic, the frequency ratio is irrational. As K increases, the Arnol'd tongues widen and eventually touch each other on the *critical surface*. Above the critical surface, the tongues overlap and the system can display chaos (see box and plate 3).

Tongue overlapping is a necessary but not sufficient condition for chaos. For example, as you can see in plate 3, it is possible to stay in a 2 : 1 ratio far

THE CASE OF THE CRINKLED TORUS

The *most* irrational ratio or winding number is the golden-mean, $(\sqrt{5} - 1)/2)$, which is the easiest place to observe quasi-periodicity experimentally. Why? Because it's the *least* likely to lock into a low denominator tongue (rationality, as it were!). The golden mean therefore is the last exit before irrationality turns into chaos. A beautiful physical experiment by Albert Libchaber—one of my favorites ever—was conducted on Rayleigh-Bénard convection in a small mercury-filled rectangular cell.[27] After an *oscillatory instability* was set up, producing a well-defined frequency of rolling motion (f_1) in the cell, Libchaber et al. injected a small alternating current of a certain amplitude and frequency, f_2. The experiment resembles someone stirring a pot of soup while heating it on a stove. Since f_1 is time independent, locking behavior can be studied by manipulating current amplitude and f_2. The entire Arnol'd tongue structure can then be beautifully mapped out and all the circle map predictions tested. By far the strongest image imprinted on my mind is near the critical surface where the two frequencies have been set up at the golden mean and the nonlinear coupling (by the current amplitude) increased. The 2-torus starts to break down, and the attractor becomes stretched and crinkled, indicating the onset of chaos (figure 4.12). Extra dimensions start to make themselves felt. Turbulence (high-dimensional behavior) is just over the horizon, and you can *feel* it in this picture.

beyond the point at which the tongues overlap. Notice, however, how the shape of the Arnol'd tongues and the overall complexity of the space depends on parameters. For instance, for fixed K and Ω, the variation of the intrinsic parameter, A, raises and lowers the critical surface, hence affecting the onset of chaos. The variation of these parameters determines the width of the stable mode-locked regions consequently delaying or accelerating irregular behavior. A familiar (by now) message appears in a different guise: *where the system lives in parameter space dictates the complexity of its behavior.*

The Devil's Staircase

A universal feature of maps such as our phase-attractive circle map is that at the critical surface the mode-locked tongues fill up the line. This is called the *Devil's staircase.*[28] The reason is that if you blow up or magnify any piece of it, you see the same thing. So, in some sense, dear pilgrim, you'll never make it to the top! Tom Holroyd did an interesting numerical experiment calculating the widths of the tongues in our map by plotting the winding number against Ω for various values of A and K. Below the critical surface, the length of the mode-locked regions is less than 1, and so the Devil's staircase is incomplete. But the smaller the steps you take the smaller the gap between mode lockings, until, at criticality, there is no longer any room for quasiperiodicity, and the staircase is said to be complete.

Figure 4.13 shows the complete staircase for three sets of different parameter values for our map. The similarity between the staircases is compelling, attesting to the universality property of these kinds of dynamical systems.

Figure 4.12 The crinkled torus observed in Libchaber's Bénard convection experiment. The winding number is close to the golden mean, and breakdown of the torus is imminent. (Reprinted with kind permission of The Royal Society)

Physicists take universality to be crucial, because without it they would not be able to predict the results of experiments in systems in which the underlying map (the dynamics) is unknown. Of course, this is usually, if not always, the case in biology and behavior.

When one examines these staircase plots in figure 4.13 one sees differences as well. In particular, the *relative* width of the tongues changes with the phase-attractive term, A, our intrinsic parameter. I find it interesting that a parameter derived from our phase-transition experiments that reflects the bistability of the phase states at $\phi = 0$ and π changes the size, for example, of the 2 : 3 ratio relative to 1 : 2.

Although it's well-nigh impossible to manipulate all the Arnol'd tongues in any biological experiment, never mind on humans, some years ago Gonzalo DeGuzman and I obtained results that are at least consistent with the mode-locking picture[29] (see below). Very similar results were obtained recently using different procedures.[30] Taken collectively, this work demonstrates the potential power of the dynamical account of multifrequency coordination. But a lot more could be done to unpack further Iberall and McCulloch's "orbital constellations."

Life at the Edge

The greatest flexibility is afforded a coordinative system when it is near the tongue boundaries where transitions to other modes are easily effected. The reason is that mode lockings in the phase-attractive map are created and destroyed through tangent or saddle node bifurcations that occur at the boundaries of Arnol'd tongues (cf. chapter 1). In other words, for motion on a

Figure 4.13 The Devil's staircase. At the critical surface, Arnol'd tongues completely fill the Ω-axis, except for a set (the Cantor set) of measure zero. The width of the mode-locked intervals is dependent on the parameters A and K. (a) $A = 0$, $K = 1.0$; (b) $A = 1.0$, $K = 0.5$; (c) $A = -1.0$, $K = 0.888 \ldots$

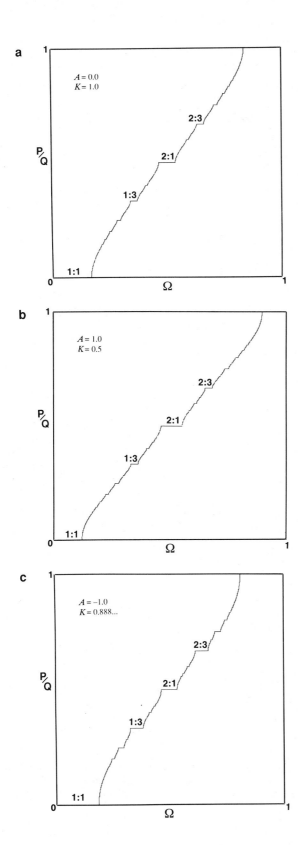

Extending the Basic Picture: Breaking Away

torus, transition pathways are by intermittency, the dynamical mechanism I proposed earlier for relative coordination. What does intermittency look like in our map, and does it really produce the relative coordination effect? For a change, let's do the analysis quantitatively with numbers.

I present an example in figure 4.14 in which I vary the values of Ω near the period 1 boundary, that is, near a $1:1$ phase and frequency-synchronized state. The boundary in this case ($A = 0$) is defined by $K = 2\pi\Omega$. For $K = 0.6$, the saddle node bifurcation occurs at $\Omega_c = 0.6/2\pi \approx 0.0455$. Figure 4.14a shows the function $f(\phi)$ intersecting the diagonal line at two points: ϕ^- and ϕ^+, where ϕ^- is a fixed point (mode locked) attractor and ϕ^+ is a fixed point repeller ($\Omega = \Omega_c - 0.03$). Initial conditions other than exactly $\phi = \phi^+$ converge to ϕ^- as $n \to \infty$. As Ω increases, ϕ^- and ϕ^+ approach each other and coalesce when $\Omega = \Omega_c$ (figure 4.14b). For $\Omega = \Omega_c + 0.01$ beyond the boundary of the Arnol'd tongue, ϕ^- and ϕ^+ cease to exist (figure 4.14c), and the system exhibits either mode lockings with higher-frequency ratios or quasi-periodic motion, depending on the exact location of Ω. If Ω is decreased, then the reverse sequence of events is observed.

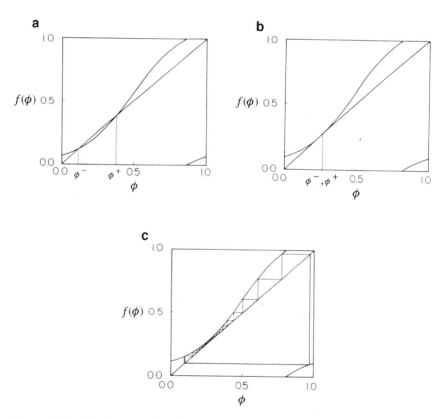

Figure 4.14 Intermittency in the phase-attractive map. Shown is the function $f(\phi)$ for three values of Ω. ϕ^- and ϕ^+ correspond to stable and unstable fixed points of the map. Notice in (b) the two coalesce at a saddle node bifurcation and then lift off in (c), giving rise to intermittent dynamics (see text).

The narrow corridor between the function $f(\phi)$ and the diagonal line in figure 4.14c induces what the French physicists Yves Pomeau and Paul Manneville call *type I intermittency*.[31] The dynamical behavior is as follows. Inside the channel, iterates of the map move very slowly, giving rise to the impression that the fixed point attractor was already in place (from the point of view of decreasing Ω). After exiting the channel, the trajectory takes large strides for a number of times before reentering the channel. That is, *phase slippage* occurs and there is no longer any mode locking, because the fixed points have disappeared. Only a faint trace of them remains.

The appearance of phase slippage means that between two channel crossings one of the oscillators gains a period: exactly the phenomenon of *relative coordination* (see again figure 4.2a). Slips in phase occur because of the unstable direction: the system escapes to explore other regions of its space before

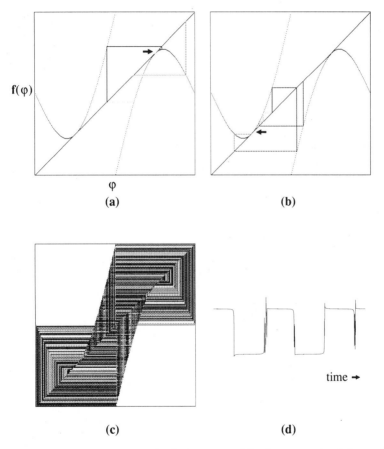

Figure 4.15   Intermittency in the chaotic regime of the phase-attractive circle map. In (a) there is a stable fixed point on the right lobe which attracts initial conditions (see arrow). (b) same as (a) except attraction is to a fixed point on the left lobe. (c) When the function is lifted off the diagonal, spontaneous switching between the fixed points occurs via a chaotic transient, as shown by the time series in (d).

wandering back into the channel (due to the 2π periodicity of ϕ), where it visits once more the remnant of the fixed point. It is just like the father walking along with his small child who, because of their intrinsically different cycle periods, must either slow down (father) or add steps (child) to keep pace with each other. The father-child system is poised near the ghost of mode-locked states (fixed points), relatively but not absolutely coordinated. The main dynamical mechanism (figure 4.14b) is the coalescence of stable (attracting) and unstable (repelling) directions in the coordination dynamics. Both stabilizing and destabilizing processes must, it seems, coexist.

When we move into the chaotic region of the dynamics the same basic mechanism is present. In figure 4.15 I show the map in the noninvertible region beyond the critical surface, where the torus breaks down. For $\Omega = 0.5$, that is, a $2 : 1$ frequency ratio, and strong coupling there is a double "hump,"

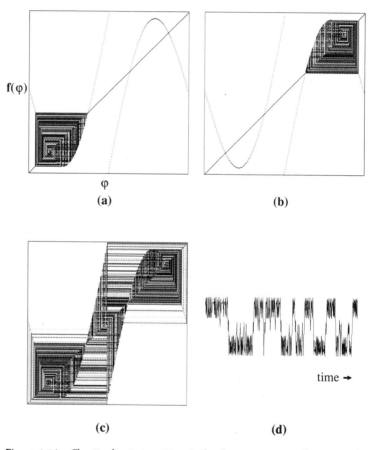

Figure 4.16 Chaotic-chaotic transitions in the phase-attractive circle map. (a) shows a chaotic attractor in the left lobe of the state space, (b) shows a different chaotic attractor in the right lobe, and (c) shows a connection between them, creating a two lobed attractor in which motion jumps spontaneously from one lobe to the other. Note in the time series (d) that motion is chaotic in each lobe.

Extending the Basic Picture: Breaking Away

each containing a stable fixed point. Which one is observed depends on initial conditions (figure 4.15a and b). When the function is lifted slightly off the line (figure 4.15c) the system escapes through a portal, jumping between the two nearly mode-locked states, via a chaotic transient. Figure 4.15d shows the corresponding time series. Such a process may well underlie the kind of perceptual switching that occurs when observers view ambiguous figures like the Necker cube (see Chapter 7) as well as the intermittent switching observed in ion channel kinetics (see chapter 8).

Finally, our map may possess two *chaotic attractors* with independent basins of attraction.[32] A random initial condition falls into one of the attractors and stays there (figure 4.16a and b). With a small parameter change, a single two-lobed attractor emerges, composed of the two original attractors connected through small portals in the phase space. A chaotic trajectory is followed inside one of the lobes until it escapes and falls into the other lobe (figure 4.16c). The time series in figure 4.16d illustrates the way the system switches randomly back and forth between these states. Although fluctuations are obviously important in biological systems, these switches are actually deterministic and not the result of noise. The slightest parameter change causes escape from the basin of one chaotic attractor to the other, reminiscent of conceptual leaps.[33]

Experimental Windows on Intermittency

Much of von Holst's work on the nature of order in the central nervous system was performed on *Labrus*, a fish distinguished by the fact that it swims smoothly using rhythmic fin movements but keeps its main body axis immobile. Von Holst referred to this preparation, in true Germanic style, as his *precision apparatus* (italics his). The time series shown in figure 4.2 are from *Labrus*. The coordination is either at (absolute) or near (relative) 1 : 1 mode locking.

I have a "precision apparatus" too, custom made and especially configured for driving and monitoring the movements of the two hands simultaneously.[34] In a series of experiments done with Gonzalo DeGuzman we manipulated the frequency ratio between the two hands. It is very difficult to require a subject to produce high-order frequency ratios on demand. My idea was to drive one finger passively using a torque motor while the other finger maintained a base rhythm. This allowed us to scan the frequency ratio in small steps (0.1 Hz) every 20 seconds or so from just below 1 : 1 to just above 2 : 1. We found beautiful evidence for intermittent dynamics in the human sensorimotor system.

A typical time series from the experiment is shown in figure 4.17. The required frequency ratio is set near 2 : 1, not 1 : 1 as in figure 4.2. The input signal used to drive the torque motor is on top. The *actual trajectory* of the driven hand is in the middle. Notice it does not exactly follow the driver. The free hand, moving at its base rhythm, is on the bottom. Enhancement of the

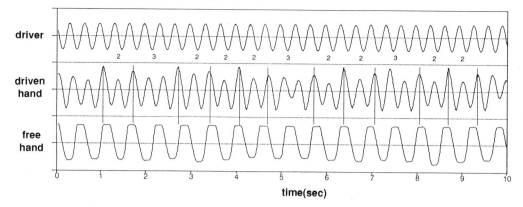

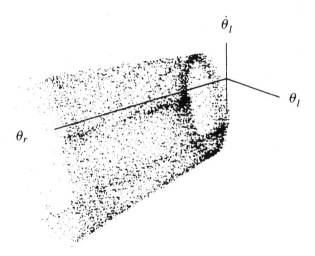

Figure 4.17 Relative coordination near a 2:1 required frequency ratio. Trajectories of the input signal used to drive the torque motor (*top*), the actual motion produced by the driven hand (*middle*), and the free hand (*bottom*). Enhancement of the peaks near in-phase regions points to the discrete nature of the coordination. Occasional but systematic slips are evident when the enhanced peak extends beyond the broadened peak of the free hand (3 steps instead of 2).

Figure 4.18 Phase space portrait of a 2:1 case in the Kelso and DeGuzman experiment. θ_l and θ_r correspond to the angular position of the left and right hands; $\dot{\theta}_l$ is the angular velocity of the left hand. the dark bands show phase concentration around 0 and π. The length of these bands is due to flattening of the right-hand trajectories at the crests and troughs.

peaks near in-phase coordination attests once more to the discrete nature of coordination. More important here, however, is that the driven hand adjusts its trajectory to sustain the natural tendency to be in phase with the slower free hand. Follow the vertical lines on figure 4.17 from left to right. Occasional slips occur when the position of the enhanced peak extends beyond the broadened peak of the free hand. Three steps instead of two! This experimental system, our little window into life, resides on the edge of the mode-locked state, not in it.

Figure 4.18 shows a 3-D plot of a 2 : 1 condition in which the position and velocity of the driven hand occupy the *x*- and *y*-axes, and the position of the free hand is plotted continuously on the *z*-coordinate. The reason a nice limit cycle is observed is because the driving torque is sufficiently large that the hand has to follow it. Two bands of phase concentration on the driven cycle show up clearly around $\phi = 0$ and π. This is because in this case the free hand visits these states for prolonged periods of time before wandering away. The length of the dark bands is due to flattening of the free hand trajectories at the crests and troughs of its motion. Here we see the same relative coordination effect produced in a different way that nevertheless attests to the mutual cooperation between the components.

When I first presented these results in 1987 at one of Hermann Haken's famous Elmau meetings (this one on neural and synergetic computers) I remarked, paraphrasing Arnold Mandell:

Neurobiological dynamical systems lack stationarity: their orbital trajectories visit among and jump between attractor basins from which (after relatively short times) they are spontaneously "kicked out" by their divergent flows.

Using the map, it is easy to plot the relative phase distributions corresponding to mode-locked and intermittent regimes of the coordination dynamics. One look at these distributions strengthens the theory that absolute and relative coordination, the two basic dynamic forms, correspond to mode locking and intermittent regimes of the coordination dynamics. The reader can prove this by iterating our map on a computer and plotting a histogram of the phases visited. That's what is displayed in figure 4.19 for parameters that are set near 2 : 1 (i.e., $\Omega \approx 0.5$). The map is iterated 10,000 times and plotted after removal of any initial transients. Three distributions are shown, with the same number of data points in each. Relative phase $[0, 2\pi]$ is on the horizontal axis expressed on the unit interval $[0, 1]$. As the frequency ratio parameter is brought closer to 2 : 1, the peaks near $\phi = 0$ (~ 0.1) and $\phi = \pi$ (~ 0.5) become higher and narrower. The system still visits all possible phase states, but less so than when it is farther away from the mode-locked state. Were one to enter the 2 : 1 Arnol'd tongue without adding a little noise, two straight lines exactly at the absolutely coordinated phase states would be observed with approximately the same number of points (~ 5000) in each. Comparison with von Holst's phase distributions in figure 4.2 is irresistible.

Anticipation . . .

The trajectories displayed in figures 4.2 and 4.17 exhibit a phenomenon that one might call, for want of a better word, *anticipation*. Notice as the vertical lines progressively strike the broadened peaks in figure 4.17 (third row) they reach a point at the edge where we *know* an extra step is going to be inserted. The system itself is clearly sensitive to the fact that if it doesn't adapt quickly, coordination and communication will be lost. Literally, in our system, it will behave irrationally. How do we understand this anticipatory effect?

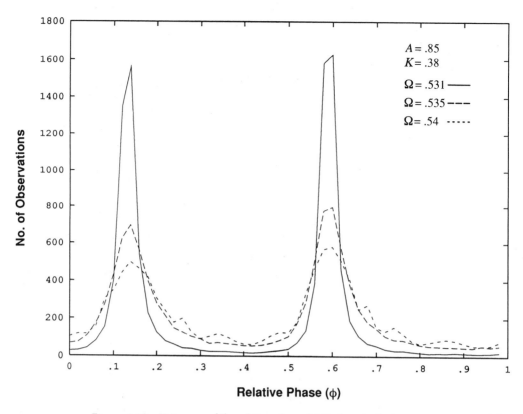

Figure 4.19 Histogram of the relative phase distributions in the intermittent regime of the phase-attractive map near a frequency ratio of 2:1 ($\Omega = .5$). Notice the peaks become smaller and more dispersed as the frequency ratio moves away from the 2:1 mode-locked state.

Referring back to figure 4.14c, note that close to the tangent bifurcation, the relative phase concentrates and slows. Phase attraction persists because the iterates are trapped in the corridor separating the function and the 45-degree line. Obviously, as the system approaches closer and closer to the fixed point, the time spent in the channel gets longer and longer (critical slowing in another guise!). This phase gathering has an anticipatory quality about it. Even though the motion is quasi-regular, it is easy to predict where and when the mode-locked state is going to reveal itself.

A good way to see this is through the corresponding bifurcation diagrams shown in figure 4.20. The bifurcation parameter, K, in this case, is varied along the x-axis and the relative phase, ϕ, is on the y-axis. The fuzzy area corresponds (mostly) to quasi-periodic motion because the frequency ratio between the components is irrational. By the way, the gaps inside the fuzziness are higher-order, very thin, mode-locked regions. Notice, however, that the progressive darkening anticipates the upcoming stable solution (the single line), which indicates that coordination is trapped or mode locked 1:1 in figure 4.20a or 2:1 in figure 4.20b.

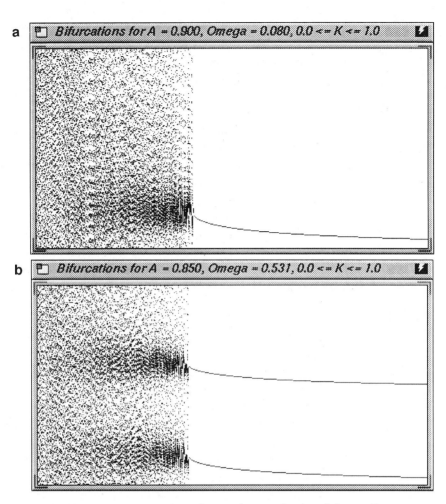

Figure 4.20 Anticipation? Bifurcation diagram showing the phase behavior when crossing the Arnol'd tongues at (a) 1:1 and (b) 2:1 for fixed A and increasing K. Progressive darkening "anticipates" upcoming stable solutions (single lines), which indicate that coordination is trapped or mode locked.

The main point is that the system spends more time near a particular phase as it approaches a critical point, giving rise to an enhanced phase density that specifies the locus of the upcoming state. Critical slowing (the darkening effect) is thus a predictor of upcoming transitions. Somewhat facetiously (but not entirely), I refer to this aspect of the coordination dynamics as an anticipatory dynamical system (ADS).[35] An ADS stays in contact with the future by living near critical points.

POSTSCRIPT

For me, there can hardly be a more powerful image of the central thesis here —that materially complex biological systems may exhibit low-dimensional, but complex dynamical behavior—than the demonstrated tendencies for humans to exhibit limited forms of spatiotemporal organization among themselves or their components. Although I hesitate to use the heady language of universality, nevertheless, a remarkable, possibly quite profound, connection seems to exist among physical, biological, and psychological phenomena. "Phenomenon" is often a dirty word in the softer sciences, somehow suggesting failure to come to grips with "mechanisms" and settling instead for "description." Often mechanisms for a given phenomenon are sought at lower levels in the interior workings of the system. That view may be flawed if lower levels also turn out to be governed by the same kinds of self-organized dynamical principles or generic mechanisms shown to be at work here.

In this book, I use the word *phenomenon* very much in the style of physics, to refer to observations obtained under specific circumstances including an account of the entire experiment.[36] This does not mean one cannot go beyond the specifics of the experiment to some deeper theoretical framework. The phenomena of absolute, and, to a lesser degree, relative coordination have been around a long time: their connection to concepts of self-organization in nonequilibrium systems and their expression in terms of dynamical laws is, I think, quite recent. Up to now, the early discoveries of von Holst and his generally valid descriptions of biological coordination, even though widely recognized, were not accompanied by a successful theoretical treatment. This had to wait for the concepts of synergetics to handle cooperative interactions, and the appropriate mathematical tools of nonlinear dynamical systems to formalize them.

In this chapter it has been difficult to contain my excitement at the recognition of unity between features of complex coordination phenomena that previously appeared quite unconnected. Absolute and relative coordination, the two basic dynamic forms, correspond to mode locking and intermittency, now unified in a single theory. Both absolute and relative coordination spring from a basic symmetry in the collective variable, relative phase. Symmetries run the laws of physics and are always tied to conservation principles. The mathematical language of symmetries dominates all the way from the conservation of momentum and energy in classical physics to the fundamental particles of quantum mechanics. Does a new conservation principle underlie coordination? Perhaps. The concept of stability seems equally if not more important, because, as shown here, coordinative states may possess equivalent symmetry yet not be equally stable.

It is the breaking of symmetry that gives rise to the complex of phenomena collectively called relative coordination. Phase slippage and the injection of additional steps correspond to the intermittent regime of the coordination dynamics. The main *generic mechanism* is the coalescence near tangent bifurcations of stable (attracting) and unstable (repelling) fixed points in the coordination dynamics. In this flexible, intermittent regime, systems exhibit phase attraction but not phase locking or phase entrainment. The collective variable is nonstationary, it does not converge asymptotically.

When Mozart wrote to his father about three piano concertos he had just completed, he is reported to have said, "They are exactly between too hard and too easy: the connoisseurs will find satisfaction, but the nonexperts will feel content without knowing why." That's relative coordination, in between the hard and the easy, the regular and the irregular where the creative pulse beats.

5 Intentional Dynamics

How does mind get into muscle? How can intentionality—considered *the* nonphysical feature of biological systems—be understood in terms of our physically inspired theory of biological and behavioral self-organization? Over the years, many have argued that the methods of doing science on inanimate objects are entirely inadequate for doing science on animals that have brains and possess intentionality. The most recent advocate of this view is Gerald Edelman, winner of the Nobel Prize for physiology or medicine in 1972.[1] Physics, as Edelman sees it, bears no relationship to intentionality because it ignores both the psychological and biological evidence necessary for understanding the problem. Thus physics is not concerned with biological structures, processes, and principles though it obviously provides the basis for them. According to Edelman "exotic physics" (his words) such as quantum mechanics and the general theory of relativity aren't going to tell us very much about the organization of organisms, especially those rather special properties we call consciousness and intentionality. On the other hand, if one accepts the theory and empirical evidence I have presented in the first part of this book, the physical concepts of self-organized pattern formation in non-equilibrium systems (synergetics), and the mathematical tools of nonlinear dynamics (including an explicit treatment of fluctuations) already provide a foundation for understanding certain essential features of organisms in relation to their environment, namely, behavioral coordination, stability of function, pattern flexibility, and change.

We have come to this understanding by way of the methodological strategy of phase transitions, of identifying collective variables or order parameters and their dynamics in the absence of any specific parametric influence. That is, the pattern dynamics has been shown to emerge as the result of *nonspecific* changes in control parameters. To clarify, using our "workhorse" example of bimanual coordination, once an initial condition is established (move your fingers back and forth in the same direction), frequency of motion is nonspecific with respect to the identified order parameter, relative phase, that characterizes the various patterns. Without knowledge of spontaneous coordination tendencies and their dynamics, it is difficult to understand what is modifiable by the environment, by learning or, as we'll see in this chapter, by intention. I call this knowledge the *first cornerstone of biological self-organization*.

To come to grips with issues of learning and intentionality, I argue that it is necessary to incorporate another kind of forcing, namely, from *specific parametric influences*. To fix ideas again, this means that in our hand movement example, instead of frequency moving the fingers through different phasing patterns, the fingers are instead required to produce a *specific* phasing pattern.

Such specific information may arise as a result of an intention, an environmental requirement or a particular task to be learned. Understanding how specific parametric influences of this kind contribute to the pattern dynamics constitutes what I call the *second cornerstone of biological self-organization*. Such forcings will turn out to be meaningful and specific only in terms of their influence on the identified order parameter or collective variable. The upshot of this is a theory of *intentional dynamics* in which two languages normally considered irreconcilable, those of intentions and dynamics, are cast in a single unified framework.

The issue of intentionality in complex systems is, of course, a slippery one. One of the problems is the big discrepancy between functional definitions of intentional behavior and how intentional behavior is typically studied. And the way it's studied, I don't have to say, colors the way we try to understand it. In this chapter I aim to provide some insights into very basic questions, such as what can be intended, how are intentions constrained by the organization of the nervous system, and what do intentions really do? To place these questions in some perspective, a little background may be helpful. First I discuss the issue of goal-directedness in biology from a general point of view. Next, I consider intentionality in light of some evidence from psychology and the brain sciences. Then I will tell my story of intentional dynamics, how it came about, and what I think it all means.

GOAL-DIRECTEDNESS IN BIOLOGY

The doyen of biologists, J. B. S. Haldane,[2] used to say that teleology (from the Greek *telos*, "goal as final cause") is like a mistress to a biologist; you can't live without her but you don't want to be seen with her in public. Goal-directedness, it is claimed, distinguishes the animate from the inanimate; more crudely, the living from the dead. It's fair to say that biology and the concept of goal-directedness are uneasy bedfellows. In fact, biologists have taken great pains to discredit the slightest hint that some immaterial vital force outside ordinary physics and chemistry underlies seemingly purposive behavior. Yes, biological organization is itself a directed process. Yes, organisms are organized things. But what is the source of organization?

Modern biology offers a two-part response. The first part, in my opinion, has been drowned out by the second, perhaps irrevocably. This, I believe, will turn out to have been a grave mistake. The reason, perhaps, is that the first part seems so obvious, it is taken for granted. Living things are open, non-equilibrium systems.[3] They exchange matter and energy with their surroundings. Without material flowing into the organism, without metabolism to break these materials down and synthesize them for its own needs, an organism could not survive. Reproduction is predicated on this fact. As emphasized here and by many others, ordinary matter under open, nonequilibrium conditions exhibits self-organization, the creation and evolution of patterned structures. But in biology, at least so far, processes of self-organization in open systems have received short shrift. Certainly, reaction-diffusion mechanisms of the Turing type (see chapter 1) have been mentioned in discussions of embryological development and biological form (how a cell becomes a finger or a toe), but for the most part it's a brief and passing tip of the hat.[4]

Why are cooperative phenomena away from (thermal) equilibrium almost totally ignored as sources of biological order? I believe the answer is due, in part, to ignorance of current trends in physics and mathematics. As 1980 Nobel laureate in chemistry Paul Berg, and Maxine Singer remark, "Biologists and geneticists in contrast to physics [sic] do not depend much on abstract concepts nor does their understanding require any mastery of mathematics."[5] Similarly, another Nobelist, Ragnar Granit, argues, "The study of the brain is hardly in much need of mathematics."[6] Passing fancy with previous mathematical approaches such as cybernetics and the communications sciences, which promised much but delivered little, no doubt have contributed to such attitudes among biologists. The issue, however, is not premature mathematization but awareness (or lack of it) of physically based principles of pattern formation and self-organization in open, nonequilibrium systems. If not the new laws to be expected in the organism that Schrödinger anticipated, then certainly some of the key concepts.

For mainstream biology, the chief source of biological organization is not its openness, but the fact that organisms are controlled by a program. Yes, most biologists admit that organisms belong to the general class of open systems. But what is really crucial is that all organisms possess genetic material that allows them to reproduce and pass on copies of themselves to their descendants.[7] In this dogma, it's the gene that controls development, transforming cells into complex organisms. According to the eminent biologist Ernst Mayr, the organism's goal-directedness—its teleonomic character—is due specifically to a *genetic program*.[8] Organisms have this in common with some humanmade devices (e.g., guided missiles) and this is what distinguishes them from inanimate nature. All that we need to know, in Mayr's view, is that a program exists that is causally responsible for the goal-directedness of living organisms; how the program originated is quite irrelevant.

If Mayr speaks for most of modern evolutionary and molecular biology, and I think he does, I must express my discomfort with his view. It seems to me that to describe the gene as a program sending instructions to cells belittles the complexity of the gene and its amazingly intricate organization.[9] What do genes actually do? The don't *program* anything: they synthesize proteins, and not even directly, as it turns out. Whereas it is often said that DNA produces proteins, the opposite is, in fact, true. Proteins, enzymes, that is, produce DNA.[10] The vast majority of any organism's characteristics is recognized even by the most ardent reductionists to arise from complicated interactions among proteins. Moreover, genetic *rearrangement* occurs, not as some sporadic event tied to random mutation, but as a result of normal processes. Genetic material is thus not fixed like a program, but is fluid and dynamic, small pieces of DNA moving from place to place influencing gene expression. Although the full implications of Barbara McClintock's heralded discovery of transposable elements ("jumping genes") are enigmatic at present,[11] there are strong hints they are important in gene repair and reconstruction. Thus, the very integrity of the genome may rest on the flexibility of DNA. Transposition is just one hallmark of the complexity of genetic organization. Far from being equated with a program, a set of instructions controlling development, the gene, at least to me, looks more and more like a self-organized, dynamical system. Programs, after all, are written by people who send a list of instructions to a machine to tell it what they want it to do. Who or what programs the genetic program? Vitalism and infinite regress (who programs the programmer that programs . . .), close friends of the biologist's mistress teleology, are waiting in the wings.

For life, the system must be open and not at equilibrium. From this openness emerged specific molecular configurations (e.g., DNA) that guarantee the continuity of life. But it takes more than DNA to build a living organism. As has long been recognized, Darwinian selection presupposes the existence of self-sustaining structures such as the gene. It does not explain how that particular configuration was selected from the primordial soup.[12]

We know from synergetics, the theory of pattern formation in nonequilibrium systems, that it is always the longest-lasting mode that survives the competition in complex systems (cf. chapter 1). In contrast to artificial machines, the gene is far more likely to be a self-organized, functional unit, a metastable, dynamical form that relies on the system's openness for its creation, integrity, and self-maintenance.

Yet a new aspect may require expansion of theories of self-organization in the biological realm. Whereas control parameters in self-organized, dynamical systems are nonspecific, moving the system through its collective states but not prescribing them, the gene may reflect specific parametric influences on biological processes such as development.[13] As Lewis Wolpert points out,[14] there are no genes for arm or leg per se, but specific genes do become active during the formation of an arm versus a leg. But unlike Wolpert, I think we should avoid talking, as he does about "*programmes* for development," thereby mixing the metaphor of computations and the physics of self-organization. Just because a specific gene is turned on at some point in the developmental process does not mean that the organism computes itself from its DNA. To understand biological development, I would rather build on the natural laws of pattern formation in nonequilibrium systems than employ the special rules and concepts of relatively recent artifacts such as the computer.

In large part, the second half of this book deals with specific parametric contributions to self-organized, pattern-forming dynamics. Rather than playing the role of a program sending instructions, intentionality is viewed as an integral part of the overall orchestration of the organism. Formally, an intention is conceived as specific information acting on the dynamics, attracting the system toward the intended pattern. This means that intentions are an *intrinsic* aspect of the pattern dynamics, stabilizing or destabilizing the organization that is already there. Before we flesh these ideas out, let's examine briefly some of the ways intentionality is treated in the behavioral and brain sciences.

The Will to Act: The Bohr and Preselection Effects

A lovely story is told about the physicists in Niels Bohr's institute in Copenhagen when they were developing and extending Bohr's theory of the atomic nucleus.[15] Theoretical physics is a creative, intensive, and laborious business. To find relief, the physicists would sometimes take themselves off to western and gangster movies, which Bohr himself loved. During gunfights, Bohr, who was extremely perceptive, noticed that the cowboy or gangster who drew first always lost. This outcome had nothing to do with who was the bad guy and who was the good guy. Bohr's young collaborators were extremely skeptical and decided to put his hypothesis to experimental test. Toy pistols were obtained and mock duels set up. Invariably, Bohr's observation proved to be correct. A new theory was formulated, prophetic in a way, of a policy that came to be called nuclear deterrence. In the theory, when the two cowboys face each other in a showdown, all they can do is talk, because both know that the one who shoots first will die.[16]

How is it that the will to act, from the setting up of the intention and its eventual realization by the muscles, takes longer than the *reaction* itself? One reason is that planning, the conversion of a complex multivariable system with many redundant biomechanical degrees of freedom into a controlled action, involves the entire nervous system at many levels. Physiological indicators of task-related processes variously named "anticipatory postural adjustments," "preparatory set," "gating," "gain control," and "selective attention," are hence ubiquitous.[17] Just ask a waiter with a tray full of glasses. Take one away without him seeing and you might upset the tray. When he does it himself no such problem arises. So-called feedforward stabilization is available in the latter but not the former case.

Whereas in the Bohr effect[18] there appear to be detrimental consequences of willing an action, there can also be benefits. For example, if a blindfolded human makes a volitional movement of his or her own choice and some time later is asked to reproduce that movement, performance accuracy is far better compared with situations in which the subject is passively moved by the same amount, or moves him- or herself to a stop defined by the experimenter. George Stelmach, Steve Wallace, and I named this superior memory performance the *preselection effect*.[19] It was and is one of the most reliable, well-studied, but least understood effects in the behavioral literature.

Brain Correlates of Intentional Action: The Bereitschaftspotential

One has to believe that the Bohr and preselection effects are strongly related to events in the brain that occur before any overt movement. The earliest sign of cortical activity starts about one second before the onset of movement and is characterized by a slow negative potential shift in the electroencephalogram (EEG), which records electrical activity of the brain (figure 5.1). This is called the *Bereitschaftspotential* (BP) or *readiness potential* and was discovered by the German neurologist Hans Kornhuber and his Austrian co-worker Lüder Deecke in the early 1960s.[20] Even for a finger movement of one hand, the BP is present in both hemispheres over parietal and precentral regions of the cortex. A progressive sharpening of the potential then occurs until, about 60 msec before movement onset, a motor potential (MP) is seen in the hemisphere opposite the hand that executes the movement. The onset and topography of this MP suggest that it reflects the activity of primary motor cortex. Of course, considerable evidence also indicates that subcortical structures such as the cerebellum and basal ganglia are active before the final motor cortex discharge.[21]

Is the readiness potential the cortical correlate of the intention to act? As a physiological indicator, it appears to be, but in my opinion this does not tell us very much. The reason is that the BP itself is generated by a complex of spatiotemporal activity patterns among several neural structures. Neurologists and neuroscientists have hypothesized about which particular structures contribute to the BP, but have yet to come to grips with the neuronal operations and interactions among the anatomical structures that create the BP and determine its temporal evolution. New brain imaging techniques and computational methods are going to be important in answering such questions (see chapter 9). For the moment, however, a great deal of attention has been given to one particular area of the brain, the supplementary motor area (SMA) where the largest BP is typically observed.

Origins of Intention?

Does the Bereitschaftspotential in the SMA reflect the intention to act? Anatomically, the SMA is situated within the superior frontal gyrus of the cerebral cortex, and has direct and indirect connections to many other cortical and subcortical structures, including the spinal cord. Thus, it seems to be in a perfect position to influence both brain and spinal mechanisms involved in goal-directed action.[22] Single unit activities recorded from SMA neurons are

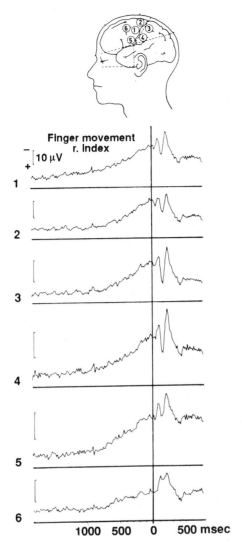

Figure 5.1 The readiness or Bereitschaftspotential recorded by scalp electrodes in response to rapid voluntary movements of the right index finger. Zero time is the onset of movement, the potentials being back-averaged from that point. (From reference 20. With kind permission of the author and Springer-Verlag GmbH & Co. KG.)

known to *precede* discharges in other areas of the brain before voluntary movement.[23] Moreover, measures of regional cerebral blood flow in humans show enhanced activity in SMA even when subjects are asked to *think through* a movement, but not actually perform it.[24].

It is well known that patients with lesions in the SMA reveal a pathological tendency to perform symmetrical, in-phase movements. For example, a person with a bilateral lesion of the frontomedial cortex who is asked to close one hand while simultaneously opening the other closes both hands in phase with each other.[25] The Vienna group led by Lüder Deecke and Wilfred Lang have recently produced very exciting results that relate the SMA to *task complexity*, in particular, the kinds of bimanual activities we studied and theoretically modeled in previous chapters. They placed electrodes on the vertex, C_z, or a bit frontal to the vertex, FC_z (around the top center of the subject's head), to pick up the electrical activity of the SMA. Then they asked students of the Vienna School of Music to perform difficult rhythms between their hands. Every pianist knows how hard it is to play quavers with one hand and triplets with the other. Playing quavers on both hands, however, is easy. Lang and colleagues report an astonishingly large activation of nearly 20 μV, according to Deecke, the largest they have seen in many years of brain research.[26] And where is this extraordinary activation located? In C_z, directly over the supplementary motor area. The SMA appears to be involved not only in initiating action, but also in its spatiotemporal coordination, especially when both sides of the body are involved.

To summarize, conventional wisdom says that goal-directedness in biology is due to a genetic program. Recent studies of brain activity prior to action suggest that goal-directedness in the human brain is the result of a single neuroanatomical structure, the SMA, that determines the right moment to start a voluntary act.[27] The reasoning behind both views is actually quite similar. Indeed, one explanation may be seen as perfectly continuous with the other. The neuronal program initiated by the SMA is a *somatic* program laid down during development under control of instructions from the *genetic* program. In both views, what gives living things their teleonomy is a program, a construct borrowed from a human-made artifact. In chapter 9 I'll offer a very different view of Sherrington's "enchanted loom" (the brain) and its relation to behavior. There I will try to come to grips with the immense complexity of brain patterns as they emerge in space and time when humans are confronted with specific tasks. But for the moment, let's consider how the present approach comes to grips with informational or intentional aspects of biological systems.

THE SECOND CORNERSTONE OF BIOLOGICAL SELF-ORGANIZATION: INFORMATIONAL SPECIFICITY

Walker, your footsteps are the path, the path is nothing more;
Walker, there is no path. You make the path by walking.
—Antonio Machado

Coordination Dynamics Is Informational

Like mind and matter, the concepts of information and dynamics have long been held distinct and separate. Usually, they are taken to refer to *fundamentally different but alternative modes of describing* complex systems. Information is assumed to be discrete and symbolic; dynamics is continuous and rate dependent.[28] An intention, for example, may be said to harness physical laws of motion in some mysterious way, but is nevertheless independent of those laws.[29] Intentions and dynamics thus appear to be irreconcilable. If Newton rules biology, as most people think, intention as cause of motion lies outside scientific explanation. Mind simply cannot get into motion. The two are logically incompatible.

But look what is done here. Instead of treating dynamics as ordinary physics using standard biophysical quantities such as mass, length, momentum, and energy, our *coordination or pattern dynamics* is informational from the very start. The order parameter, ϕ, captures the *coherent relations* among different kinds of things. Unlike ordinary physics, the pattern dynamics is context dependent: the dynamics are valid for a given biological function or task, but largely independent of how this function is physiologically implemented. Thus, if we accept that the same order parameter, ϕ, captures coherent spatiotemporal relations among different kinds of things, and the same equations of motion describe how coordination patterns form, coexist, and change, it seems justified to conclude that order parameters in biological systems are functionally specific, context-sensitive *informational* variables; and that the coordination dynamics are more general than the particular structures that instantiate them.

Notice, coordination dynamics is not trapped (like ordinary physics) by its (purely formal) syntax. Order parameters are *semantic*, relational quantities that are intrinsically meaningful to system functioning. What could be more meaningful to an organism than information that specifies the coordinative relations among its parts or between itself and the environment? This view turns the mind-matter, information-dynamics interaction on its head. Instead of treating dynamics as ordinary physics and information as a symbolic code acting in the way that a program relates to a computer, dynamics is cast in terms that are semantically meaningful. The upshot of this step, which, I stress is empirically motivated, is that intentions do not lie outside self-organized coordination dynamics. They are an intrinsic, inseparable part of them. Intending and doing are but two aspects of a single behavioral act.

What Intentions Do

The thesis developed here to explain the enormously complex nature of interactions within an organism and between organisms and their environment is that new kinds of coupling forces have to be introduced whose observables relate to coordination. Such coordination variables (order parameters) capture the (often long-range) informational coupling among the components of the system. Evidence reviewed in previous chapters confirms the hypothesis that self-organized coordination dynamics are, at their very roots, *informational*. All this evidence comes from experiments that manipulate control parameters in a nonspecific fashion, research that constitutes the first cornerstone of self-organized coordination dynamics.

How, then, does intentionality fit in? The key point, mentioned briefly in the introduction, is that once the relevant pattern variables and their dynamics are identified, it is possible to talk rigorously about *what* it is that is changed (or indeed is changeable) by specific parametric influences such as an intention. Such specific information is expressed in terms of the same order parameters or collective variables that characterize observed spontaneous coordination tendencies. Intentions, in other words, are written in terms of the very order parameters that characterize the coordination activity of the nervous system. Theoretically, intentional information has no meaning outside its influence on the order parameters. The full coordination dynamics—spontaneous coordination tendencies perturbed or forced by specific parametric influences (here, intention)—is an open informational unit, not a closed physical system acted on from the outside.

What might the operational consequences of all this be? One prediction is that intentions should be constrained by properties of existing coordination tendencies or patterns, in particular, their relative stability. For example, I may want to produce an antiphase pattern of coordination between my hands, say, at ten cycles per second, but the unstable nature of that pattern at that rate limits my intention. Or, I may want to produce a phasing pattern of, say, 72 degrees between the hands, but without a great deal of learning, that pattern cannot be stabilized (see the next chapter). On the other hand, intentional information should be capable of stabilizing (within limits) and destabilizing already existing intrinsic patterns such as in-phase and antiphase. Let's see if any of these ideas can be grounded in fact.

INTENTIONAL BEHAVIORAL CHANGE

The statement that an intention is not arbitrary with respect to the underlying order parameter dynamics it modifies is not merely some radical claim or philosophical commitment. It can be shown to work in the scientific sense of an explicit mapping between theory and experiment. Which came first, the reader inquires, the theory or the experiment? Well, they sort of co-evolved as a result of a collaboration among three of us.[30] So let's start with the intuitions that led to the experiment that led to the theory that led to ...

As I mentioned before, despite the fruitfulness of biological and philosophical approaches to intentionality, nevertheless a discrepancy exists between the functional definition of intentional behavior (e.g., as a belief or desire)[31] and how such behavior is studied experimentally. A goal-directed behavior is one that is undertaken precisely because it tends to produce the desired result. But to what degree does an organism's existing organization constrain its intentions? Put another way, what can be intended? In general, such questions are not easy to answer because the levels of description (intentions and behavior) are so far apart, so incongruent.

This is precisely where the strength of the present approach reveals itself (I'll address its limitations later on). Recall our previous work on hand movements demonstrating intrinsic coordination tendencies and spontaneous transitions between them. Since the intrinsically stable patterns are already attractors of the coordination dynamics, we can assume that they can be intended. But remember also that the antiphase pattern is inherently less stable than the in-phase pattern. This suggests that the relative stability of the two attractors may influence the process of *intentional switching* between them. For example, since the antiphase pattern is intrinsically less stable than the in-phase pattern, the system should be able to switch faster from the former to the latter than vice versa.

A further intuition is that an intention can change or alter the dynamics of these intrinsic patterns, for example, stabilize or destabilize them. In my earlier work described in chapter 2, I showed that it is possible, through deliberate conscious effort, to stabilize the antiphase mode, thereby delaying spontaneous transitions. This was not possible, however, without some cost, for example, in terms of amplified phase fluctuations and the amount of energy required to sustain the antiphase mode at higher frequencies. From such results, one may construe intention as specific information acting on or perturbing the basic (intrinsic) coordination dynamics.

Note once more that the conceptual distinction between the basic coordination dynamics and specific intentional information is meaningful only if the former are identified under conditions that do not involve intentional change (i.e., where patterns evolve spontaneously). Indeed, it is this essential feature of the approach that enables predictions about the process of intentional behavioral change to be possible in the first place.

A nice way to visualize the consequences of this theoretical perspective is through the potential pictures of the original HKB model. On the top left of figure 5.2 is the basic coordination dynamics, $V(\phi)$, in the absence of an intention to change behavior. The two intrinsic attractors at in-phase ($\phi = 0$ degrees) and antiphase ($\phi = 180$ degrees) are shown. On the top right is shown the potential, $V_\psi(\phi) = -c_{intent} \cos(\phi - \psi)$, corresponding to an intentional force of a given strength, c_{intent}, in which the solid line represents an intended pattern that is in phase ($\psi = 0$ degrees) and the dotted line an intended pattern that is antiphase ($\psi = \pm 180$ degrees). Note that in this figure the two intended patterns are equally stable because intrinsic coordination tendencies are absent.

The full coordination dynamics—the superposition of intrinsic and intentional contributions ($V(\phi) + V_\psi(\psi)$)—is displayed in figure 5.2 (bottom). The system now has only one minimum at the intended pattern, reflecting the fact that each pattern can be established by a specific intention. Intentional information, in other words, breaks the symmetry of the coordination dynamics. We see, in addition, that the little ball will travel much faster down the steeper slope (from $\phi = 180$ degrees to $\phi = 0$ degrees) than vice versa. We also see that adding an intentional force to the less stable of the two patterns at $\phi = 180$ degrees will stabilize it (dotted line in figure 5.2, bottom). Not only that, but one might imagine that a very strong intentional effort may be able

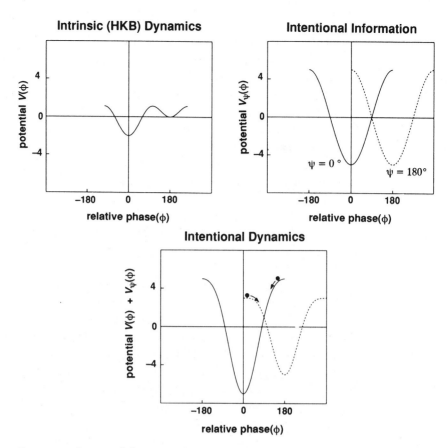

Figure 5.2 Intentional dynamics. (*Top left*) The HKB intrinsic dynamics as a potential, $V(\phi)$. (*Top right*) Intentional information as a potential, $V_\psi(\phi)$, in the absence of spontaneous coordination tendencies. A single parameter, c_{intent}, determines the strength of the intention to produce one pattern or the other. *Bottom.* The full intentional dynamics, $V(\phi) + V_\psi(\phi)$. The little ball travels much faster down the steeper slope at intended phase $\psi = 0$ than $\psi = 180$ degrees, consistent with experimental observations.

to stabilize the antiphase pattern beyond the critical transition frequency at which it typically (i.e., spontaneously) loses stability.

How might these joint predictions—first, that faster switching should occur from less stable to more stable patterns than the converse, and second, that intentions can change the dynamical stability of the patterns—be subjected to experimental test? Scholz and I devised an experiment in which the subjects' task was initially to cycle the fingers either in phase or antiphase. Subjects were paced for ten cycles by a metronome, which was then turned off. Instructions were to continue cycling at the set frequency until an auditory tone came on, indicating that the subjects should switch as fast as possible to the other mode of coordination. Using interactive computer displays of the trajectories and the usual calculations of the relative phase between them, the length of the actual transient (or switching time) between the two modes of coordination was measured. This is the total amount of time from where the

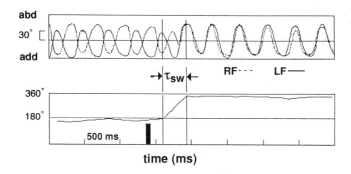

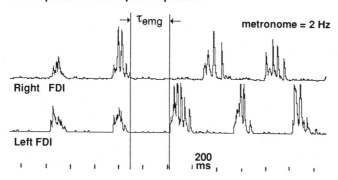

A. in–phase to anti–phase pattern

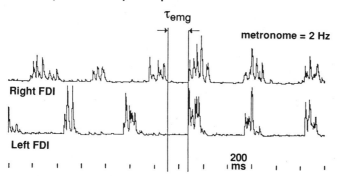

B. anti–phase to in–phase pattern

Figure 5.3 (*Top*) Switching time estimate from kinematics, including the time series of the right (RF) and left (LF) index fingers and their relative phase. Solid bar = switching pulse, instructing the subject to switch from one pattern to the other. *Bottom.* Switching time estimate from the EMG of the right and left first dorsal interosseus (FDI) muscles, showing a voluntary switch from (A) in-phase to antiphase and (B) antiphase to in-phase.

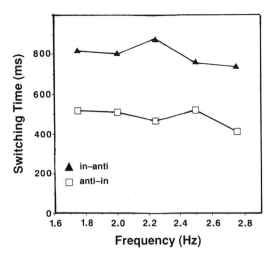

Figure 5.4 The experimentally obtained switching time (mean duration of transient from one mode of coordination to the other) averaged over subjects and experimental trials as a function of the required oscillation frequency (Hz). Note that it takes much longer to switch from the more stable in-phase pattern to the less stable antiphase pattern than vice versa.

relative phase, ϕ, first diverged from the previous mean state ($\phi \approx 0°$ or $\phi \approx 180°$) to when the relative phase stabilized ($\pm 15°$) for two consecutive cycles. These measures were corroborated independently by EMG readings (figure 5.3).

The results were unequivocal. Intentional switching from the antiphase mode to the in-phase mode was almost twice as fast as the other way around (figure 5.4). This makes perfect sense; in theory, the intention has to act only as long as is necessary to get the system into the basin of attraction of the intended mode. Then, because the hill is steeper for the in-phase mode (the slope of the potential is greater, see figure 5.2), the system runs there much faster than the other way around.

Schöner calculated the theoretical distribution of switching times for the two directions of switching. These distributions are shown in figure 5.5 (top); note the sharper peak and shorter tail for switching to the more stable pattern (in-phase, solid line). The actual distribution of switching times for switching from in-phase to antiphase and antiphase to in-phase is shown in figure 5.5 (bottom). Comparison between the Schöner-Kelso theoretical model and the Kelso-Scholz experimental results is quite striking, both qualitatively and (within the inherent bounds of accuracy) quantitatively.

More generally, all subjects were able to intentionally maintain the antiphase pattern, as well as intentionally switch from the in-phase to the antiphase pattern even at frequencies well beyond each subject's *spontaneous* transition frequency. Thus, the system is able to intentionally sustain a pattern that is intrinsically unstable. This is because, in our theory, an intention, as a crucial part of the dynamics, is able to change the stability of the intrinsic attractors (see figure 5.2). Thus, intention acts to parametrize the intrinsic

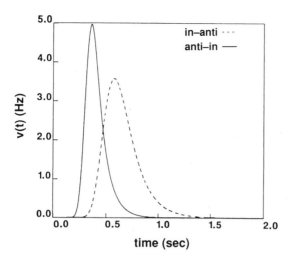

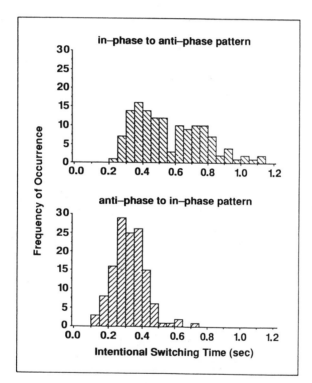

Figure 5.5 *(Top)* The probability density of switching time for both directions of switching as obtained from the Schöner-Kelso theory of intentional behavioral change. The only adjusted parameter, c_{intent}, was chosen to account roughly for the observed order of magnitudes of the mean switching times. The solid line is the antiphase to in-phase distribution; the dashed line is the in-phase to antiphase distribution. *(Bottom)* The experimentally obtained frequency distributions of switching times collapsed over subjects and trials for both directions of switching.

dynamics, stabilizing one pattern and destabilizing the other and vice versa. *Yet the presence of the intrinsic pattern dynamics—preferred coordination tendencies —is always felt.*

Naturally, intentionally sustaining intrinsically unstable patterns has its costs. Were this the usual information processing or cybernetic closed-loop account of voluntary behavior, one might say the errors (deviations between the intended and actual phase states) are much greater for antiphase patterns. Here, however, the level of fluctuations actually aids intentional switching from less stable to more stable states and hinders transitions in the opposite direction. Intentional behavior, in other words, is strongly influenced by the stability of the patterns.

We can generalize this approach to other levels of description and other kinds of movements as well. Of course, my hope is that the conceptual framework of intentional dynamics can be adopted quite generally. One benefit of having a specific theoretical model is that it allows us to understand features of our experimental results that may not have been otherwise obvious. For example, due to our knowledge of the individual oscillators (see

anti–phase to in–phase pattern

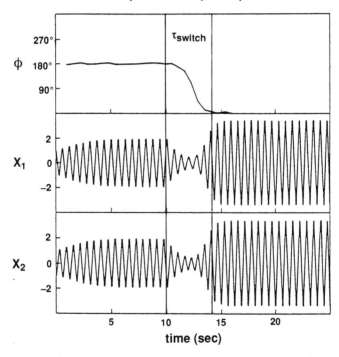

Figure 5.6   Intentional dynamics at the component level. (*Left*) Antiphase to in-phase switching. At $t = 10$ seconds the intention to move in-phase comes on and the oscillators switch within approximately 3 seconds. (*Right*) In-phase to antiphase takes longer. At $t = 10$ seconds the intention to move antiphase becomes active. The oscillators then switch to an antiphase pattern within about 5 seconds.

chapter 2), intentional switching among coordination modes can also be modeled on the component level. It turns out that only a linear coupling term is required to perturb the oscillator dynamics intentionally. When the intended pattern, ψ, is zero, this coupling stabilizes the in-phase pattern; when it is π or 180 degrees, it stabilizes the antiphase pattern.

Figure 5.6 (left) shows a typical computer simulation that starts close to the antiphase condition, to which the model oscillators relax quickly. At $t = 10$ seconds, an intentional force specifies the in-phase mode. Within approximately three seconds the oscillators switch to in-phase. Figure 5.6 (right) shows the opposite in-phase to antiphase switching at the same parameter values. Just as in our experiment, switching in this case takes considerably longer.

No attempt has been made here to fit experimental results by playing around with parameters. Rather, the simulations show the main qualitative effects of intentional switching at the oscillator level of description. On the one hand, the influence of the relative stability of the intrinsic patterns on the intentional switching process at this level is clear. On the other, it is obvious that an intentional perturbation of a given strength can stabilize one coupled dynamic pattern and destabilize another, just as theory would have it.

in–phase to anti–phase pattern

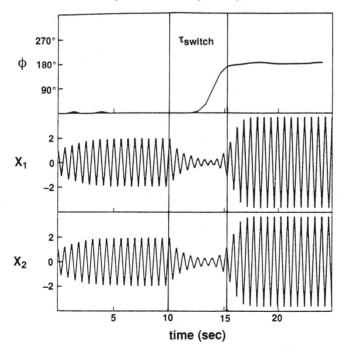

Figure 5.6 (continued)

RELATED VIEWS: TERMITES, PREDATOR-PREY CYCLES, AND QUANTUM MECHANICS

What have termites and predator-prey capture got to do with intentional dynamics? Perhaps not much, but they may share some intriguing, although fairly abstract relations. Let's see what they are. Termite architecture—the building of a nest by a group of social insects—is a fascinating example of self-organization. A team of insects, only locally informed of each other's activities and quite ignorant of the overall societal goal, exhibit collective intelligence by their remarkable ability to construct nests, pillars, arches, and domes. Belgian physical chemist J. L. Deneubourg played a major role in understanding the mechanisms behind the problem-solving abilities of social insects.[32] It turns out that insect societies, with no brain or central executive, have developed decision-making capabilities that depend only on communication among individuals and physical constraints of the system. This kind of decision making, such as the selection of the nearest food source, functions without any symbolic representation whatsoever. There is no blueprint, no a priori program for the insects' coordinated activity: no single individual knows the alternatives or contains an explicitly programmed solution. Only by coordinating their activity is a collective decision (unconsciously) reached.

Bugs are not people, however, but it has been noted that, just as insects go through a set of modes representing qualitatively different kinds of behaviors, so a person weaves a path among different modes in daily life.[33] This was one of the main themes discussed in the previous chapter. Remember in Chapter 4 how Iberall and McCulloch describe a set of action modes that they claim are connected with the internal states of the organism, such as sleeping, eating, drinking, copulating working, envying, and so forth. Over time, within the complex of possible patterns, the system begins to adopt those patterns that suit it. As Iberall and McCulloch remark, "The system does not drift through life aimlessly. It is unstable. It selects both foci of behavior and orbits. These 'suit' the individual. They involve more and more routines that become sub-cortical. *They begin to form a patterned field in the brain* (italics mine). People become doctors, drunkards, woman(man)-chasers, intellectuals, politicians." Sheer poetry. But nothing, I might add compared to their discussion of the "great flaring maturational instabilities" (e.g., at adolescence, when "the genital interface explodes on the scene").

Extrapolating from the insect example and the notion of action modes, Kugler and colleagues hypothesize that each mode is constrained by a unique number of attractors. They propose an internal field within the organism that itself is organized by the layout of attractors. Such attractors are viewed as goal states that act to constrain the system's internal degrees of freedom. Changes in the number and configuration of goal states *qua* attractors result in a change in the field that causes qualitative shifts to new action modes.

This story has, I think, a ring of truth to it, and bears some resemblance to our analysis of intentional dynamics. It is most necessary, however, to identify what the relevant degrees of freedom in the system are, how the hypothesized attractors are defined, what their layout corresponds to, and so forth (cf. figure 5.2). To posit an internal field is to presuppose that one can calculate its variation in space-time. This remains a future task for the above hypothesis. Once (and if) these requirements are met, it may then be possible to predict something and devise experimental tests.

Notwithstanding the challenge that understanding intentional goal-seeking behavior presents, similar kinds of problems confront René Thom's notions of pregnant forms.[34] In Thom's theory, pregnant forms carry biological significance for the animal and include such interactions as prey for hungry predators (and predators for their prey) and the readiness of one sexual partner for another. Recognizing these forms produces a stream of hormonal, emotional, and behavioral effects on the perceiver designed to attract or repulse the inductive form. Unlike the closely related Gestalt notion of *Prägnanz*, Thom uses the term "pregnance" (remember, he's a Frenchman), not to characterize the morphology of the stationary state itself, but more dynamically in terms of those factors that *trigger* or *regulate* pathways *toward* stationary states. His difficult book *Semiophysics* contains a number of fascinating and imaginative ideas ranging from interpretations and reinterpretations of Pavlovian conditioning to embryology.

Thom makes contact with our theory, in which intentional information acts on or perturbs the intrinsic dynamics, with his notion that the pregnant form triggers actions that aim to obtain satisfaction or avoid pain. The difference is that Thom models the trigger character of pregnant form by the *tunnel effect*,

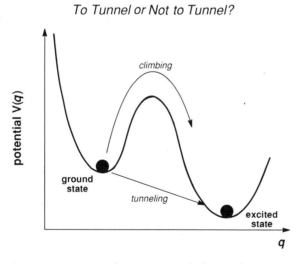

To Tunnel or Not to Tunnel?

Figure 5.7 To tunnel or not to tunnel? Illustrated is quantum mechanical tunneling out of a trap through a barrier. In QM, a particle (little ball) has some probability of penetrating a barrier or wall that certainly stops a classical particle. Tunneling, according to Thom, occurs due to the apparition of a pregnant form. Also shown is the particle jumping over the barrier, which takes more energy.

well known in quantum mechanics. Thom conceives an animal as normally resting in a ground state, a local minimum surrounded by basins of attraction that possess even deeper minima corresponding to excited states (figure 5.7). The perception of a pregnant form (say a prey) creates a tunnel effect, switching the animal into an excited state. Once the prey is caught and eaten, satisfaction lifts the basin of attraction of the excited state and eventually annihilates it. The animal then returns to the ground state until sometime later, when hunger strikes again, and the basin of attraction surrounding the excited state is recreated.

What's the take-home message from this discussion? I admit that not much seems to be gained by introducing the concept of pregnance and pregnant forms to explain goal-directed behavior. Yet the attempt to treat such concepts in terms of dynamical systems, rather like the action modes of Iberall and McCulloch, is intriguing. Indeed, one might construe my efforts here as an attempt to clothe the action mode idea in terms of identification of relevant variables, control parameters, and so forth. What appeals to me about Thom's formulation, and the reason I mention it at all, is the notion of tunneling. I have often asked myself why a biological system should have to climb over a barrier in order to switch state. The transitions I have been talking about in this book do not involve much energy at all. Outside of gait changes in horses (and perhaps even these under natural conditions) switching is *informationally* based. The couplings between things are informational, not force mediated in the conventional sense.[35] In the theoretical models described so far, parameters (nonspecific and specific) act to deform or raise and lower basins of attraction surrounding (nonequilibrium) steady states. The formulation is not of the quantum tunneling type, but the *effects* are the same. Do macroscopic, informationally based neurobehavioral dynamical systems tunnel through barriers? Photons, the elementary particles that emanate from your light bulb, tunnel. Do brains? (See box.)

SUMMING UP

A dog, as the great Sherrington remarked, not only walks; it walks to greet its master.[38] What inroads, if any, have we made on the issue of intentionality? Ask yourself the question, if a dog on a leash is tied to a tree, what free will can it exert? The answer is only in the space of a circle whose radius is defined by the length of the leash. This represents a fixed, rigid *physical* constraint on the dog's intentions. What I have tried to demonstrate here is the significance of functional dynamical constraints with respect to intentional behavioral change. The stability of the intrinsic phase dynamics influences whether a given intention can be carried out or not, and how quickly. To some degree coordinative stability restricts *what* can be intended, although not, one hastens to add, what can be imagined.

Here I have shown how the two languages of intentions and dynamics, usually considered irreconcilable, may actually be captured in one unified

QUANTUM MECHANICS

It would be disingenuous of me not to acknowledge that this question is not merely a rhetorical one. In the last couple of years a number of papers have appeared on this very issue, namely, how quantum mechanics (QM) may account for mental influences on the brain. In one, the theoretical physicist Friedrich Beck and the eminent neurophysiologist and Nobel laureate Sir John Eccles argue that mental intention triggers the probability of *exocytosis* in selected cortical areas such as supplementary motor neurons.[36] Exocytosis refers to the (probabilistic) release of neurotransmitter into the synaptic cleft, the summed activity of which results in depolarization and the discharge of an impulse by a cortical cell. Beck and Eccles model this hypothesized trigger mechanism by a QM tunneling process in a potential. They propose that by momentarily increasing the probabilities for exocytosis, mental intention couples the large number of probability amplitudes of the synaptic boutons into coherent action.

In contrast, using an ontology due to Heisenberg, physicist Henry Stapp argues that every conscious event is an actual event (a quantum jump) that selects large-scale, metastable patterns of neuronal activity in the human brain.[37] Whereas Beck and Eccles' view of the relationship between mind and matter is dualistic, Stapp argues that physical events in the brain and psychic events in the mental world are images of each other under a mathematical isomorphism. It remains to be seen if either or neither of these views bears fruit in the sense of experimentally testable predictions. But one cannot discard them summarily.

picture. Specifically, an intention acts in the same space of order parameters as that in which preferred coordination tendencies—action modes if you like—are observed and measured. Intentional information defines an attractor in the space and is meaningful to the extent that it attracts the system to an intended behavioral pattern. At the same time, intentions are constrained according to the intrinsic attractors of the coordination dynamics. The ability to generate a particular pattern is influenced by the relative stability of the available patterns. *Circular causality* shows up yet again: intentions parametrize the dynamics but are in turn constrained by the dynamics.

An important limitation of the theory of intentional dynamics is that it is only when the dynamics of the relevant collective variables are known (e.g., by the phase transition strategy) that theory has predictive value. Of course, this is as much of a challenge as it is a limitation to understanding intentional systems in which the relevant informational variables and their dynamics are not known, but have to be found. Newtonian mechanics is not much help at this higher level, and it is still unclear what use QM can be. As shown here, extending the concepts and tools of synergetic self-organization seems rather more fruitful.

For over 100 years psychologists, physiologists, and neurologists have used reaction time to infer the nature of cognitive processes before overt action occurs. Little attention has been paid to the actual behavior itself, which typically involves a key-pressing task. The tool employed here—switching time—shown to reveal dynamic constraints on intentionality may now be put to more general use. For instance, in cases where the behavioral pattern dynamics are not known and phase transitions are difficult to find, switching time measures may provide information about the relative stability of the underlying patterns. Such methods may offer a new view, even a paradigm, for understanding behavioral dynamics.

On a more general, philosophical note, and returning to where we began, intentionality is typically taken to demarcate mental from physical phenomena. Purpose, according to some,[39] although an essential part of biology is more a point of view, analyzable in neither chemical nor physical terms. Presumably that is why the influence of the computer has been so strong. The concept of program—genetic program, somatic program, motor program— seems like the only resort when it comes to matters of goal-directedness in biology. And the guys in the trenches, doing molecular genetics and molecular neuroscience, for the most part don't care. But they should.

When physics and biology are torn asunder on the issue of purpose and intentionality, an enormous void is created. Here I have argued that an *appropriate* physics of dynamic pattern formation and cooperative phenomena in open, nonequilibrium systems is a step toward filling this void. For the most part, not the evolutionary biologists (e.g., Mayr), the developmental biologists (e.g., Wolpert), the geneticists (e.g., Berg and Singer), the neuroscientists (e.g., Edelman and Granit), or the physicists (e.g., Penrose, Hawking, and Weinberg) appear to be aware of this appropriate physics when they discuss the topic of intentionality and consciousness. Obviously, the strategy employed here is motivated by the physical concepts of nonequilibrium phase transitions (synergetics). Identifying order parameters for self-organized behavioral patterns proves to be the crucial step in establishing a theory-experiment relation, making it possible to define precisely those aspects of behavior that modify intention and that can be intentionally modified. The result is that, far from severing mental and physical phenomena, the *language* of our admittedly primitive analysis links them together inextricably. Planning and execution are but two aspects of a single act.

6 Learning Dynamics

What we have to learn to do, we learn by doing.
—Aristotle

Learning is the process of acquiring skill. It involves a change of behavior through practice or experience. The ability to learn is a chief characteristic of living things, endowing the organism with a means to escape its limited built-in behavioral repertoire. Yet the topic of learning is plagued by a variety of definitions, approaches, theories, and methods. Traditionally, learning was psychology's baby, but nowadays much of the action is in the neurosciences and related fields. How does your brain change when you learn and remember? Or, more important, at least to my mother-in-law's friend Celia, why do you forget the older you become? Answers to these questions are being sought at many different levels within the nervous system, from studies of molecular changes in a single neuron all the way up to how neurons communicate with each other.[1]

My tack in this chapter is a bit different. I aim to answer a small set of questions concerning how ordinary human beings learn new skills. *What changes with learning?* What is the *nature of change* due to learning and how do we explain it? What factors govern the *rate* of learning? And the question on every parent's lips (especially those whose children attend public schools): How on earth does learning occur at all?

I am comfortable with the fact that different kinds of learning are likely realized by a wealth of cellular and molecular mechanisms. Rather than try to decipher these, my goal is to seek *generic principles* of adaptive change on the assumption that some of them might be at work in other systems and at other levels as well (see chapter 8). I make no attempt to be exhaustive in my treatment of the various approaches to learning. All I can provide is a broad and rather personal sweep. On this point, I beg the professional reader's forbearance, if not forgiveness. I hope that any novelty contained in the present perspective, or insights provoked by it, will compensate for the lack of comprehensive treatment of other people's work.

The main thrust of this chapter is to change the way we view the issue of learning, at least at the level of coordinated behavior, and to set the agenda for future developments. Another goal is to understand the nature of change itself, to clarify the mechanisms that determine whether change is slow or fast, smooth or abrupt. Finally, I want to establish a conceptual connection between dynamic processes that occur on vastly different time scales, as in continuing behavior, learning, development, and evolution. My experimental focus on these issues, is, of course, how humans learn new perceptually specified coordination patterns. In short, the problem of perceptual-motor learning.

ISSUES IN LEARNING

Many studies of learning, indeed, most traditional approaches, select arbitrary responses for people to learn. Even then, concerns are usually raised about how instinctive patterns (the built-in repertoire presumably possessed by the organism) interfere with the learning of such arbitrary behaviors. Everyone knows the individual isn't a tabula rasa, but it is difficult—usually impossible—to characterize the initial global state of the learner. So what do we do? The usual solution is to equate for different initial states (individual differences) by having subjects learn as novel a skill as possible. Then we feel comfortable about averaging their performance over learning attempts or trials and generating the familiar learning curve. In such approaches, the individual is just a statistic. Everybody is treated the same. Any differences due to experience, maturation, ancestry, or what the subject had for breakfast are canceled out. The organism, to put it bluntly, is treated like a machine whose task is to associate inputs and outputs. Any autonomously active, intrinsic organization within the organism or between organisms and their environment, although present, is swept under the rug.

The same kind of criticism can be levied at cellular and molecular approaches to learning that rely, say, on classical conditioning.[2] In classical conditioning, first described by Russian physiologist Ivan Pavlov, a previously ineffective stimulus paired and appropriately timed with an effective stimulus eventually produces a large (or different) behavioral response. From this point on, analysis of mechanisms proceeds at ever more intricate levels down to second messengers, receptor binding at the cell membrane, and so forth. All terribly interesting and important. However, the basic question of *why* a particular stimulus is effective in producing a response in the first place, or why some stimuli are effective and others are not, is crucial to understanding learning, even though it is often ignored at the molecular level.

Similarly, a veritable avalanche of research on artificial neural networks places high priority on learning and associative memory. Much attention is focused on modifying the synaptic strengths or coupling constants among the component neurons as the main learning mechanism. In such networks, neurons (sometimes in analogy to the spin of the atoms in a glass or metallic mixture) take on the values of 1 or 0—they are either on or off. Each element becomes fixed in a direction depending on the coupling constants and the position of other spins. For a given set of coupling constants (synaptic strengths), there are many stable spin configurations. Hence, attractors of the spin dynamics can be used to model content addressable memory or associative recall. This means that all stimuli in a given basin of attraction will retrieve the same memory. Useful though such machines are, biological components are not ones and zeroes, and they often exhibit intrinsic activity patterns among themselves. These aspects of learning machines are almost entirely ignored.[3]

So what's my point? Instead of ignoring the fact that individuals bring different backgrounds and capacities into the learning environment, I argue that these should be carefully evaluated before exposure to a new task. New things to be learned must be linked with intrinsic tendencies or constraints already present in the learner at the time new material is introduced. Learning, in this view, occurs as a *specific modification of already existing behavioral patterns in the direction of the task to be learned*. In general, this means that the individual must be treated as the significant unit of analysis, because each person brings a personal history of experiences into the learning environment. Because each one possesses his or her own "signature," it makes little sense to average performance over individuals. One might as well average apples and oranges. This does not mean that putative laws and principles of learning cannot be generalized across individuals; laws wouldn't qualify as such if it were not possible to do so. It only means that the way the law is instantiated is specific to the individual. The trick, of course, is to uncover sensitive probes (procedures and operations) that render an evaluation of hypothesized intrinsic tendencies possible, always in the context, of course, of what is to be learned. When this knowledge base is tapped, specific predictions about current and future behavior become possible.

Another point about typical learning research is that the continuing process of learning is never directly assessed. Usually, some hypothetical construct located inside the head, such as a *schema*[4] or a *trace*[5], is said to be built up or strengthened as a result of the learning process. Internal changes in the schema may only be inferred through correlated changes in performance that occur with practice. Learning, in this somewhat impotent view, is a covert process forever inaccessible to observation: only the effects of practice may be seen. Needless to say, changes in performance with practice have been studied under a large variety of experimental conditions. The general picture of learning curves that emerges is well known. Practice produces a directed drift of the response in the direction of the task requirement. In plain terms, the subject's performance improves and becomes less variable. Your grandmother could have told you that.

The concepts and methods developed here lead to a rather different picture of the learning process. Far from one thing changing, typically observed as improvement in a single task, I will present evidence that the *entire attractor layout* is modified and restructured, sometimes drastically, as a given task is learned. Learning doesn't just strengthen the memory trace or the synaptic connections between inputs and outputs; it *changes the whole system*. No one knows at present what happens in a complex system like the human brain when learning occurs. Some intriguing PET (position emission tomography) scan findings claim that the overall metabolic activity of the brain drops after subjects have learned a complex task.[6] But measures of brain activity probably drop under many other circumstances, as we will see in chapter 9. At this point such observations tell us only that some global change has occurred in the brain after people have learned a task. The rest of this chapter aims to explain the *evolution* of the learning process, how and why the system changes the way it does.

THE MAIN CONCEPTS

One of my favorite definitions of skill and skill acquisition is that of the late Paul Fitts[7]:

[A] skilled response is ... highly organized, both spatially and temporally. The central problem for skill learning is how such organization or patterning comes about.

I love this definition no doubt because it resonates deeply with how I am trying to understand the problem. Skilled performance for Fitts constitutes a particular kind of spatiotemporal organization, a pattern in space-time. Thus, it is essential to have in one's possession tools and concepts that afford an identification of skilled behavior in terms of the formation and change of patterns. In the context of basic perceptual-motor learning, it is nontrivial to distinguish the difference between skilled and nonskilled behaviors in terms of coordination patterns. Ask yourself, what makes Leonardo's or Picasso's brush stroke different from yours and mine?

Although Fitts certainly didn't approach the problem of learning in terms of pattern-formation processes, his quotation says (at least to me) exactly that. Learning somehow involves the emergence and stabilization of spatiotemporal patterns. It will therefore come as no surprise that I believe the key to understanding learning lies in extending the theory of self-organization in nonequilibrium systems, in particular, to include the key concepts of *intrinsic dynamics* (spontaneous coordination tendencies) and *specific parametric influences* acting on those dynamics. Here, the latter take the form of the task to be learned. Concretely, it is important to stress that what is specific on one level of description may be nonspecific on another. For example, a specific parameter at the component level of our coordination system (e.g., frequency) is nonspecific at the collective, coordinative level. Similarly, at the collective level of description, naturally occurring or experimental manipulations may invoke specific (e.g., required relative phasing) or nonspecific (e.g., frequency) constraints.

The nature of the attractor layout prior to learning must be established to know *what* has been modified or *what* has been learned. This fundamental requirement follows from arguments along exactly the same lines that I developed in the previous chapter on intentional dynamics. The concept of

ON INTRINSIC DYNAMICS

My use of the term *intrinsic* is not intended to invite distinctions in terms of boundaries formed by the skin. Nor does it refer to motivational processes said to be intrinsic to the organism, though it may include them. Nor does it mean that coordination tendencies are innate, though it by no means excludes heredity. Intrinsic dynamics bear no particular relation to falling bodies (classical mechanics) or biomechanics. I use the concept of intrinsic dynamics to convey the idea that patterns may arise due to nonspecific changes in a control parameter (cf. chapter 1). Again, the intent was not to invite distinctions with contrasting terms such as *extrinsic*. Misunderstandings can be avoided (for once!) by an appreciation of the following formalism:

$$\dot{\phi} = f_{intr}(\phi) + f_{inf}(\phi)$$

where the first term on the right-hand side corresponds, say, to the basic HKB dynamics (figure 6.3, left) and the second expresses specific informational requirements in terms of a collective variable or order parameter, ϕ. The above equation may be viewed as a minimal conceptual form of the full coordination dynamics in which specific information, $f_{inf}(\phi)$ is meaningful only in terms of its influence on (relevant) pattern variables.

intrinsic dynamics offers a pathway to solving an issue that learning theorists have historically emphasized but not been able to do much about, namely, that organisms acquire new forms of skilled behavior on the background of already existing capacities. The initial state of the organism never corresponds to a disordered random network, but is already ordered to some degree. Thus, it is very likely that learning involves the passage from one organized state of the system to another, rather than from disorder to order. Schrödinger's order-order transition principle comes to mind once again.

Notice that the concept of intrinsic dynamics is not to be equated with *innate mechanisms*, but rather reflects capacities that exist at the time a new task is to be learned (see box). In this, it departs from the views of ethologists such as Konrad Lorenz and Niko Tinbergen who stressed that learning must be preceded by a study of the innately given behavioral repertoire. Remember, the term intrinsic dynamics simply represents relatively autonomous coordination tendencies that exist before learning something new. Unlike many of the fixed action patterns of the ethologist, we can quantify these coordination tendencies directly. Like the ethologist, however, I view learning very much as a molding or sculpting of intrinsic dynamics.[8]

The other key to understanding learning lies in the way specific information in the form of a pattern to be learned modifies the intrinsic dynamics. Learning is the process by which the pattern becomes memorized. We say that a behavioral pattern is learned to the extent that the intrinsic dynamics are modified in the direction of the to-be-learned pattern. Once learning is achieved, the memorized pattern constitutes an attractor, a stable state of the (now modified) pattern dynamics.[9]

In the previous chapter, intentional information was treated very much along the foregoing lines, but here there are a few additional, quite important steps. First, note that we only treated intention in terms of those patterns that were already contained in the intrinsic dynamics. That was the beauty of the formulation: we knew what patterns could be intentionally stabilized and destabilized. We did not, however, deal with patterns that were not part of the intrinsic dynamics. Obviously, this is the case in learning. Second, changes in the pattern dynamics during learning cannot be accommodated by the collective variables alone. Additional degrees of freedom pertaining to the dynamics of remembering have to be introduced. In other words, the relative strength of memorized information must evolve during the learning process.

Each of these steps involves some new concepts, but I will proceed one step at a time. As always, I will adopt the Galilean strategy of choosing as simple an experimental model system as possible that reveals these new concepts, which, it will turn out, are absolutely essential to understand learning.

THE SEAGULL EFFECT: COMPETITION AND COOPERATION

Old habits die hard. As an entry point into learning, ask yourself what happens when behavioral requirements do not conform to existing behavioral tendencies. Put another way, how are the dynamics of the intrinsic patterns modified by the environment? To answer these questions we return to our workhorse, bimanual coordination, but this time with a twist. Instead of just probing the two basic in-phase and antiphase patterns and their local stability, we study the stability of all the relative phases, or at least a fair sample of them. Thus, we obtain a view of the entire potential landscape or attractor layout. Knowing this, we can study how the landscape evolves with learning.

Two sets of experiments provide a springboard into learning.[10,11] Although subtle and important differences between them exist, the results are remarkably similar. Both deal with the concrete case of bimanual rhythmic movement paced by two visual metronomes. I'll first describe briefly the Tuller-Kelso study because it facilitates the conceptual development that I'm striving for here.

The subject's task was to flex the left index finger every time a light for the left hand came on, and to flex the right finger every time a light for the right hand came on. The set of conditions involved different lag times between the onsets of the two lights varying in steps of one-tenth of a second from perfect synchrony (no lag in-phase) to a half-second lag (or 0.5 antiphase) and back to synchrony. The cycle time for the lights was constant at 1 second. In the language I've been using throughout this book, the experiment can be described succinctly as follows. Two visual metronomes one for each finger, specified a particular temporal relationship between the hands. Frequency was fixed at 1 Hz, and the relative phase between the metronomes varied, thereby providing a continuously available environmental pattern for the subject to match.

We were interested in the difference between the *required* relative phase, ψ, and the pattern that the subject produces, the *actual* relative phase, ϕ. A

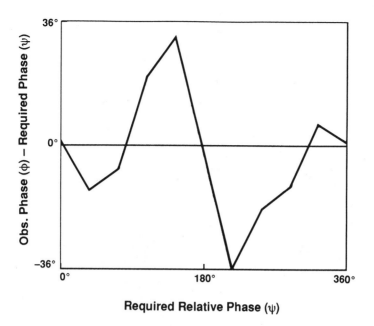

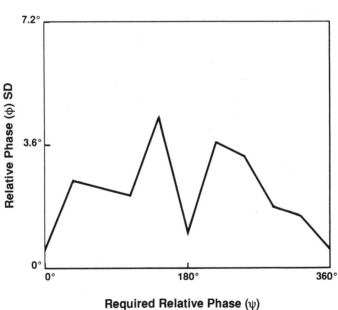

Figure 6.1 The seagull effect. (*Top*) The required relative phase as specified by the pacing lights plotted against the difference between the observed and the required relative phases. A negative number means that the required phase relation was underestimated. (*Bottom*) Standard deviation of the relative phase produced plotted against the required relative phase. Note the seagull-shaped curve (adapted from reference 10).

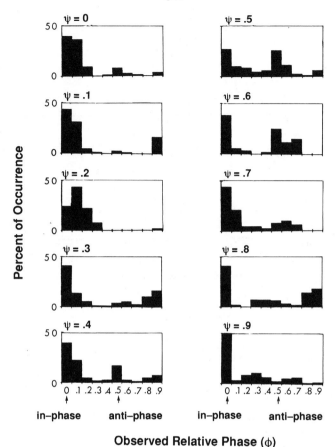

Figure 6.2 Frequency distributions of two split-brain subjects in Tuller and Kelso's experiment. Plotted is the percentage of responses produced at the relative phase required (ψ) by the pacing lights (see box for details).

representative subject's results are shown in figure 6.1 (top) in which the deviation from the required relative phase (error) is plotted against the required phasing (as specified by the visual metronomes). The bottom portion shows the variability of the subject's response as measured by the standard deviation (SD). Obviously in-phase (0 and 360 degrees) and antiphase (180 degrees) patterns are the most stably produced. This is true, by the way, for musicians and nonmusicians, as well as for people who have had the corpus callosum severed surgically to reduce or eliminate epileptic seizures (see box).

The top portion of the figure shows how these two coordination states attract neighboring states. Notice that the slope is always negative passing through 0 and 180 degrees, indicating overshoot and undershoot of nearby required phases in the direction of the closest stable phase.

Obviously, variability is greatest in between the two intrinsic patterns giving rise to the characteristic M-shaped curve that I dubbed the seagull

VP

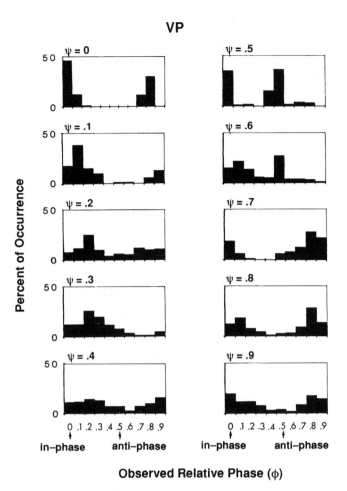

Figure 6.2 (continued)

effect. (This is what living on the coast of southeast Florida does to one. Actually, I much prefer pelicans, who are far more social and even fly in formation, but that's another story.)

The main results of our experiment are that not only is performance at the two intrinsic patterns better than all the others (low error, small variability), but that nearby phasings are attracted toward these stable patterns. Yamanishi and colleagues' results are basically identical, even though they required subjects to reproduce different relative phasing patterns from *memory*. Thus, they gave feedback to subjects during the experiment when they deviated $\pm 5\%$ from the required relative phase. After the patterns had been learned, they were elicited by pacing the subjects' fingers with the metronomes again for ten cycles, after which the metronomes were turned off. Their seagull effect is similar to that shown in figure 6.1 (bottom). The only difference is that the Tuller-Kelso patterns were produced with information from the environment available at all times; the Yamanishi et al patterns were produced from memory, with no external signals present.

ON SPLIT BRAINS

What happens to coordination when the two cerebral hemispheres are divided by surgical section of the corpus callosum? Such "split-brain" subjects are of special interest because previous research suggests that if pacing stimuli specifying each hand's timing are projected to separate hemispheres, then (because control of finger movements is strongly lateralized) the temporal patterns produced should conform to these external phasing requirements. In fact, Tuller and I found that the intrinsic patterns (in-phase and antiphase) turn out to be even more dominant in split-brain than in normal subjects. Figure 6.2 shows sets of ten histograms for two such subjects in which the relative phase produced (ϕ) is expressed as a function of the required relative phase (ψ) between the hands. Both subjects, but especially JW, show a strong dominance of the intrinsic dynamics. Magnetic resonance imaging (MRI) reveals that the corpus callosum of JW is completely split, whereas VP has some residual callosal fibers in the area of the splenium and rostrum. JW shows an especially strong bias toward synchrony at all required phases. However, he is not simply moving his fingers at the same frequency and ignoring the pacing lights, because antiphase movements appear at required relative phases of 0.4 (144 degrees), 0.5 (180 degrees), and 0.6 (216 degrees). Antiphase patterns are all but absent at required phases closer to the in-phase mode. The performance of split-brain subjects reflects the persistence of the intrinsic dynamics even when environmental demands specify other patterns. This strong reliance on the intrinsically stable patterns produces less flexible, not more flexible, coordinative behavior. Environmental information arriving in visual cortex thus cannot overcome the intrinsic patterns, which likely reflect the organization of the subcortical neural networks (see chapter 8). Viewed from the present theoretical perspective, specific environmental information only weakly, if at all, perturbs the intrinsic, order parameter dynamics of split-brain subjects.

How do we understand these results? Gregor Schöner and I modeled both of them at the cooperative (relative phasing patterns) and component (oscillator) levels. As in the case of intentional information, a relative phase required by the environment, ψ_{env}[10] or by memory, ψ_{mem}[11] acts on the pattern dynamics, attracting the order parameter toward the required pattern. Once again, note that *specific information* (from the environment or about a learned behavior) *is expressed by a variable of the same type as the collective variable or order parameter*.

The left part of figure 6.3 plots the familiar HKB potential in the bistable regime of the intrinsic dynamics. The relative stability of the two attractors at 0 and 180 degrees is reflected by the depth of each well, and the strength of their attraction by the slope of the curve. Clearly, the system will relax into one of the two attractors as long as the initial condition is in the basin of attraction of one of the modes. Note that $\phi = 90$ degrees and its symmetry partner ($\phi = 270 = -90$ degrees) are unstable or repelling fixed points of the intrinsic dynamics. All this should be old hat by now. The main idea is that the intrinsic dynamics *persist* when environmental information specifying a required behavioral pattern is present. The same applies to memory: its contents are viewed as specific information about a learned behavior. Yet

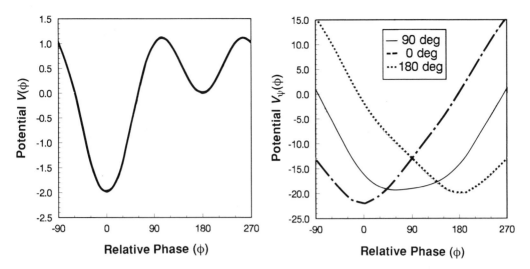

Figure 6.3 Visualization of the coordination dynamics. (*Left*) The familiar HKB dynamics without a specific environmental requirement, here a required relative phase. (*Right*) An environmental requirement is introduced that either cooperates with one of the intrinsic patterns (in-phase and antiphase) or competes with them. Note how the potential is deformed when the required 90-degree pattern competes with the intrinsically stable patterns.

even after learning, some remnant of the intrinsic dynamics is assumed to persist (one of the causes of action slips and airplane crashes?).

The right panel of figure 6.3 plots the potential when specific information requires three different relative phases ($\psi = 0$, 90, and 180 degrees, dashed, solid, and dotted curves, respectively). Notice that the new informational requirement breaks the $\phi \rightarrow -\phi$ symmetry of the original pattern dynamics. This is entirely consistent with the fact that pacing information is specific to the left or the right fingers.

Two additional concepts reveal themselves. When the required relative phase coincides with one of the stable intrinsic patterns, the minimum of the potential is where you'd expect it to be, exactly at the required relative phase ($\psi = 0$ or 180 degrees). Notice the shape of the potential. It is more articulated at in-phase than antiphase, reflecting the differential stability of the two collective states. Clearly, this case reflects the cooperation between specific informational requirements and existing coordination tendencies. On the other hand, if the required relative phase does not correspond to one of the intrinsic patterns, say $\psi = 90$ degrees (the solid curve in figure 6.3, right), competition between the two sources of information leads to a deformed potential. The minimum is pulled away from the required relative phase, implying a larger error of performance. At the same time, the minimum exhibits a wider, less articulated shape, signifying enhanced variability. In short, the extent to which specific behavioral information cooperates or competes with spontaneous self-organizing coordination tendencies determines the resulting patterns and their relative stability. The interplay between these two sources of information corresponds, in other words, to a *selection mechanism*.

QUESTIONS OF LEARNING

Viewed from the perspective of (averaged) learning curves, learning appears to be a smooth, inexorable process. But that isn't necessarily the way it proceeds. Learning, viewed as the mere strengthening of synaptic connections, tacitly ignores the presence of any meaningful relation between the things being learned and the intrinsic organization of the system doing the learning. The concepts I described above give a new slant to learning, or at least they provide some insights into these issues. In particular, the cooperative or competitive interaction between specific learning requirements and intrinsic organizational tendencies (see box) has important consequences for how learning is to be understood. Not least of these is that the entire coordination dynamics—the evolving attractor layout—can be tapped throughout the learning process. This is a primary feature of the research program that I discuss next.

The Zanone and Kelso Experiment

Pier-Giorgio Zanone is a developmental psychologist who came to work with me in 1987. We began a program of research on learning, founded on the ideas and methods described in this book. The theoretical aspects of our work were carried out jointly with Gregor Schöner, but it is the experimental approach and implications that I want to emphasize in this section. The whole idea was to build on our knowledge of spontaneous organizational tendencies in perceptual motor coordination as a means of understanding learning. Using the scanning procedure of the Tuller-Kelso and Yamanishi, et al. (TKY) paradigms, we sought to determine how the individual attractor layout evolves throughout the process of learning. By probing the attractor layout before, during, and after practice of the to-be-learned pattern we were able to test novel predictions about the nature of the learning process. I will describe these predictions in relation to the experimental results that Zanone and I obtained. But first let me describe what we did.

The subject's task was to learn a specific phasing relationship between the hands of 90 degrees. Note that this is not typically one of the intrinsic patterns. As in the TKY paradigms, subjects cycled their index fingers according to a frequency and phase relation specified by two visual metronomes, one for each finger. Two types of procedures were run on five consecutive days. In the learning trials, subjects had to perform a 90-degree relative phase for twenty seconds, after which quantitative feedback was provided about their performance. Fifteen such trials were run (in three blocks of five) each day. At the beginning and end of each daily session, as well as between blocks of learning trials, subjects performed a scanning run to probe the current attractor layout. During each scanning run, the environmentally required relative phase was increased from zero degrees (simultaneous, in-phase blinking between the two metronomes) to 180 degrees (alternating antiphase blinking) in twelve discrete steps of 15 degrees. Unlike the learning trials, *no feedback or knowledge of results was provided during scanning probes*. Without such feedback, learning does not typically occur. Seven days after the last learning session we brought subjects back to the laboratory and asked them to *recall from memory* the 90-degree pattern. We followed this by a scanning probe to see what the attractor layout looked like.

Do people learn this task? Of course they do. After the first couple of days performance hovers around the to-be-learned pattern of 90 degrees. Do they remember the pattern? Yes, indeed. Seven days after the learning phase is over, performance is pretty much unchanged. These are the kinds of questions that are typically asked by what I call *product-oriented* approaches to learning. Literally thousands of practice and learning studies are of this kind. But to answer questions about the nature of the learning process, we have to establish the relationship between concepts and data. I do this next in the context of the questions I raised at the beginning of this chapter.

What Changes with Learning?

The *entire attractor layout changes with learning*, not simply the coordination pattern being learned. We know this because the scanning probes, our way of quantifying each individual's attractor layout are dramatically different before and after learning. Figure 6.4 shows two subjects' attractor layouts on the first and last days of practice. The results of the scanning probe of the recall session are also shown. The top curves show the systematic error between the environmentally required relative phase and the actually produced phasing (delta RP) as a function of the required phasing. The main point to notice is that performance is modified in conditions other than that of the learning requirement of 90 degrees. That is, with learning, the relative phase of 90 degrees also becomes attractive for neighboring conditions. Look, for instance, at subject MS on day 5, where this effect is unbelievably powerful. Required phasings of less than 90 degrees are overshot, whereas those of greater than 90 degrees are undershot. There is a negative slope all the way across the required phasing interval between 45 and 180 degrees intersecting the zero axis at 90 degrees. In other words, the learned pattern at 90 degrees has become a strong attractor with a large basin sucking in many nearby patterns.

Compare also the difference between the scans on days 1 and 5 for subject TM. In the first probe one observes a humped curve as a function of the required pattern: error is lowest at zero degrees and 180 degrees, the two intrinsically stable patterns. But by day 5 the learned pattern is attracting nearby conditions, although not quite as strongly as for MS. Notice that her probe on day 1 is a little more wrinkled than that of TM. She already shows

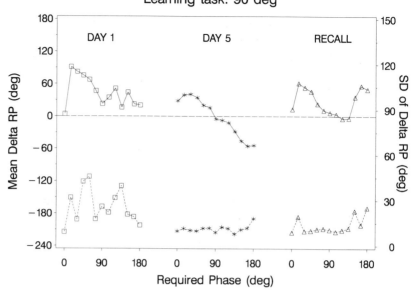

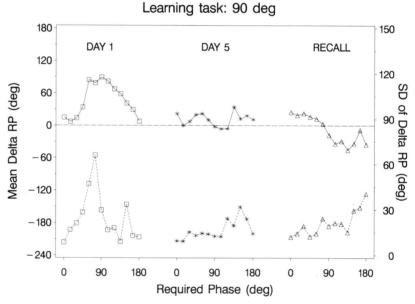

Figure 6.4 Individual attractor layouts before (day 1) and after (day 5) practice, as well as during the recall session. The upper solid graphs plot the mean delta RP (actually produced minus required relative phase) as a function of the required phase. The bottom dotted graphs show the variability of phasing behavior as measured by the corresponding standard deviation (see text for discussion).

hints of reasonable performance at 90 degrees, even before any learning has taken place. If anything, in her case practice seems to have stabilized the 90-degree pattern even further.

These two examples should be sufficient to illustrate my main point: *learning changes not just one thing, it changes the entire system*. Not just one association or connection is being strengthened, although that occurs also. Other connections are being altered at the same time. In the case of MS, even the intrinsically stable patterns exhibit signs of destablization. This is a far cry from the acquisition of habits and associations through repetition that have tended to dominate theories of learning in one form or another. Moreover, each subject possesses his or her individual "signature" or attractor layout. This means that the individual learner, not the group or the species, must be the significant unit of analysis.

What Form Do Changes Due to Learning Take?

The reader can already intuit that the attractor layout may change *qualitatively* with learning. This means that learning can take the form of a phase transition. In the case of TM, after five days of practice, not only are the two intrinsic patterns attractive states of the dynamics, but the 90-degree pattern is as well. The attractor layout has switched from being bistable to multistable, the learned pattern now becoming part of the coordination dynamics. Moreover, such qualitative modifications of the attractor layout persist over one week. The right-hand graphs of figure 6.4 show that the picture for the recall probe still exhibits the main features of the last learning probe. To a certain degree such a result provides an answer to the question, Why does learning persist? The answer is that the learned pattern has become a stable attractive state of the underlying coordination dynamics.

Whereas qualitative changes occur over the time scale of learning, theory also predicts *instabilities during the process itself*. The idea is that as the environmentally required relative phase is learned, the strength of memorized information continuously increases, destabilizing the weaker of the two initial attractors. One of Schöner's simulations of our theoretical model is shown in figure 6.5. The top panel shows the evolution of the memory variable which converges to the to-be-learned pattern. The middle and bottom panels show the evolution of the actual phasing pattern when the initial conditions are near antiphase and in-phase, respectively. Obviously, the coordination dynamics is initially bistable: two solutions exist for different initial conditions. The middle and bottom panels show that two instabilities during learning are possible. The first occurs when the antiphase pattern switches to near in-phase. After this instability, notice the current behavior is still quite far away from the requirement because of the persistent influence of intrinsic coordination tendencies. As learning proceeds and the memory variable gets stronger, a second instability occurs, a new attractor is created close to the requirement.

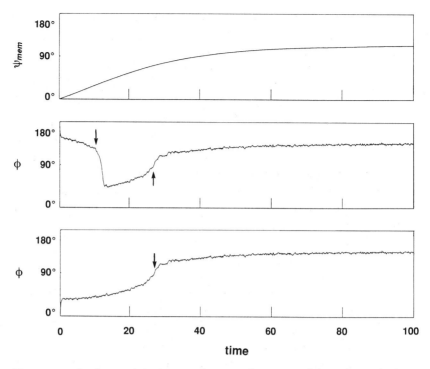

Figure 6.5 Simulation of the learning dynamics showing instabilities during the learning process. The upper panel shows evolution of the memorized relative phase, ψ_{mem}. The middle and bottom panels show the performed relative phase when the initial conditions are near antiphase and in-phase. Two instabilities are predicted to occur during learning (see text).

Figure 6.6 (left) shows the same instabilities but in the form of theoretical distributions of relative phase, plotted as a function of learning time.[12] The distribution is bimodal initially, reflecting the underlying bistability of the attractor layout. At the first instability, the distribution becomes unimodal. Then, an abrupt shift of the peak position accompanied by a broadening of the distribution signifies the second instability. Zanone and I were fortunate to detect the predicted first instability during the learning process. Our experimental results are shown on the right in figure 6.6 and can be compared with the theoretical results on the left. The performance during learning trials is plotted as a histogram of relative phase for each day. Remember, the pattern to be learned is 90 degrees.

On day 1 the distribution is clearly bimodal due to the fact that the subject switches between two coexisting patterns, one close to the learning task and one closer to the antiphase pattern. On the following days, the distributions have become unimodal, centered about the learned relative phase. This bistable to monostable change shows that an instability occurred during learning, the intrinsic attractor close to antiphase losing stability. Note also the progressive sharpening of the distribution over time, indicating the stabilization of performance at the learned pattern. In short, one of our main theoretical predictions is confirmed. *Learning may take the form of a phase-transition process that involves stabilization of the required pattern as an attractive state of the coordination dynamics.* Changes in behavior with learning, traditionally measured as simple improvements in performance of the learning task, are instead the outcome of modifications of the entire underlying dynamics.

What Mechanisms and Principles Govern Learning?

Whether some tasks are learned more easily than others (e.g., in terms of rate of learning and performance efficiency) depends on the extent to which specific parameters (e.g., a to-be-learned pattern) cooperate or compete with existing organizational tendencies (intrinsic dynamics). Suppose the intrinsic dynamics were not known before the learning task is introduced, as is nearly always the case. Suppose that one person learned the task quickly and another learned it much more slowly. Without a theory of individual differences (here revealed by identifying existing coordination tendencies before exposure to the new task), the responses of both subjects would typically be grouped together, and one would be none the wiser about why learning was rapid in one case and not the other.

Our choice of $\psi = 90°$ in the Zanone-Kelso experiment was theoretically motivated by the fact that 90 degrees is an unstable fixed point of the intrinsic pattern dynamics separating the two basins of attraction for $\phi = 0°$ and $\phi = 180°$ (see figure 6.4). Thus, the task as originally conceived was to make an unstable fixed point stable through the process of learning, thereby promoting qualitative change in the attractor layout by the mechanism of competition. Of course, if by chance or design the task requirement happens to be close to the existing dynamics, learning is likely to be rapid because of cooperation. The bottom line is that the learning process—the nature of change itself—can't be understood as the result of the competitive or cooperative interplay between task parameters and intrinsic tendencies unless the latter are known beforehand.

TRANSFER AND GENERALIZATION: SYMMETRY AGAIN

What good is it to learn something if you can't transfer the knowledge to other situations? Transfer requires understanding and linkage between the things learned. Connections, perhaps at a rather abstract level, have to be established. It's no use, for example, knowing Newton's equations of motion if one doesn't understand the basic physical concepts that allow these laws to be applied to everyday events. It turns out that even in our little model system of bimanual coordination something rather abstract and, I believe, important is going on.

Figure 6.7 presents individual attractor layouts before and after learning from another study that Zanone and I did using the same concepts and methods that I described before. I won't go into all the details because the

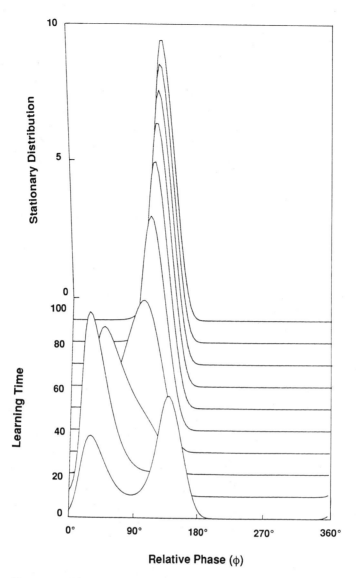

Figure 6.6 Phase transition in learning. (*Left*) Theoretical distributions of performed relative phase as a function of learning time. The distribution is initially bimodal, reflecting the bi-stability of the dynamics. At the first instability, the distribution changes to a unimodal one whose peak is still displaced from the learning requirement at 90 degrees. The second instability occurs later and is seen as a shift in the peak position accompanied by a broadening of the distribution. (*Right*) Experimental distributions of performed relative phase as a function of learning time. Histograms of relative phase are plotted for each day of learning. On the first day the distribution is bimodal due to switching between two coexisting patterns, one close to the learning requirement of 90 degrees and the other closer to antiphase. As learning progresses, the distributions become unimodal and sharper as predicted.

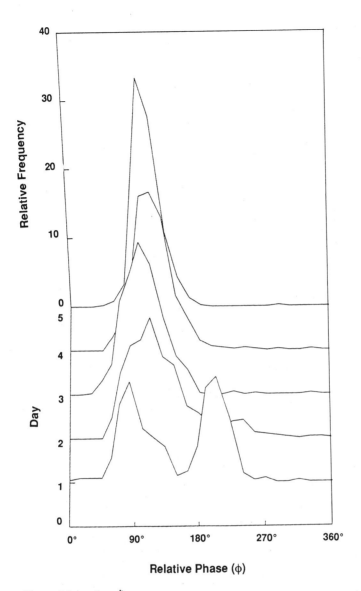

Figure 6.6 (continued)

pictures pretty much speak for themselves. Two differences with the previous work are already obvious. First, the scanning probe encompasses the full range of phases between zero and 360 degrees. This means that the right hand leads the left in the interval between zero and 180 degrees, and the left leads the right between 180 and 360 degrees. Second, the learning requirement is determined individually for each subject. Why? Because in the case of JG (top) the 90-degree phasing pattern is not part of the initial intrinsic dynamics, whereas in other subjects, such as BF (bottom), it is. After learning (right graphs) JG exhibits a novel attractive state at the learned pattern of 90 degrees, reflected

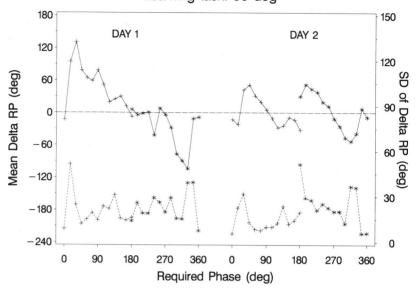

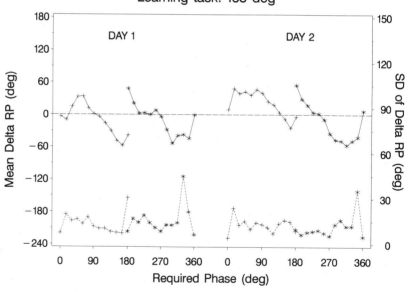

Figure 6.7 Individual attractor layouts before (day 1) and after (day 2) learning showing automatic transfer of learning (see text for details).

by the negative slope and low variability about that value. This is a non-equilibrium phase transition in learning, similar to that described before. Most remarkable is that a new attractor also appears at 270 degrees, although that pattern *was never practiced* at all. Not only is the 270-degree pattern stable, but it also attracts neighboring patterns. This constitutes clear evidence of *transfer of learning*.

Similarly, the picture for BF (bottom) is also fairly symmetric. Initially 90 and 270 degrees are stable patterns of the coordination dynamics, so this subject practiced a 135-degree pattern. By the end of day 2 (right graphs) the learned pattern has clearly become part of the attractor layout, *but so has its unpracticed symmetry partner*. Transfer of learning, in other words, occurs automatically. A future point of theoretical interest is that another route to learning is identified through these data. That is, the initially stable patterns at 90 and 270 degrees (day 1) are gradually shifted toward the required phasing pattern of 135 degrees (day 2) without any qualitative change in the coordination dynamics. Thus, it may be that when $\phi = 90°$ is already a stable pattern, the resulting changes due to learning do not take the form of an instability but are smooth and continuous. Here then is a subtlety: the distance in collective variable space between behavioral requirements and intrinsic coordination tendencies appears to be a factor determining the nature of change.

More important, I think, than evidence for transfer itself and the form that it takes, is what these results tell us about what is learned by the brain in coordination tasks. The fact that transfer occurs spontaneously to the same timing relation but a different ordering (left-right) between the hands suggests that a very abstract kind of learning has occurred. *What is learned is a phase relation that is apparently quite independent of how it is instantiated.* At the heart of transfer of learning lies a symmetry of the collective variable, relative phase. I find this a bit ironic, given that learning *breaks* the symmetry of the original HKB dynamics.

MOTOR EQUIVALENCE

The idea that what is learned is an abstract phase relation between co-ordinating components speaks to one of the great mysteries of human action—the ability to achieve the same goal in very different ways. People, for example, can write their signatures small or large, or even with a pencil attached to their nose or big toe. The material they use to do this—the various parts of the brain and the body, not to mention postural and task contexts—is very different in each case. So what's the mechanism? Of interest, many years ago it was found that people might write their names big or small using various and sundry muscle and joint combinations, but the time taken was roughly the same.[13] This observation fits with our identification of the phase relation as an order parameter. Remember, phase is a relative timing variable, an abstract relational quantity that is capable of being realized by many different effector systems (cf. chapter 3). If phasing is what is learned, then the issue of generalization of learning from one different coordination system to another is open to empirical challenge. All that it is necessary to show is that the pattern to be learned constitutes an attractive state of both coordination systems (e.g., arms vs. legs) after practice on just one of them. Such a result means that not only does learning involve changes in the coordination dynamics of the system that practices the task but in other coordination systems as well. This would constitute solid proof that the degree of abstraction by the brain is very high indeed. Most significant, in my view, is that common dynamical mechanisms and principles underwrite motor equivalence, transfer, and generalization of learning.

SOME OTHER KINDS OF LEARNING

Does the same bag of tricks apply to other kinds of learning such as learning to read and write or learning to talk? (I'll talk about learning to walk in the next section on development.) Frankly, I don't know. Science always chooses specific cases on which to build theoretical understanding, and some concrete cases are bound to be better than others. Bimanual coordination happens to be a good bet for a Galilean equivalent for biology and behavior. It may not be considered a standard task, say for psychologists, but neither, one could argue, was rolling a ball down an inclined plane. Yet, in principle at least, I don't see why the same ideas can't be used for other kinds of learning. Whether they can or not, of course, lies in the details.

I remind the reader of recent work in speech perception and production demonstrating phase transitions from syllables such as "ape" to "pay" as speaking rate is systematically increased (chapter 3). The transition is invariably from a vowel-consonant (VC) form to a consonant-vowel (CV) form, and not vice versa. It is interesting, once again, that it is the phase relation among glottal and oral articulators that distinguishes VC and CV syllables, both in production and perception. Such CV forms are also known to predominate in infant babbling. Babies say "ma, ma," not "am, am."

Recent work in child language development seems to show that the phonetic characteristics of early words are highly similar to each individual baby's babble.[14] Indeed, infants apparently attempt to produce an adult sound only when it matches or is close to one of their own babbling sounds. Access to the world of language, it seems, depends on these preferred perceptual-motor patterns. As a first pass, it seems to me that what we call babbling corresponds to a kind of intrinsic dynamics that can be modified by species-specific information, the sounds of adult words, acting on those dynamics. As I have stressed all along, such information is only meaningful and specific to the extent that it attracts the system to the required (perceived, learned, memorized, intended) coordination pattern. It would be very exciting if phase transitions were found to occur during language learning.

1

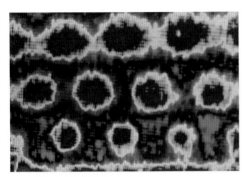

Plate 1 Hexagonal convection pattern in a layer of silicone oil arranged as a honeycomb. (Courtesy of M. G. Velarde).

Plate 2 A computer-enhanced image of an experimental Turing structure. (Courtesy of P. de Kepper.)

Plate 3 Arnol'd tongues (color coded) for the phase attractive circle map theory of coordination. The pattern of locked tongues is shown as a function of the frequency-ratio, Ω (x-axis, scaled between 0 and 1) and the coupling strength, K (y-axis, scaled between 0 and 3). *Red, blue, green*, etc. correspond to $1:1, 1:2, 1:3, 2:3, 1:4, 3:4, \ldots$ mode-locked regions, only a few of which are shown. Black areas are either quasi-periodic or, above the critical surface where the tongues overlap, chaotic. The intrinsic parameter, A is varied from left to right ($A = -1; A = 0; A = 1$). Note the thinning and broadening of the tongues and the period doubling route to chaos.

2

3

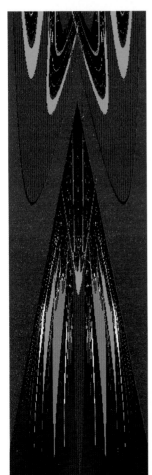
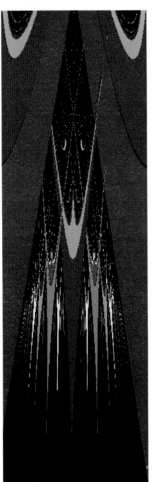
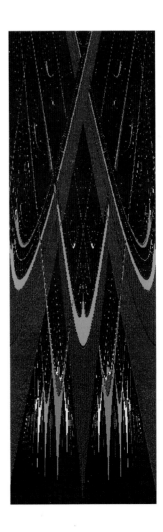

4

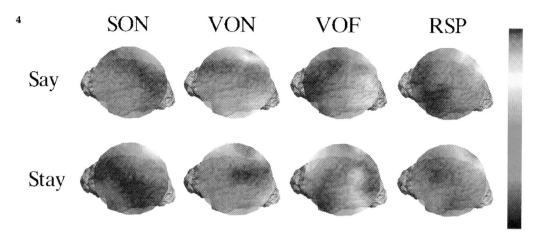

SON   VON   VOF   RSP

Say

Stay

Plate 4 Topographic distributions of brain activity underlying the perception of "say" versus "stay" super-imposed on the cortical surface. For ease of viewing, the magnitude of brain activity is compared at stimulus onset (SON), vowel onset (VON), vowel offset (VOF), and response onset (RSP). Differences between the perception of "say" versus "stay" *for the same acoustic stimulus* show up strongly just after vowel offset.

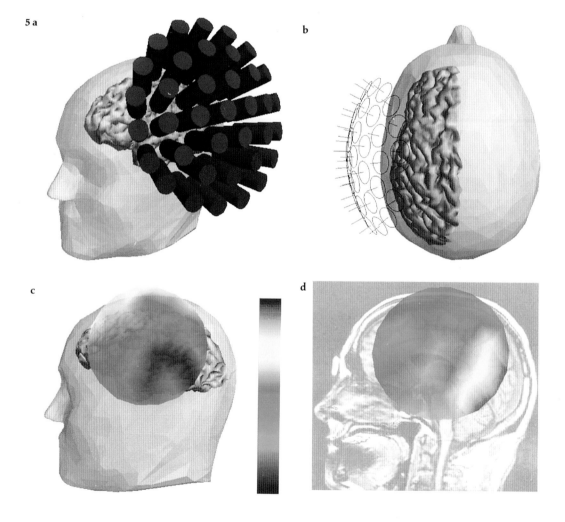

5 a   b

c   d

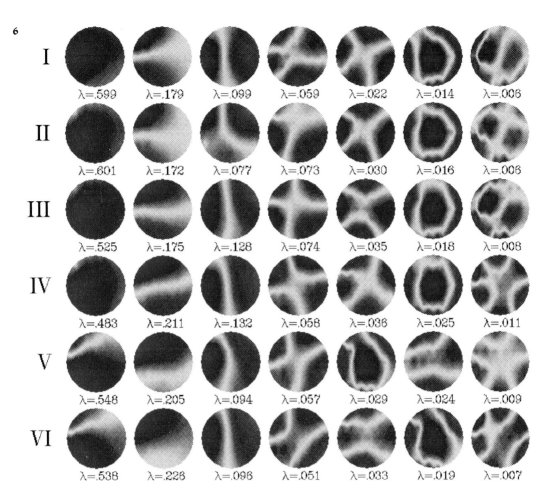

I
λ=.599 λ=.179 λ=.099 λ=.059 λ=.022 λ=.014 λ=.006

II
λ=.601 λ=.172 λ=.077 λ=.073 λ=.030 λ=.016 λ=.006

III
λ=.525 λ=.175 λ=.128 λ=.074 λ=.035 λ=.018 λ=.008

IV
λ=.483 λ=.211 λ=.132 λ=.058 λ=.036 λ=.025 λ=.011

V
λ=.548 λ=.205 λ=.094 λ=.057 λ=.029 λ=.024 λ=.009

VI
λ=.538 λ=.226 λ=.096 λ=.051 λ=.033 λ=.019 λ=.007

Plate 6 Spatial modes on the different plateaus of the Juilliard experiment. The transition occurs between plateaus III and IV. The first and second modes dominate but there is coherent spatial behavior in at least six modes.

Plate 5 (a) Reconstruction of the subject's head and location of SQUID sensors. (b) Construction of cortex model using MRI. Slices were taken in the coronal plane with a spacing of 3.5 mm. The location and orientation of each SQUID sensor is superimposed. (c) Example of magnetic field activity detected by SQUIDs and displayed on the cortical surface, and (d) on a sagittal MRI slice.

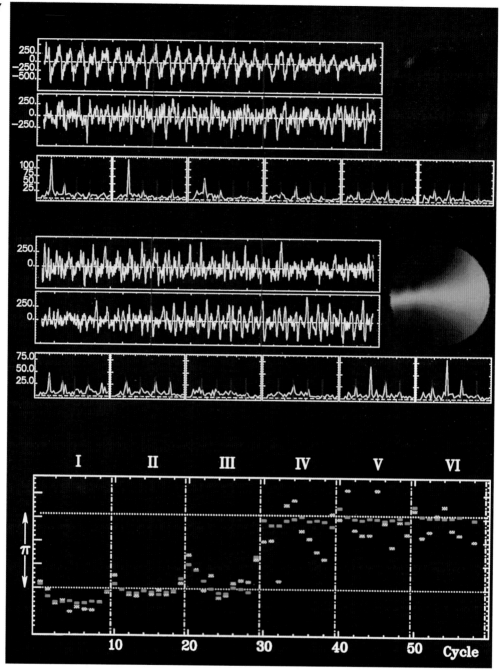

Plate 7 Dynamics of the first two spatial (KL) modes that capture about 75% of the power in the signal. (*Top right*) Dominant KL mode. Amplitudes and power spectra on frequency plateaus 1 through 6. (*Middle right*) Second KL mode and corresponding amplitudes and power spectra. (*Bottom*) Relative phase of the behavior (blue squares) and the amplitude of the top mode (gold squares) with respect to the stimulus. Note the qualitative changes in all three displays around the beginning of plateau IV (see text for details).

BEHAVIORAL DEVELOPMENT

I already speculated a little about how the concepts and principles used here to understand learning of perceptual-motor patterns apply to other kinds of learning that occur in developmental time. The entire framework of self-organization, how new or different forms emerge spontaneously due to instabilities, seems tailor made for basic questions of development. Human beings want to know how the minds of children grow and develop, how they come to know the world around them, how they become socialized, and so forth. They want to be sure that their child is developing normally, and to understand better what they can do to enhance and nurture this development. When new forms do arise, the first words, reading, mathematical ability, catching, kicking, throwing, whatever, we wonder at the processes that gave rise to them. What changes in the organism and in the child's relation to the environment have occurred? What's the nature of change? Where do these new behavioral patterns come from? These and other fundamental questions are precisely the ones that I've addressed in this book, and especially in the present chapter. Such issues, of course, occupy the attention of scientists in many disciplines.

Although "dynamic systems" concepts are part of the intellectual heritage of developmental theory, they have in my view promised much and delivered little. Recent examples, prominently published in psychology's main theoretical journal, *Psychological Review*, are typical. One uses logistic growth curves to describe various aspects of cognitive and language skill development.[15] The other resurrects catastrophe theory to explain the stagelike properties of cognitive development.[16] In both cases, the linkage between models and data is far from transparent. A more frank assessment is that as scientific theories go, these efforts are pretty barren. The words of van der Maas and Molenaar are revealing:

Catastrophe analysis requires knowledge of the mathematical equations of the transition process ... [but] this variant is not yet feasible as we do not know the precise dynamical equations of the processes governing cognitive stage transitions.[17]

They proceed to talk about the "intricate problems" of catastrophe modeling that concern the actual fitting of models to data.

Later the same authors remark, as they say, "succinctly," "Catastrophe theory only applies to dynamic nonlinear systems that can be characterized by a so-called potential function and that conserve energy."[18] Given this totally unrealistic constraint, one wonders what gave the authors (never mind their reviewers) pause. Living, developing things are assembled and sustained under open, nonequilibrium conditions. I find it astonishing that so-called dynamic systems approaches to development using logistic functions and catastrophe theory have proceeded with an absolute minimum of hard data about the *actual* dynamics of the situations they seek to describe and understand.

An exception to this criticism is work in the area of motor development, such as why and how infants walk when they do.[19] Esther Thelen has studied a variety of rhythmic stereotypes in human infants and other animals. In the early 1980s she spent a summer working with me to learn dynamical systems methods for treating kinematic data (baby kicks) and to become more familiar with the concepts of synergetics. The outcome of that summer is nearly a decade of detailed analysis on the ontogeny of infant locomotion.[20]

To motivate empirical study, Thelen and Beverly Ulrich import lock, stock, and barrel the same concepts that emerged from our analysis of bimanual coordination: identify the collective variable (e.g., the phasing of alternating steps); characterize its stability (perturbation analysis); identify transition points (e.g., abrupt shifts in phasing); and identify potential control parameters (e.g., the speed of a treadmill that the baby steps on, loading of the limbs, etc.). As here, they stress "dynamic accounts at many levels of analysis" and aim to clarify the role of different subsystems that contribute to shifts in motor (body build, practice, muscle strength, etc.) and cognitive (memory capacity, perceptual abilities, social factors, etc.) task performance. Behavioral development is notoriously noisy and nonstationary. Thus, as Thelen notes,[21] it may not be possible to establish a very precise mapping between data and theory in immature humans, but it won't be for the lack of trying.

What is much more important, in my view, is the perspective of behavioral development as a self-organized process that has emerged from her detailed experiments. As Peter Wolff remarks:

[A] rigorous application of the proposed strategy [described briefly above] with an appropriate respect for the complexity of dynamic systems can lead us to radically different methods of developmental investigation and radically different theoretical perspectives on the processes of novel pattern generation. By this route, the perspective opens the way for experimentally investigating phenomena that were previously inaccessible because of our theoretical pre-conceptions.[22]

EVOLUTION AND MORPHOGENESIS

Between May and August of 1876, Charles Darwin wrote the main text of his *Autobiography* in which he tells the story of the long and tedious intellectual journey that culminated in his *Origin of Species*. The autobiography is a brief and revealing book in which, as his granddaughter remarks in the preface, we can watch Darwin's diffidence slowly give way to scientific assurance. Although he never wavered from his faith in natural selection as the motive force of evolution, doubts arose in his mind toward the end of his life about the contributions of other forces. One particular passage is especially intriguing:

Everyone who believes, as I do, that all the corporeal and mental organs (excepting those which are neither advantageous or disadvantageous to the possessor) of all beings have been developed through natural selection, or the survival of the fittest, *together with use or habit* [my emphasis] will admit that these organs have been formed so that their possessor may compete success-fully with other beings, and thus increase in number.[23]

No short sentences for Darwin. I am not going to comment very much on Darwin's words except to say that to me they convey clearly that learning plays a role in evolution. Not just structure but behavior itself is subject to natural selection. In fact, if one takes the word "together" seriously, Darwin puts learning (due to practice and training) pretty much on the same footing with natural selection. Habit, after all means acquired behavior.

I therefore find it tantalizing that the concepts and principles described in this chapter for a process that evolves on the time scale of learning might also apply to processes on other time scales. I have argued (and shown) that pattern selection, an essentially Darwinian concept, plays a central role in learning. Here, the pattern selected arises as an a posteriori fact of competi-tion or cooperation between specific information (environmental, intended, learned) and the intrinsic dynamics of the system. Thus, selection and self-organization go together like bread and butter. Indeed, the language of selec-tion is in precisely the same terms as the underlying pattern dynamics. What is usually missing, of course, is a clear characterization of the intrinsic dynamics—the basic forms of coordination—on which specific information is deemed to act. In self-organized coordination dynamics, selection appears as an emergent consequence rather than an a priori agent of change.

Is it not conceivable that a similar kind of selection principle might exist in the development of organisms? Stuart Newman has drawn attention to a number of what he calls "generic physical mechanisms," such as the coupling mechanism of reaction and diffusion identified by Turing, surface tension, gravity, phase separation, and so on, that contribute to morphogenesis and pattern formation in animals and plants.[24] This is an old theme, stretching back at least to the English naturalist D'Arcy Thompson, revered for his work, *On Growth and Form*. Biological form, in Thompson's famous words, "is a diagram of forces." Whether cell, tissue, shell, bone, leaf or flower, "God always geometrizes ... in obedience to the laws of physics."[25]

Whereas Thompson eschewed selection,[26] Newman recognizes the impor-tant contribution of genes in stabilizing successful generic forms as well as destabilizing them, bringing about radical changes in body plan. It is tempting to suggest an analogous selection principle for development of multicellular organisms to that advocated here in which selection arises from cooperative-competitive interactions between specific (genetic) parametric constraints and intrinsic (generic) dynamics. The latter, I should stress, is not to be simply equated with ordinary physics and chemistry, but includes coordination phenomena of the kind I have described. Notice that in such a view the gene in no way constitutes a program for development. Rather, it's a participant playing a parametric role in a self-organizing, synergetic process. Stuart Kaufman in his book *Origins of Order* sees as a "fundamental problem ... how to think about the relation between selection and the natural forms generated by different classes of developmental mechanism."[27] The present view of selection as an a posteriori consequence of nonequilibrium pattern formation processes, extended here to include a role for both specific and nonspecific parameters, seems to be a step in the right direction.

Returning to evolution itself and specifically the origin of species, I find exhilarating recent work that suggests that it is the *dismantling* of genotypic constraints, described by words such as "internal balance" and "genetic homeostasis," that effects speciation.[28] Environmental stress, such as the evaporation of Lake Turkana in Africa, creates developmental instabilities, causing rapid changes in mollusc sequences. Accompanying these points of rapid taxa change, the breakup of genetic homeostasis, is significant elevation in phenotypic variance.[29] If we view Lake Turkana's drought as an environmental influence and the cohesion of the genotype as intrinsic dynamics, speciation itself looks very much like the result of a self-organized pattern-formation dynamic.

It is tempting to construe the enormous phenotypic variance preceding the emergence of new species as a striking reflection of critical fluctuations in the theory of nonequilibrium phase transitions. This does not say that the conventional neo-Darwinian dogma in which new species (forms) arise by accumulation of slight successive variations in the genotype is wrong. But it does suggest that self-organizing processes play a significant role as well.

SUMMARY AND CONCLUSIONS

What processes underlie the ability of human beings to learn new skills? Every major learning theorist knows that organisms, even neonates, are not tabulae rasae but bring different backgrounds and capacities into the learning environment. People enter the learning situation with a certain degree of pre-organization that constrains the form that learning takes. The problem is how to assess these constraints, which are not static, but evolve dynamically with the learning process. Although very different in origin, the present approach is continuous with others, notably "biological constraint" and "behavior systems" perspectives.[30] These theories view constraints as partially innate, species-specific, and generally reflective of evolutionary processes that shape the organism's ability to survive. For me, whether coordinative constraints are innate or due to previous experience is not a crucial distinction. What is absolutely essential is that the information to be learned must be structured in relation to existing constraints that can be identified and measured. Stimuli are not just arbitrary bits of information to associate with responses.

Through perceptual-motor learning I have introduced new concepts and methods that allow a quantitative evaluation of existing constraints (patterns and their dynamics) before learning a new skill, as well as how these dynamics evolve in time as new tasks are learned. Predictions regarding the learning process follow from knowledge of the coordination dynamics that is specific to the individual learner. Thus, an advantage of the present approach is that it overcomes a troublesome problem for most learning theories, namely, how to treat individual differences explicitly. In this account, without information about the individual coordination dynamics, each subject's "signature," we would be none the wiser about why learning is rapid in one case and not the other. Somewhat paradoxically, generic principles of learning emerge by studying the individual, not the group or the species as the relevant unit of analysis.

The originality and relevance of the dynamic pattern approach rests not only on identifying functional constraints but in showing, both in theory and experiment, that the entire attractor layout changes with learning, not simply the particular behavioral pattern being learned. Learning may take the form of instabilities or phase transitions depending on the relation between what is to be learned and the organism's existing coordination tendencies. Evidence of such qualitative change not only validates dynamic pattern theory, but renders observed gradual changes interpretable as well. As the landscape evolves, some collective states are eliminated, some created, still others stabilized.

Competition and cooperation (cf. chapter 1) are the hypothesized mechanisms governing learning, essentially determining the behavioral outcomes at any point in time. Both, as I've shown, are measureable. Competitive mechanisms operate when extrinsic requirements do not coincide with a stable state of the current pattern dynamics. In contrast, when they do coincide, or nearly so, cooperative processes run the show. It is not fortuitous, in my opinion, that such processes come into play at more microscopic levels; for example, in the stabilization and elimination of neuronal connections. In developmental time, too, selection of behavioral patterns that are best tuned or adapted to the specific environment can be interpreted as the natural outcome of the interplay between existing intrinsic and environmental constraints.

In the last decade or so it has become clear that learning and memory are heterogeneous phenomena at cellular and molecular levels.[31] The mechanisms for cellular and molecular changes during learning are diverse across the animal kingdom, just as Darwin would have it. I find the possibility all the more intriguing, therefore, that the dynamical mechanisms and principles proposed in this chapter may apply to adaptive change on different space and time scales. This may not help us identify which neurotransmitter or receptor facilitates synaptic changes involved in learning, but it may provide insight into how human beings learn.

7   Perceptual Dynamics

The mind cannot fix long on one invariable idea.
—John Locke

What is it you see when you look at something? What is it you feel when you touch something? What is it you smell when you sniff something? What is it you hear when you listen to something? Questions such as these have puzzled philosophers and psychologists throughout the ages. For many, the problem of perception is all about how properties of the world come to be represented in the mind of the perceiver. In this chapter I will be rather less concerned with the *contents* of perception than with the *dynamics of perceiving*, an issue that I consider to be largely ignored by perceptual theory. Looking, touching, and sniffing are obviously dynamical processes. Listening, on first blush, appears less so. But what is the nature of the dynamics? Is the language of collective variables, attractors, control parameters, and so forth, valid? Do the concepts of stability, instability, transitions, and the like apply? Can perception be understood as a synergetic, pattern-forming process, this time perhaps in the brain itself? In short, is the process of perceiving—like those of coordinating, intending, learning, and developing—subject to dynamics?

My goal is not to propose a new theory of perception, but rather to draw attention to certain perceptual phenomena and findings that seem to demand interpretation in terms of dynamic pattern concepts. My hope is that a more coherent perspective might emerge in a field that is not short of theories and facts.

One powerful theory of perception, stemming from the pioneering research of James Gibson and his school of ecological psychology, places its primary emphasis on informative events that occur in the environment that are relevant to the activities of organisms.[1] A competing theory associated strongly with Wolfgang Köhler and the Gestalt school, stresses not the richness of stimulation in the world per se, but the achievement of the nervous system in organizing it.[2] My view is that to the extent that the latter notion underestimates meaningful information contained in higher-order patterns of stimulation, this is regrettable. Similarly, to the extent that the former ignores the role of the brain in perceptual organization or, in the extreme, even rejects the entire possibility of intrinsic organizational tendencies in perception, that is a pity as well.

Any attempt to unite two such seasoned theoretical positions is doomed to failure. Yet I believe both share essential dynamical features. On the one hand, I will show that the relevant variables that capture the mutual relationship between animals and their environments are meaningful quantities that exhibit dynamics in terms of stabilities, attractor layout, critical points, and so on. On the other, I will present evidence that the neural organization supporting perception also has dynamics that to some degree is independent of stimulus structure, higher order or otherwise.

The advantages of casting perceptual processes in terms of dynamics, beyond the predictive power obtained by such a step, are several. First, the laws underlying organism-environment couplings concern meaningful informational events and are largely unknown. Second, and relatedly, the dynamics of perceptual organization has seldom been formally specified or subject to rigorous experimental test. Third, as we will see, the dynamical perspective opens up entirely new ways of inquiring what perception is, including the role the human brain plays. Finally, neither of these theories is especially explicit about the processes—neural, dynamical, or computational—underlying perception at the scale observed.

The ecological perspective eschews computational processes in favor of an underelaborated "resonance" mechanism of information pick-up or detection.[3] The Gestaltists viewed the brain as a thermodynamic system converging toward equilibrium in an energy function. In light of recent developments in the fields of artificial neural networks and computational neuroscience, several scholars have expressed optimism about the correctness of the Gestalt picture.[4] But neither proposal, I will argue, is up to the facts it seeks to explain. I will provide an alternative view that respects the perceptual dynamics at both brain and behavioral levels. Indeed this will be our first step toward establishing the nature of the linkage between events on different scales.

THE BARRIER OF MEANING: PERCEPTUAL DYNAMICS I

The brilliant mathematician and polymath Stanislaw Ulam is known, among other things, for his work with Edward Teller on thermonuclear reactions that made the H-bomb possible. In his spare time, Ulam thought a little bit about the problem of how we perceive the world around us. He absolutely hated the idea—promoted by cognitive psychologists and artificial intelligence researchers—of *representing* properties of objects (or the world, in general) in the mind.[5] A representation is something that stands for something else; it is some kind of resemblance of the world outside stored inside the head. According to representational views, things are not as they seem. What we are aware of is not our world, only our representations of it. A representation is like a little person inside the head making sense of the meaningless sensations that impinge on the eye (or the ear, or the hand, or the nose). For

Ulam, when you perceive intelligently, you never perceive an object in a physical sense. What you perceive is a *function*, a role that is tied inextricably to *context*. The key to your door is not some disembodied image registered in the brain; it is only a key in the context of what it is used for. Its functional role is subject to change according to the needs of the organism. For example, some people, but regrettably not I, are very good at using keys as beer bottle openers. The point is that context is everything. Remove context and meaning goes with it.

In a similar vein, but entirely independently, James Gibson coined the term *affordance* as the relevant entity of perceiving. Affordances, such as whether or not an object is graspable or a surface can be walked on are directly perceived properties of the environment that permit certain actions on the part of an animal. Perceiving is not strictly speaking *in* the animal or an achievement of the animal's nervous system, but rather is a process in an animal-environment *system*. For Gibson and those who follow his approach, perceptual systems evolved so as to be able to pick up information specific to affordances. The radical claim of this ecological view is that affordances are not mediated: they are detected unaided by inference, memory, or representation. This theory is grounded on information, invariant properties of optic, haptic, acoustic structure that are relevant to an organism's action capabilities. Information is out there and available to a suitably attuned organism.[6]

Of course, organisms vary in their level of attunement. Information is not the same to a child as to an adult, or to a novice versus an expert. It is not an absolute physical quantity transmitted from a sender to a receiver. Information can be of very high order indeed. Consider Father Zossima, the priest in Dostoyevsky's *The Brothers Karamazov*. For so many years, people had bared their souls to him that, "In the end he had acquired so fine a perception that he could tell at the first glance from the face of a stranger what he had come for, what he wanted and what kind of torment racked his conscience" (p. 30). Perceiving, then, might be viewed as a highly developed skill; so highly developed, it seems, that by observing someone's behavior we can see into the person's state of mind.

Optic Flowfield Dynamics

Although the research program begun by Gibson has many dimensions,[7] I will focus on just a few major highlights where his ideas have been pinned down quite precisely. I will cast the problem in a way that it can be appreciated in terms of self-organized dynamic patterns, especially the concepts developed in the last two chapters. Later on I will consider the kinds of neural mechanisms involved.

To a first approximation, the problem of perception is a kind of mirror image of the problem of action. Very large numbers of degrees of freedom are involved on both sides of the coin. One goal of this book is to understand the ways nature has devised to compress billions of (potential) degrees of freedom into a few functionally relevant macroscopic quantities. On the perception side, when you consider the light rays to the eye, the retinal mosaic and the neural processing structures involved, the dimensionality is huge. Yet, somehow, despite this vast dimensionality (or because of it), organisms successfully perceive the world around them and behave in relation to it, usually in appropriate ways.

Gibson developed the idea of the optic flowfield as a relevant macroscopic description of the light to an eye (note, *any* eye and *any* visual system) that is specific to the layout of surfaces and the activity of a moving point of observation. Ask yourself what a person or an animal moving through a cluttered environment has to know. It has to know something about its own motion, where it's heading, and how fast. It has to know something about the layout of the environment, the regularity of the ground, and the obstacles in its way. And it has to be able to detect other moving objects and its relation to them. All these properties, amazingly enough, are *specified* (mark, not represented) by information in the optic flowfield. Remember the chase scene through the woods in George Lucas's *Star Wars?* That's pure optic flow.

Consider again, for the sake of illustration, the observations made by Lee and Reddish on diving gannets (cf. chapter 2). When these birds have located a prey (a nice fish or even a school) they plummet into the water from their cruising height, reaching speeds of up to fifty-five miles per hour. A split second before touching the surface they rapidly stretch their wings back for a streamlined "spearhead" entry. Now, in principle, many of the usual physical variables, such as height, velocity, acceleration, and so on, could be used by the bird to monitor and regulate its behavior. Lee and Reddish found, however, that the gannet reliably timed its movements in terms of only one optical parameter, tau, $\tau(t)$, even when diving from quite different heights. From an enormously detailed *microscopic* description a single *macroscopic* variable emerges to which the gannet is remarkably sensitive (see box for details of how this parameter is extracted from the geometry of the optic flowfield).

How might we understand the gannet's (qualitative) behavior in terms of coordination dynamics? The first step is to identify an order parameter. A good candidate on the kinematic level is the wing position coordinate, x, and its velocity, $\dot{x}$. In the present context under the current boundary conditions (Joe Gannet is adequately hungry, weather conditions, time of day, etc., are appropriate) the bird's wing posture can take on two stable states, the initially open cruise position and the stretched final position for the dive. We can map these states onto attractors of the wing coordinate dynamics. Environmental information is captured by the expansion rate of optical structures in the flowfield. As Lee and Reddish found, $\tau(t)$ reliably specifies the timing of wing folding. We can include $\tau(t)$ as an informational parameter in the dynamics in that one state (initial wing posture) is stabilized at low expansion rates, but the other (spearhead posture) is stabilized at high expansion rates. As the gannet dives, an environmentally induced behavioral change—a *phase transition* in

HOW TIME-TO-CONTACT IS SPECIFIED BY THE GEOMETRY OF THE OPTIC FLOWFIELD[8]

For ease of description, the eye is conceived schematically as a flat retina, orthogonal to the instantaneous direction of movement. Extension to spherical (3-D) coordinates and flow patterns on a real retina of any shape can be made easily.[9] The eye is located at time t at a height $Z(t)$ and assumed to be moving vertically downward with velocity $V(t)$, in this case, toward the water below. Surface texture elements, such as ripples, reflect light to the eye, that passes through the nodal point of the lens and projects an expanding optic flow pattern on to the retina (figure 7.1).

That is, all optic texture elements spread outward along radial flow lines emanating from O. Any single image element P' at a distance $r(t)$ from O moves outward with optic velocity, $v(t)$. Time-to-contact is specified as follows. From similar triangles,

$$Z(t)/R = 1/r(t),$$

differentiating with respect to time

$$V(t)/R = v(t)/r(t)^2,$$

eliminating R by division

$$Z(t)/V(t) = r(t)/v(t) = \tau(t).$$

Hence, time-to-contact under constant closing velocity is specified by the optical parameter $\tau(t)$. This conceptually simple quantity is a meaningful invariant that can be detected by *any* visual system and used to guide locomotory activities as different as a gannet diving for fish, a fly extending its legs for landing,[10] human long-jumping, playing table-tennis, etc.[11]

our terminology—will occur when τ reaches a critical value.[12] Note that the behavioral dynamics are stable under perturbation of the flowfield (like a gust of wind) or of the behavioral variables.[13]

It turns out that the global morphology of optic flow contains other macroscopic parameters that organisms (and machines) can use to regulate their behavior. All of these *motion invariants*[14] have units of 1/time, and all of them can be used to steer cars (or robots), avoid obstacles, detect brinks, and so forth. Their significance is that they are directly available in the optic flowfield without recourse to indirect measurements of distance, speed, or acceleration of approaching objects. Of importance, optic flow parameters demarcate different kinds of situations for a moving point of observation such as an organism. For example, the rate of change of τ, $(\dot\tau)$, specifies *harshness* of contact, how hard you are going to contact an environmental surface or part of it, should current conditions of motion persist. Above a certain critical value $(\dot\tau > -0.5)$ contact will be soft; below it $(\dot\tau < -0.5)$ contact will be hard. This makes sense: the negative value means you have to decelerate quickly enough or you'll crash. In other words, $\dot\tau$ effectively distinguishes different states of affairs.

Of course, the time to contact strategy depends considerably on a constant velocity approach and the ability of the system to decelerate appropriately.

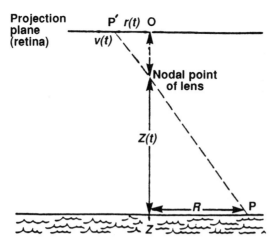

Figure 7.1 How time-to-contact is specified in the optic flowfield according to Lee and Reddish. Reprinted with kind permission of *Nature* and the authors.

Any constant $\dot{\tau}$ approach between $-1.0 \leqslant \dot{\tau} \leqslant 0$ will result in arrival at a target with zero velocity. However, some approaches, say below $\dot{\tau} < -0.5$, require an exponential increase in braking that may be physically impossible.[15] Harking back to the famous Bénard instability example of pattern formation in chapter 1, $\dot{\tau}$ is an analogue of the Rayleigh number but for *optical* not fluid flows.[16] Not only do $\tau(t)$ and $\dot{\tau}$ provide continuous information for the modulation of activity insuring its adaptive stability, they also are capable of effecting bifurcations to different modes of behavior.[17]

I can't leave this discussion without mentioning one of the most compelling demonstrations of the salience of optic flow in humans. It also allows me to make a key point about the perception-action relation in terms of *pattern-forming dynamics*. Figure 7.2 shows one of the famous swinging room phenomena. The two traces represent the movement of the room (bottom) which is being oscillated back and forth by only 6 millimeters every 4 seconds or so, and the movement of the person's trunk (body sway) inside the room (top). Only the walls of the room move, not the floor. Clearly the optic flow, global contraction and expansion, induces an oscillation in the sway of the standing subject. One can imagine that were the person walking forward in a room whose walls and ceiling were moving in the same direction, but faster, that the resulting global optical contraction would cause the person to perceive that he is actually walking backward! Such is, in fact, the case.[18]

But figure 7.2 is included not just as another testament to the power of optic flow to regulate movement. Rather, it points to an intimacy between visual (room) and motor (postural sway) variables that takes the form of an action-perception *pattern*. And guess what the key informational variable is that captures the coupling between these two components? Our favorite order parameter relative phase, of course.

Sad to say, the measure of (in)stability employed in this classic study was the mean speed of the trunk, a variable referential to only the actor compo-

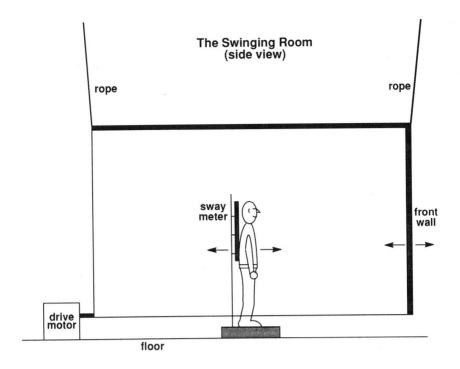

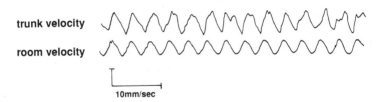

Figure 7.2 (*Top*) Schematic of the swinging room paradigm. The swinging room is a large box open at the bottom and at one end. To provide visual structure, floral wallpaper is hung on the front and side walls, and the ceiling. The room is suspended above the floor on ropes. The subject stands inside, and postural sway is measured as the room swings back and forth. (*Bottom*) Recordings obtained in the swinging room. The top trace is the measured velocity of the subject's trunk. The lower trace is the velocity of the room. Note the strong phase coupling between environment (room) and organism (human postural sway). (Adapted from reference 19.)

nent of the action-perception system.[19] Unfortunately, no control parameters (frequency and amplitude of swinging room?) were systematically manipulated so that the coordination dynamics of this action-perception system might be uncovered. But the relative phase sure looks like the relevant macroscopic quantity that uniquely specifies the relation of animal to environment in this case. Obviously, all the methods and measures (fluctuations, relaxation times, switching times) described in previous chapters could help quantify the coupling between vision and action components.[20]

The key conceptual points are twofold: first the dynamic concept of stability (this time of an action-perception pattern) is once again at the core of understanding the animal-environment relation; second, we see in the moving room example how an enormously detailed microscopic description on the perceptual side and an enormously detailed microscopic description on the action side, are reduced to a potentially simple linkage characterized by only one task-specific macroscopic collective variable.[21]

On the Neural Support for Direct Perception

Proponents of ecological psychology, following Gibson, advocate a theory of direct perception unmediated by neural representations. Direct perception, Gibson says, is what one gets from seeing Niagara Falls rather than a picture of it. Thus, perception is epistemically direct, not *mediated* by pictures, whether retinal, neural, or mental. (See also my earlier comments about Ulam.) Direct perception is the activity of obtaining information from the ambient (optic, haptic, acoustic, olfactory, gustatory) array. Short shrift is given to the idea that the environment can be known objectively by awareness merely of brain states. Rather, the nervous system is viewed as "functioning vicariously," not as an efficient cause of perception, (i.e., as a processor or producer of percepts).[22]

As advocates of ecological psychology have noted, this does not mean that the nervous system does not *support* perception, only that this support function is not one of processing visual information. Although the support metaphor has yet to be elaborated, I find it especially interesting that various animals have specialized groups of neurons that respond selectively to objects approaching on a collision course, such as the nucleus rotundus in the brain of the pigeon[23] or the medial superior temporal (MST) area of primate visual cortex.[24] These neural structures, in my view, support some of the essential facts of seeing emphasized by Gibson and his school. This does not, on the other hand, necessarily imply any *awareness* or analysis of brain states by a Cartesian ghost in the machine. I look to the next generation of students of Gibson's approach to embrace fully the neural basis of perceiving. In my opinion, they will miss out badly if they don't. As Gibson himself contended, "there is no other avenue [than the nervous system] for contact with or knowledge of the environment."[25]

Affordance Dynamics

According to Gibson, the basic affordances of the environment are perceivable, usually directly, without much learning. As Gibson said, affordances point two ways, to the environment and the observer. Pass-throughability (openings that one can pass through), climbability (places that one can climb on), and swimability (substances that one can swim in) are all affordances. Affordances have to be measured relative to the organism because they are specific to the organism. Ordinary physical units won't do. To capture a climb-upable place or a step-down-able place necessitates intrinsic measurement scaled to the animal's dimensions and action capabilities, not some external, user-independent metric.

In this section, my aim is not to evaluate the status of Gibson's theory. Current perceptual theory takes some affordances, such as graspability, climbability, traversability, and so forth, as noncontroversial. Other functional properties, for example, those tied to cultural or societal convention, seem to constitute affordances of another kind, perhaps mediated by cognitive inference.[26] Be that as it may. My mission here is to show that affordances are subject to dynamics—stabilities, critical points, transitions—the whole bailiwick.

William Warren's remarkable research deals with the affordance of climbability.[27] We all know that some stairs are easier to climb than others. At least in my experience, some, such as those that spiral up to the crown of the Statue of Liberty or the dome of Saint Peter's in Rome, are so steep that they switch us from bipeds to quadrupeds, clambering up them on all fours. What causes these behavioral transitions? Fear, says the reader! Warren has a potentially different account.

Warren reasons that if you take the affordance idea seriously, somehow you have to express the relation between environmental variables and action variables in a common language. One way to do this is through *dimensionless numbers*. For example, in the case of stair climbing, the environmental variable might be the height of the riser, and a relevant climber dimension might be the length of a person's leg. This makes intuitive sense. A taller person can more readily climb a taller step (ignoring other factors such as strength and flexibility) than a shorter person. Scaling riser height, R, to leg length, L, yields a simple number, $\Pi = R/L$ (both in units of length, and therefore dimensionless). Independent of different *absolute* values, Π is a simple expression of the fit between stairway and climber.

Using a biomechanical model of stepping, Warren calculated the critical value at which bipedal climbing should switch to four-legged clambering ($\Pi_c = 0.88$). This occurs when the riser height becomes too high for the climber to get his center of gravity over the support foot, which provides a base for raising the next leg. The point is that regardless of the climber's height, the critical riser height should be a constant proportion of the climber's leg length. Were this Π number continuously changed as a control parameter, a bifurcation or phase transition should occur, causing a qualitative change in action mode.[28]

Can people perceive this critical point, whether a stairway is climbable or not (without having to resort to all fours, a most ungainly act)? Two groups of subjects of different stature, one rather short (5'4" or 163 cm) and one rather tall (6'4" or 193 cm), were shown slides of stairs varying in riser height and asked to judge whether the stairs were climbable or not just by using the subjects' normal stepping patterns. When the perceived riser height was measured in inches (i.e., as an absolute measure), the two groups were statistically very different. However, when the same data were expressed in units of leg length as a dimensionless number, the two groups were identical. And what was the value of Π_c? Exactly $\Pi = 0.88$, as predicted by the biomechanical model.

What factors underlie these judgments? Warren, like a good Gibsonian, reasoned that the perceived difficulty of stair climbing must be related to the observer's own action capabilities. Is a person perceptually sensitive to how costly (in terms of oxygen consumption) stairs are to climb? And, given a choice, does a person choose stairs that correspond to the minimum energetic cost? And is this also scaled to the dimensions of the climber? The answer is yes on all counts. Warren used an adjustable, motor-driven stairmill and measured the energy consumption of climbing stairs of different heights. The minimum energy expenditure, indicative of an optimum riser height, was once again scaled to the height of the climber, $\Pi_0 = 0.26$. When subjects were asked to judge which stairway could be climbed most comfortably, these judgments—only when scaled to leg length—were precisely the same as the optimum Π value obtained by metabolic measures of actual stair climbing.

These experiments of Warren's were crucial in establishing the reality of affordances. Perceptual categories—what's climbable and what isn't—correspond to critical points in an action-perception system. Perceptual preferences—what's easiest to climb—are based on optimality criteria. Both critical and optimal points are *scale independent*. Since this seminal work, invariant relations between critical action boundaries and pertinent action components have been found in such activities as sitting, grasping, and reaching.[29]

Does all this imply that people are walking around with critical ratios stored in their brains, ready to be retrieved when the circumstances require it? Although much more work remains to be done on this issue, I very much doubt it. I'm reminded of von Holst's study in which he systematically amputated the legs of a centipede, leaving only three pairs intact.[30] Regardless of how large an anatomical gap was left between remaining legs (up to five segments), the centipede, which normally walks with adjacent legs about one seventh out of phase, assumed the gait of a six-legged insect. It's ridiculous to suppose that the centipede stores all possible representations of coordination patterns in anticipation that some innovative experimenter (or child) might perform an amputation. In chapter 9 I'll refer to some recent and quite fascinating results that show phenomenally rapid reorganization in the brain after amputation, learning, or even relatively minor changes in task conditions.

Is there an alternative way to interpret affordances? Contemporary cognitive science, for the most part, asserts that mental processes involve computations defined over internal representations. In the case of climbability judgments, the cognitive system forms a mental model of the stairs and simulates stair climbing. Based on these simulations, various hypotheses are evaluated and a decision is made about whether the stairs are climbable or not. Warren's answer is very different: optical information is available about the relationship between relevant environmental properties and properties of the observer's own action system that specifies category boundaries. There is no recourse to computational processes in a representational medium.

What the information is, in the case of affordances, seems to me an open question. Affordance studies have not actually studied action-perception as a dynamic pattern-formation process. Strong hints about candidate control parameters (e.g., the Π ratio) are present, but as yet no attempt has been made to identify order parameters, which, as I showed in chapters 3 and 4, are *informational quantities*. Moreover, the dynamics of affordances—how they are modified by growth and development, learning, intention, and other contingencies—have received limited empirical attention. Nice descriptions exist regarding infants' abilities to detect whether surfaces can be traversed or not.[31] But the actual behavioral patterns produced have not been subject to dynamic pattern strategy and analysis. Herein lies a challenge to ecological psychology. Affordances appear to involve dynamic perception-action couplings, but thus far have been treated statically (e.g., the R/L ratio in Warren's work). Bifurcation studies in which control parameters are manipulated, and system relevant collective variables and their dynamics identified, are necessary to supplement what so far is a functional description.

One aspect is clear, however, due in large part to Gibson's genius. The optic array is literally alive with structure. Forms in the array corresponding to surfaces in the environmental layout undergo accretion and deletion. Disturbances in the array specify the creation and destruction of objects. The perceptual system is enormously sensitive to flowfield geometry and its dynamic properties (fixed points, critical values, etc.). We humans and all seeing creatures are just as immersed in the optic flowfield as a rock is in a river. Just as vortices, eddies, ripples, and splashes arise and vanish in the total process of fluid motion, so too does the optic flowfield reveal meaningful events—brinks, passages, openings, collisions—as we intentionally move about in it.

A Brief Summary

Let's take a quick breather and summarize where we are so far. Any description of the world is high dimensional if not infinite. When organisms move in it, however, or when their surroundings move in relation to them, they are subject to stimulation that (in the case of vision) constitutes a flowing optic array that is chock full of global forms or patterns. When you move backward it consists of *sinks* in which velocity vectors contract and flow to a point. When you move forward, the point to which you are walking is a *source* from which all vectors undergo magnification. In short, this very high-dimensional world is compressed into low-dimensional patterns that constitute information for an active, perceiving organism. Similarly, to characterize an organism's behavior in terms of functional synergies or coordinative structures (chapter 2 and ff.) is to appreciate that a very high-dimensional system exhibits low-dimensional spatiotemporal dynamics.

When we put the world and the organism together, as in Lee's moving room experiment, we see that their *coordination* may take the form of a low-dimensional pattern that once again is subject to dynamics. The themes unifying the biophysics of stimulation, the relation between stimulus properties and perception, and the coupling between perception and the action capabilities of the organism, are self-organized pattern formation and nonlinear dynamics. The details in each case lie in identifying what the relevant pattern variables are and the exact nature of their dynamics.

THE BARRIER OF MEANING: PERCEPTUAL DYNAMICS II

A complete theory of behavior, or at least any serious move in that direction, must deal not only with events outside the organism but also with events inside the organism. Intrinsic coordination tendencies and their dynamics constitute, as we've seen, one of the main cornerstones of the approach thus far, allowing us to understand better how the environment, learning, and intention sculpt behavior. Are there also intrinsic determinants of perception, and do these too exhibit dynamics? That is to say, is the perceptual system itself subject to equations of motion that characterize its (self-)organization? If so, how do we find them?

Perceptual organization is not a new idea, of course, although perceptual self-organization and dynamics might be. Perceptual organization lies at the heart of Gestalt psychology. *Gestalt* is German for "pattern" or "configuration." According to Gestalt theory, perception of an object's form or a piece of music is an emergent property, intrinsically meaningful in its own right. Gestaltists maintain that perceptual experience can't be broken down into primitive sensations or pictures on the retina. Rather, parts of a scene or an object are organized into wholes based on rules of grouping such as proximity, similarity, closure, and good continuation (figure 7.3).

Gestalt psychologists such as Wertheimer and Köhler took phenomena of the kind shown in the figure as self-evident demonstrations of autonomous order formation in perception and the brain.[32] Explanations of these phenomena unfortunately did not go much beyond the effects themselves. Here I will try to establish a closer connection between perceptual experience and specific theoretical models based on the concepts of self-organized dynamical systems.

I will confine analysis principally to two situations. The first pertains to circumstances in which perception *switches* under systematic changes in stimulus parameters. The stimulus is assumed to play a nonspecific role as a control parameter. Such nonspecific information might favor one perception over another, but the essential point is that the information in the stimulus is not enough to account for what is perceived. The evidence and theory I will present concerns how we *categorize* objects and events in a world that is

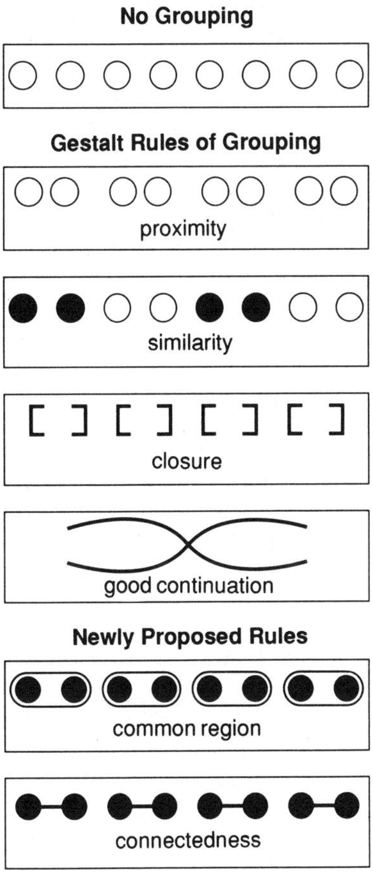

No Grouping

Gestalt Rules of Grouping

proximity

similarity

closure

good continuation

Newly Proposed Rules

common region

connectedness

Figure 7.3 Gestalt grouping rules. On the bottom are two new rules proposed by perceptual psychologists Rock and Palmer. One refers to an observer's tendency to group elements that are located in the same perceived region (common region rule). The other (connectedness) refers to the tendency to treat any uniform region such as a spot or a line as a single unit. (Adapted from reference 2, with kind permission of the authors.)

subject to continuous change. Categorization of visual, language, and speech events will emerge as a dynamical process in which stability, instability, hysteresis, and so forth turn out to be properties of perceiving, not of the stimulus per se.

In a later section of this chapter I'll consider a second set of circumstances in which the *physical stimulus is kept constant, but perception undergoes spontaneous change.* Such is the case in reversible or ambiguous figures (figure 7.4). Descendants of the Gestalt school such as Michael Stadler and Peter Kruse have emphasized this research paradigm as a means of demonstrating the self-organizational activity of the brain. They remind us of Köhler's prophetic remarks over fifty years ago:

Stationary visual percepts, a tree, a stone, or a book are as a rule extremely reticent as to the nature of the neural events which underlie their existence.

 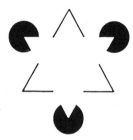

Figure 7.4 Some well-known examples of ambiguous or reversible figures. The contours on the Kanizsa figure (lower right) are illusory.

We may hope to learn more about brain correlates if we turn to instances in which percept processes seem to be in a more active state. This is the case with reversible figures.[33]

New evidence presented here leads to a rather more flexible and fluid view of perceiving than previously envisioned, in which the underlying neural dynamics are *intermittent* (cf. chapter 4). Rather than residing in attractors of a neural network the system dwells for varying times near attractive states where it can switch flexibly and quickly. Faced with ambiguous stimuli, the brain will be revealed as a twinkling metastable system living on the brink of instability. Notice that my intent is not to use illusions to shore up the usual claim that veridical perception necessarily requires some kind of cognitive enrichment of basically ambiguous stimuli. Rather, it is to use ambiguity as a way to uncover the dynamics of the perceptual process.

On Categorization

Scientists are prone to put other scientists in boxes or categories. In physics, the distinction between a theorist and an experimentalist is historically well established and quite clear cut. In psychology and biology, the category boundary is a bit fuzzier. A theoretical physicist colleague told me that he once suggested some experiments to a practicing neurobiologist and was told, in not very polite fashion, to go off and test them himself. (How, with a paper and pencil?) As one who hates to be categorized, I enjoy taking old ideas and experimental paradigms and reshaping them into something new or different. "Old wine in new bottles," as an editor of a respected journal once categorized one of my papers. (I'll forever categorize him as an editor.) Categorization is, of course, a serious business. How we categorize things is a big puzzle that philosophers and psychologists are paid to think about.[34]

Here we will view categorical perception as a dynamical process, even though it has not been characterized as such. Quite rightly, much research deals with establishing necessary and sufficient criteria for assigning things— family members, colors, chairs, and so forth—into categories. One of the reasons that perceptual categorization is often cast as a static process is because it is studied that way (see also earlier comments on affordances). Yet multistability of the perceptual dynamics would appear to constitute a reasonable theoretical basis for categorization, with categories determined by the stability of attractive states (qua perceptual patterns), and category boundaries separating different basins of attraction. Just as the different gaits of a horse constitute categorically different coordinative patterns, when we look at an animal locomoting, we can readily recognize its behavior as a walk, trot, or gallop.

Harking back to chapter 3, it will be remembered that Johansson point light displays of people walking, running, limping, and so forth can be recognized by a computer algorithm in which the stored or learned patterns are defined in terms of the amplitudes and relative phases among the moving points.[35] This is a specific case in which the relevant information for recognizing dynamic patterns lies in the very order parameters or collective variables that characterize the pattern's generation. The basic issue, then, is how we sort a continuously changing signal into an appropriate category, whether it be of objects, emotions, or events. Let's have a look at how relatively simple visual stimuli and speech sounds are categorized.

Categorizing Visual Stimuli

Glass Patterns   Random dot moiré patterns are one of the most eye-catching phenomena in visual psychophysics. You can create them yourself. Splash some paint on a piece of paper (as speckled as you can) and make a transparency of it. Then superimpose one on the other and apply a slight transformation, say a rotation of one relative to the other. Watch as a circular pattern emerges and then dissolves as the rotation is increased. Other patterns are also possible. Translation produces a streaky wood appearance with the grain oriented in the direction of the dots, uniform dilation generates a radial pattern, and so forth (figure 7.5).

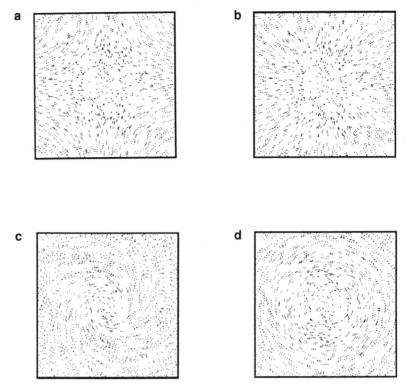

Figure 7.5 Four basic archetypal Glass patterns: (a) saddle, (b) node, (c) spiral, and (d) vortex. They may be generated by evolving a simple set of equations (see text for details).

The physical simplicity of Glass patterns, named after their originator, Leon Glass, and the near immediacy of their detection belie their psychological and neurophysiological complexity.[36] Glass and others, such as Bela Julesz, proposed that perception of moiré fringes is based on the human visual system's ability to extract information about local correlations from different regions in the visual field when forming a global percept. There is also some neurophysiological evidence that cells in extrastriate areas of monkey visual cortex including V2, V4, and the inferotemporal cortex are selective for highly complex stimuli such as concentric, radial, and spiral forms.[37]

Study of the formation or destruction of Glass patterns has been limited to static displays. Yet these patterns afford a wonderful opportunity to expose the underlying dynamics of perceiving. For example, the evolution from a random pattern to an ordered pattern can be generated by allowing the following pair of differential equations to evolve in time:

$$dx/dt = ax + cy,$$

$$dy/dt = dx + by.$$

Depending on the specified parameters (see table 7.1), four basic archetypes may be generated, as shown in figure 7.5. By systematically varying a parameter, or by simply allowing the equations to evolve in time, we may obtain

Table 7.1 Parameters specified to generate four basic archetypes

| Parameters | | | | Archetype |
a	b	c	d	
1.0	− 1.0	0.0	0.0	Saddle
1.0	1.0	0.0	0.0	Degenerate node
1.0	1.0	− 2.0	2.0	Spiral
0.0	0.0	− 2.0	2.0	Vortex

insights into transitions between perceived disorder and order (and vice versa), multistability (differential perception of the same physical stimulus configuration), switching from one perception to another, hysteresis (perceptual persistence despite parameter changes that favor an alternative), and other perceptual effects.

An intriguing aspect about the displays shown in figure 7.5 is that they give a strong impression of motion. In actual fact, the global forms correspond to components of optic flow that hint at their common mathematical origin. For example, figure 7.5b shows the radial velocity field that would be induced by an observer moving straight toward a rigid plane. I think it would be terribly interesting to know if the neuronal populations in parietal cortex that fire selectively for radial motion also fire for static displays that only give the impression of motion.[38]

A former graduate student Elizabeth Horvath and I studied Glass patterns under three conditions: two involved sequential presentations of patterns that change from disorder to order and from order to disorder, and one involved the random presentation of patterns.[39] Each pattern was displayed on a computer screen for half a second or so, followed by an interval during which the subject had to indicate by clicking a mouse button whether the pattern was random or ordered. Then the pattern was advanced one frame (referred to as position in time) and the perceptual judgment task repeated. Three distinct effects were observed for the sequentially ordered patterns (figure 7.6).

The first is called *critical boundary*; it shows that perception depends only on the pattern displayed, not the order of display. This effect only occurred on 7% of trials. The second perceptual effect was hysteresis; a larger position in time was necessary to perceive a change from disorder to order than from order to disorder. Hysteresis, direction dependence based on previous experience, constitutes strong evidence for nonlinearity and multistability in visual perception. It was by far the most dominant perceptual effect and occurred in 72% of all trials. However, we also observed cases in which a switch from disorder to order occurred at a *smaller* position in time than when the patterns evolved from order to disorder. This phenomenon is called *enhanced contrast*, implying that the boundary between patterns shifts to enhance perceived differences. It occurred 21% of the time.

Shortly, I'll describe a theoretical model that accommodates all these patterns of categorization and makes a number of testable predictions. But first

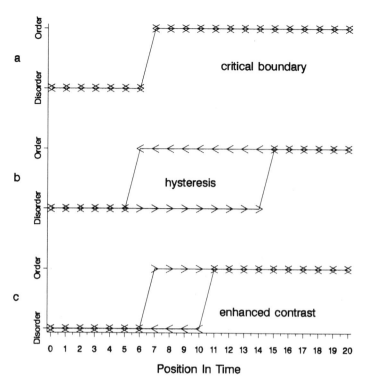

Figure 7.6 Three distinct ways in which observers perceived Glass patterns: (a) critical boundary, (b) hysteresis, and (c) enhanced contrast. Arrows indicate the direction in which the stimulus configuration was changed (see text for details).

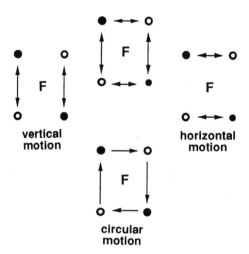

Figure 7.7 The motion quartet paradigm. One pair of diagonal lights is alternated with the other, giving rise (mostly) to horizontal or vertical motion. F is a fixation point on which the subject focuses. (Adapted from reference 32 with kind permission of the authors and Springer-Verlag.)

let's see how general these results are by studying an entirely different set of circumstances called apparent motion.

Apparent Motion   Take an evening walk in Times Square or Piccadilly Circus and you'll be bathed in displays of apparent motion. A small-scale example called the motion quartet paradigm[40] is shown in figure 7.7. It goes like this. Four point lights are positioned as the corners of a rectangle (although other forms such as circles or Coca-Cola signs can also be used). Apparent motion is set up when one pair of lights corresponding to diagonal corners is alternated in time with the second pair. If the delay time of alternation is appropriately tuned, the display is perceived in two different ways: parallel vertical and horizontal motion. Very occasionally, a subject might see motion traveling around the rectangle. The perception of other patterns can be induced by changing which lights are activated and the timing between them. But here I'm going to concentrate on horizontal and vertical motion paths, in the language of perception called *stroboscopic alternative motion*.

One can see clear parallels here between our work on coordination and what the Gestaltists call *perceptual organization*. A coordination pattern, after all, is defined as a reproducible relationship among the components of a system that is stable. Similarly, perceptual organization refers to the establishment of a relationship (or lack of it) between components. The perception formed is characterized by its stability, which we all know by now is a dynamical concept. Although the term *multistable perception* has often been applied to ambiguous displays of the kind shown earlier,[41] in my view, multistability has been used mostly in a metaphorical fashion. Only recently has it been possible to connect perceptual processes—how perceptions are formed and changed—with the concept of stability in dynamical systems. Ambiguous stimuli may well be multistable in a dynamical sense, but this is true only if stability can be established as an attribute of perceiving.

Inspired somewhat by these ideas, my friend and colleague perceptual psychologist Howard Hock set out to test them using the motion quartet paradigm.[42] It is well known that the stimuli underlying perception of vertical or horizontal motion can be biased toward one or the other by changing the aspect ratio—the ratio of height to width—of the rectangle. If the rectangle is tall and skinny, this favors the perception of horizontal apparent motion; short and fat favors vertical motion. By changing the aspect ratio as a control parameter, Hock et al showed that perception switches from one state to the other.

Monostability, the existence of a single perceptual state, was observed only at the extremes of the stimulus series. *Bistability*, the coexistence or simultaneous availability of two perceptual states, was established in two ways: perception of either vertical or horizontal motion was shown to be possible for the same aspect ratio; and sudden, spontaneous changes in perception occurred most often at the category boundary separating the two states. When the control parameter was gradually increased or decreased, the locus of perceptual switching depended on the direction of change of the aspect ratio. When the vertical:horizontal aspect ratio progressively increased, the switch from vertical to horizontal motion did not occur until the aspect ratio became quite large. Similarly, when the aspect ratio decreased, the switch back to vertical motion wasn't observed until the ratio became relatively small. Thus, hysteresis involved the persistence of a given perception despite settings of the control parameter that would otherwise favor the alternative.

Are these effects really of dynamic origin? That is, are they linked to the concepts of stability and instability? Two key experimental facts from a beautiful series of seven experiments provide an affirmative answer. The first comes from a clever manipulation: by fixing the control parameter just short of a value that typically caused a perceptual transition and letting the display idle at this value, Hock et al observed a very high rate of spontaneous perceptual change. The probability of switching, in other words, increases at the category boundary, a clear reflection of loss of stability in the initially established perception.

The second key result concerns the dynamic basis of the hysteresis phenomenon. Perception is predicted to depend not only on the direction of stimulus change, but its rate of change as well. One can understand this intuitively. If the control parameter is changed slowly as we approach the bistable regime, there's always a likelihood of spontaneous switching due to fluctuations in the perceptual system. If we move the system through the bistable regime faster, however, it has much less time to exhibit spontaneous changes. Thus, the size of the hysteresis region (the full range of bistability) should depend on the rate of parameter change. Just as predicted, Hock and colleagues found that the size of the hysteresis region was reduced when the rate of stimulus change was slowed. Dynamically, this reflects the presence of competing intrinsic tendencies: persistence under gradual parameter change favors the initially established perception, but slowing the rate of parameter change enhances spontaneous change.

In summary, Glass patterns and motion quartets are just two of many largely untapped possibilities that allow one to manipulate stimulus parameters systematically and study corresponding perceptual dynamics. Such experiments clearly demonstrate spontaneous perception of order from a disordered stimulus array at a critical parameter value; a transition from perceiving one pattern to perceiving another that is due to loss of stability; multistability, the same stimulus configuration being perceived in two or more different ways; and hysteresis, the pattern initially perceived persists and is dependent on the direction of parameter change. That multistability, transitions, and hysteresis are properties of perceiving attests to the fundamental role of pattern forming dynamics in visual perception.

Not so long ago, hysteresis in perception was equated with unexplained set or "recency" effects, things to be avoided. And not so long ago, intrinsic determinants of perception were viewed by many as outside the reach of science because they could not be accounted for in terms of stimulus analysis alone.[43] But nowadays, even illusory contours typical of the famous Kanizsa triangles (see figure 7.4) have been shown to produce neural responses in extrastriate visual areas of the monkey brain.[44] Soon we will see that hysteresis and multistability, far from being recalcitrant anomalies, open up exciting possibilities for detecting brain events in humans corresponding to perception and perceptual transitions.

For now, let's turn to another situation, the perception of speech, that has led to a theoretical model that captures all known patterns of categorization change within a unified dynamical account. This outcome is due to a theme that I've pushed all along in this book, namely, how important it is for skilled experimenters to work closely with theoreticians. In this case the main players are the experimental psychologist Betty Tuller and the theoretical physicist Mingzhou Ding, with strong support from our graduate student Pamela Case, and me as a hanger on.[45]

Categorizing Speech

In typical speech-perception studies, listeners judge stimuli as belonging to the same speech category despite large variations in acoustic parameters that the experimenter manipulates. Only when the value of the manipulated parameter reaches a critical boundary does the percept change, and then it typically does so abruptly and discontinuously. Usually the stimulus set is presented to listeners in random order with the aim of eliminating short-term sequential effects known to modify category boundaries. Rather than employ procedures designed to *eliminate* possible dynamical effects, Tuller et al. studied them in their own right, the hypothesis being that they reflect inherent properties of the speech perception system. We used a say-stay stimulus continuum, but varied an acoustic parameter sequentially (i.e., as a control parameter) to explore the temporal evolution of sound categorization and change. The manipulated parameter was the amount of silence after the initial fricative "s" in the syllable. Typically, when silence duration is long, people perceive "stay" not "say." Sequentially ordered patterns involved silent gap duration *increasing* in 4-msec increments from zero to 76 msec and then decreasing back to zero; and silent gap decreasing in 4-msec steps from 76 to 0 msec and then increasing back to 76 again. The subject's task was to identify each stimulus as either "stay" or "say."

Once again, three distinct patterns of results were observed. Critical boundary occurred only 17% of the time across subjects and conditions. Hysteresis was far more common, occurring in 41% of runs across subjects and conditions. Enhanced contrast, namely, boundary shifts that act to enhance the contrast between speech categories, occurred on 42% of runs across subjects and conditions. How can we understand all three patterns of category change, and what factors determine the preference of one form over another? For this we have to find a theoretical model.

Theoretical Modeling of Categorical Effects

Our model is based on mapping stable perceptual categories onto attractors of a dynamical model. In this, we followed essentially the same phenomenological strategy that Haken, Bunz, and I used in our theoretical treatment of coordinated movement. For a single perceptual category, a local model containing a fixed point is adequate. However, if several categories are stably perceived, a nonlinear dynamical model has to be found. Task variables (e.g., deformation of a vector field in Glass patterns, the aspect ratio of motion quartets, acoustic parameters in speech) are presumed to act as parameters on the underlying dynamics. Figure 7.8 shows the attractor layout for the proposed perceptual dynamics:

$$V(x) = kx - x^2/2 + x^4/4,$$

where x is the perceived form, and k is the control parameter specifying the direction and degree of tilt for the potential, $V(x)$. The long-term behavior of x is captured by stable fixed points (the minima shown in figure 7.8). For ease of visualization, the figure shows the potential for several values of k. With

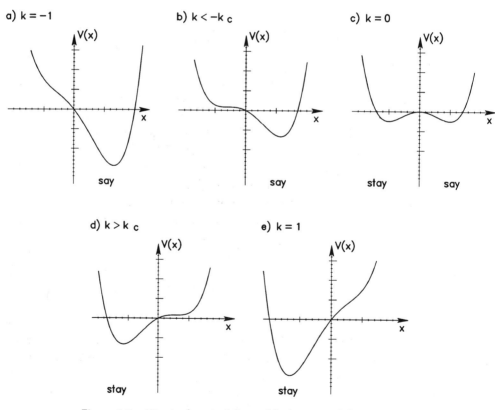

a) k = -1

b) k < -k c

c) k = 0

d) k > k c

e) k = 1

Figure 7.8 Attractor layout of the model of perceptual dynamics proposed by Tuller and colleagues. Here, x is the perceived form and k is the control parameter, which is a function of an experimentally manipulated acoustic parameter.

$k = -1$, only one stable point exists corresponding to perception of a single word. As k increases, the potential landscape tilts but otherwise remains unchanged in terms of the composition of attractor states. When k reaches a critical point, however, an additional attractor appears by a saddle node bifurcation in which a point attractor (with $x < 0$) and a point repeller (maxima in figure 7.8) are simultaneously created. The coexistence of two possible perceptions continues until $k = k_c$, where the attractor corresponding to one percept ceases to exist by a reverse saddle node bifurcation, leaving only a single stable fixed point in the system. Further increases in k only deepen the potential minimum corresponding to the alternative.

Any accurate portrayal of a real world problem must take into account the influence of random disturbances, sources of which in a perception experiment may correspond to factors such as fatigue, attention, boredom, and so on. Mathematically, spontaneous switches among attractive states occur as a result of these fluctuations, modeled as random noise. For a given attractor, the degree of resistance to the influence of random noise is related to its stability, which, in general, depends on the depth and width of the potential well (basin of attraction). As k is increased successively in figure 7.8, the stability of the attractor corresponding to the word initially perceived decreases (the potential well becoming shallower and flatter), leading to an increase in the likelihood of switching to perceiving the alternative category. This implies that perceptual switching is more likely with repeated presentations of a stimulus near the transition point than with repetition of a stimulus far from the transition point. Such predictions have, in fact, been confirmed in experiments.[46]

To account for the three patterns of categorization observed (critical boundary, hysteresis, enhanced contrast), let's examine the dependence of the parameter, k, on the experimentally manipulated parameter of silent gap duration. Without going into details, the tilt of the function may be shown to depend on the initial conditions, a parameter, λ, that is linearly proportional to the gap duration, n, which is the number of perceived stimuli in a given run and, ε, a parameter that represents cognitive factors such as learning, linguistic experience, and attention. Note that the introduction of criteria stemming from cognitive processes is not without precedent; for example, attention and previous experience play a significant role in synergetic modeling of perception of ambiguous visual figures,[47] and contribute to factors that determine adaptation level in Helson's classic work.[48]

When n and/or ε are sufficiently small, the tilt of the potential is dependent only on gap duration and the initial configuration. Figure 7.9a and b shows three regions corresponding to different states of the system in the $\varepsilon - \lambda$ parameter space in the first half of each run when n is small, and in the second half of each run when n is large. White regions indicate the set of parameter values for which a stimulus has but a single possible categorization in the represented portion of the run. Shaded regions indicate the set of parameter values for which a stimulus may be categorized as either one form or the other

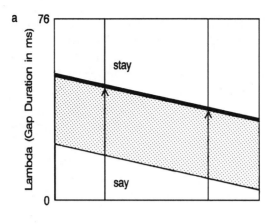

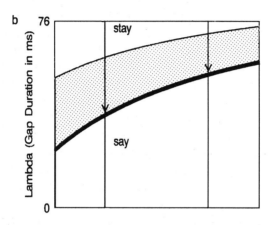

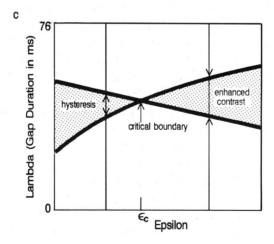

Figure 7.9 The phase diagram or parameter space for the perceptual dynamics of speech. Arrows indicate the direction in which the acoustic parameter (gap duration) is manipulated. Shaded regions in (a) and (b) indicate regions where perception of *both* "say" and "stay" is possible. In white regions only one perception or the other is predicted to occur. Superposition of (a) and (b) in (c) accommodates the hysteresis and enhanced contrast effects that are observed experimentally.

(the bistable region) and thus represent the condition from $-k_c$ (the lower border of each shaded region) to k_c (the upper border of each shaded region).

Consider the initial condition with initial $k_0 = -1$ and the parameter λ (gap duration) increasing. As λ increases, the stimuli are categorized as "say" for any value of ε as long as the $\varepsilon - \lambda$ coordinate remains below the shaded region. Within the shaded region, the stimuli are still categorized as "say" despite the percept becoming progressively less stable. As λ continues to increase, perception switches from "say" to "stay" at the upper boundary of the shaded region (heavy line), after which "say" is not a perceptual possibility. Note that for different values of ε, switches in perceiving occur at different durations of silent gap.

In the second half of the run, λ is decreasing and the resulting division of the $\varepsilon - \lambda$ plane looks somewhat different. This portion of the figure should be read from top to bottom, from large silent gaps to small ones. As λ decreases, the stimuli are categorized as "stay" for any value of ε as long as the $\varepsilon - \lambda$ coordinate remains above the shaded region. Again assuming the absence of a perturbing force, the subject continues to categorize the stimuli as "stay" within the bistable (shaded) regions despite perception becoming less stable. As λ continues to decrease, the lower boundary of the shaded region (heavy curve) marks the switching from perception of "stay" back to "say".

In figure 7.9c the boundaries at which switching occurs in figures 7.9a and b are simply superimposed because each describes a different segment of a run. The line with negative slope represents category switching in the first half of the run, and the curve with positive slope represents the switch back to the initial category. Their intersection yields a critical value of ε, ε_c, for which a critical boundary would be observed in the "say-stay" continuum. For $\varepsilon > \varepsilon_c$, the system exhibits enhanced contrast. In the region $\varepsilon > \varepsilon_c$, λ is smaller for the negatively sloped line than for the positively sloped curve. For $\varepsilon < \varepsilon_c$ (e.g., the vertical line to the left of ε_c), the classical hysteresis phenomenon is obtained: λ is larger for the line than the curve.

The main qualitative features of the observed data are thus captured in the model. Moreover, the parameter plane looks essentially the same when the effect of random fluctuations is explicitly considered. The effect of such fluctuations is to enhance the likelihood of observing switches to the other category as the boundary is approached rather than crossed. Regardless of whether an individual's response patterns displayed hysteresis, enhanced contrast, or a critical boundary, stimulus repetition maximized the probability of category change near the boundary.

Our model accommodates all three perceptual patterns observed and spawns further experiments. The potential landscape is seen to deform with variations in a parameter, with the rate of deformation depending on the number of perceived repetitions, as well as an experience factor with the stimuli. For example, near ε_c, small fluctuations in ε are more likely to lead to a change in pattern than when the system is far from ε_c (see figure 7.9c). The observable consequence is that if a subject shows both hysteresis and enhanced contrast, the overlap and underlap regions should be small. On the other hand, if the subject shows only hysteresis or only enhanced contrast, the overlap or underlap region should be large. This was, in fact, the case.[49]

A second prediction concerns the influence of cognitive factors, as the experiment proceeds in time. If a subject shows at least two of the three possible response patterns, the model predicts a smaller probability of hysteresis in the second half of the experiment compared with the first half. This prediction was also confirmed by Tuller et al.

A third prediction is that spontaneous changes in perception should occur only in the bistable region. One way to test this is to repeat the last stimulus in a run, thereby maximizing the opportunity of seeing spontaneous changes. In principle, as a category loses stability, the shallower minimum in the potential should have a less restraining influence on the fluctuations, and changes in category should occur earlier in a stimulus sequence with repetition than with single stimulus presentation. Remarkably, this prediction was also confirmed experimentally.

In short, traditional approaches to understanding cognitive and perceptual processes emphasize static representational structures that minimize time dependencies and ignore the potentially dynamic nature of such representations. The present experiments, ranging from vision to speech perception, are a striking testament to the basic concepts of self-organized dynamics, such as control parameters, transitions, attractors, multistability, instability, and hysteresis. Of course, no effort has been made here to distinguish particular cues involved in these processes. Instead, the aim is to uncover generic mechanisms underlying the dynamical nature of categorization and change. The evidence suggests that the same mechanisms apply across particular cues, modalities, and categories, thus unifying a broad set of data and providing a coherent basis for predicting contextual effects.

A Glimpse of the Categorizing Brain

The prevalence of hysteresis creates a unique but thus far unexplored situation for studying the neural processes underlying perception and cognition. For example, in the case of speech, the subject may hear *two different words for the identical acoustic stimulus*, depending on the direction of change in the silent gap duration parameter. Thus, the idea is to use or exploit the nonlinear effect of hysteresis to dissociate brain processes pertaining to the physical (acoustic) *stimulus* from brain activity patterns specific to phonetic *perception*.

Here, I present some recent results using a multisensor SQUID (superconducting *qu*antum *i*nterference *d*evice) array to monitor neural activity patterns in the human brain.[50] I'll go into more details about this technology in chapter 9. For now, all we have to know is that SQUIDs pick up signals generated by intracellular dendritic currents in the brain directly, without having to stick electrodes in the brain or on the skull. Because the skull and scalp are transparent to magnetic fields generated inside the brain, and because

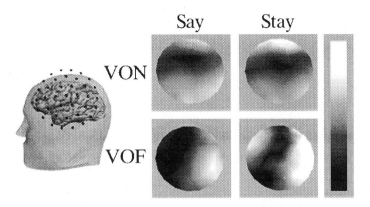

Figure 7.10 How does the brain perceive categories? (*Left*) A phantom view of the subject's head showing the position of the SQUID array relative to skull and cortex. The cortical surface is reconstructed from magnetic resonance images. (*Right*) Topographic distributions of brain activity just after vowel onset (VON) and vowel offset (VOF).

the array is large enough to cover a substantial portion of human neocortex, this new research tool offers a noninvasive means to evaluate the brain's pattern-forming capabilities and their relation to cognition.

The SQUID array was placed over the left auditory evoked potential field, which was determined empirically before the start of the experiment. Figure 7.10 (left) shows the location of the montage (small circles) with respect to the subject's head and cortex (reconstructed from magnetic resonance images). The speech-categorization paradigm was very similar to that described before. A "say-stay" continuum was presented, with the gap duration sequentially increasing and decreasing. During the interval between stimulus presentations, the subject's task was to press one of a pair of microswitches to indicate which word was heard. Analysis of responses showed that at a gap duration of 32 msec, 50% of the stimuli were perceived as "say" and 50% as "stay." For each of the thirty-seven sensors, we averaged the raw signal across tokens with the same response, then calculated the root mean square amplitude of the averaged waveforms. Figure 7.10 (right) shows a sequence of topographic distributions of the field amplitude corresponding to the perception of "say" and "stay" for the same acoustic stimulus (plate 4).

It seems quite clear that differences in patterned brain activity for perception of the two words are minimal up to the offset of the vowel. However, a few milliseconds after vowel offset, a large difference in field amplitude between the two percepts is observed in the posterior part of the array, despite the fact that the physical stimulus is identical. By the end of the epoch the fields are again the same.

These results are encouraging and appear to open up a wide range of exciting possibilities, not just in speech but in other modalities as well. The methodology allows a clear separation between the neural analogues of the acoustic (or any other) signal, and what the brain is doing when people perceive. Although preliminary, to our knowledge this is the first clear evidence of brain activity patterns that are specific to changes in phonetic percept. More generally, studying the brain (e.g., with large SQUID arrays, multichannel EEG, functional MRI, etc.; see chapter 9) under conditions where the input is constant but perception switches appears to be a powerful paradigm for understanding the neural basis of perceptual organization. Such a paradigm shifts the focus to intrinsic (dynamical) determinants of perception and goes beyond static approaches based on the processing of stimulus features.

Related Theoretical Models: Stochastic Resonance, Neural Networks, and the Synergetic Computer

The dynamic theory represented in figures 7.8 and 7.9 is quite simple and intuitive, but the effects it models and predicts are far from trivial. Other models do exist, however. Some, like ours, are at the phenomenological level (stochastic resonance); others use artificial neural networks, such as Anderson's brain state-in-a-box approach,[51] Williams and colleagues' cooperative neural network,[52] and Carpenter and Grossberg's adaptive resonance theory.[53] Still another model, based on the parallelism between pattern formation and pattern recognition,[54] is the synergetic computer that Ditzinger and Haken used to account for a large number of perceptual effects seen experimentally.[55] The thrust here, of course, is to stick to the noble desideratum that theory make contact with experiment. This is not always the case in modeling work. Nevertheless, let me say a few brief words about each class of model, drawing attention to some similarities and differences between them. Then I want to move on to some new experimental results that suggest a slightly different conceptual picture, and relate this to proposed mechanisms for perception that are grounded in recent, rather exciting neurophysiological findings.

Stochastic Resonance   The basic physical idea behind stochastic resonance, developed and elucidated most clearly by Frank Moss, is of a multistable, nonlinear system subject to combined stochastic and periodic forces.[56] Sound familiar? Intuitively, imagine that our double well potential (figure 7.8) is rocked gently to and fro. The amplitude of this modulation is always smaller than the barrier height between the minima, so without noise the system will not switch state. With a little noise, however, transitions occur that to some degree are coherent with the periodic modulation. With too much noise, this coherence disappears and the system's response becomes random. Between these noise limits, however, an optimum noise intensity maximizes coherent switching. Thus, the genesis of the term stochastic resonance.

Stochastic resonance has been demonstrated in certain artificial physical systems, but what has it to do with biology, never mind perception? Recent experiments on single mechanoreceptor neurons of the crayfish show that weak signals can be enhanced by applying an optimal level of external noise.[57] The idea is that stochastic resonance can facilitate information transmission by amplifying weak signals as they flow through the nervous system. External noise, in other words, might help the crayfish detect weak signals that it otherwise might not. Stochastic resonance, however, is a *self-generated* optimization process with noise as an intrinsic parameter. So, exciting though the notion is, there is as yet little or no evidence that the nervous system itself controls or optimizes its noise levels for the purpose of transmitting information. But it well might. After all, as I've shown, fluctuation enhancement is a key feature of nonequilibrium phase transitions in sensorimotor behavior and learning. Noise in that case is beneficial for probing the stability of coordinated states and discovering new ones.

My reason, however, for bringing up stochastic resonance in the present context is an interesting experiment[58] that used Fisher's[59] man-girl reversible figure as an example of a noisy bistable system (figure 7.11). Starting on the man's side, cast your eyes from left to right, and your perception should switch to that of a woman. Start on the woman's side, and perception should switch back to the man, but at a different point. Fisher's man-girl figures have always been taken as classical evidence for hysteresis and multistability. Notice that unlike the carefully controlled Glass patterns and motion quartets, many aspects of the stimulus configuration change haphazardly from one figure to the next.

That criticism aside, Chialvo and Apkarian first trained subjects to rank the images from 1 (man) to 17 (girl). Then an initial arbitrary image was selected and presented to the subject who ranked it as before. The resulting answer was then added to a random number (noise of uniform distribution) and/or subjected to sine wave modulation to determine the next image. In other words, perception of the images was studied as an iterative process. Examination of the statistics of dwell times in one or the other of the stable states showed peaks at the predicted modulation period. Increasing the noise amplitude for fixed modulation parameters increased the signal to noise ratio, reminiscent of stochastic resonance and noisy modulated bistable dynamics.

Stochastic resonance is a fascinating model of information transmission in sensory and perceptual processes that deserves a close look, although a lot of

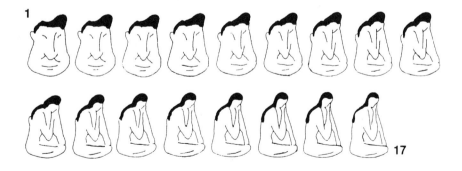

open questions remain. In particular, noise in the central nervous system has to be shown to be controllable and modulable in ways useful for perception and cognition. In the present context of specific models that apply to specific processes, I wonder about other perceptual effects such as critical boundary and enhanced contrast, demonstrated to be characteristic of (unmodulated) noisy multistable systems. These have to be modeled as well, as I've shown. Other paradigms might also be tried, such as those described earlier in which the stimulus conditions are better defined and controlled. Also, stimuli that are intrinsically unstable and can be made more so, such as Necker cubes (see below), could be subjected to weak periodic modulation. On a more biological front, isolated neurons in the optic nerve show conductances that are modulated by a slow circadian rhythm,[60] increasing at dusk and decreasing at dawn. Is this pacemaking capability in single neurons noise enhanced? An implication of stochastic resonance is that neural systems have evolved to make use of their own internal noise, just as the theory of self-organized pattern formation (synergetics) would have it.

Neural Networks   Neural network models are computer simulations in which the units, "neurons", are supposed to correspond functionally to real biological neurons. Each neuron has many inputs, some of which excite it and others inhibit it. The neuron generates an output if the weighted sum of all the inputs exceeds a certain threshhold. Although it's much too broad a categorization (no pun intended), the neural nets I'll talk about here can roughly be classified as bottom-up or top-down. There is no way I can do justice to all (or even any) of these models; the intent is just to give the reader a flavor. Both top-down and bottom-up approaches recognize the striking anatomical parallelism and the highly interconnected nature of neurons in the cerebral cortex.

In the bottom-up approach, neurons are coupled together at synapses. Strength of coupling is proportional to the product of presynaptic and postsynaptic activity and can be changed through learning, according to Hebb's rule, named after the Canadian psychologist. Hebb's learning rule is at the core of associating arbitrary input and output patterns.

In Anderson's brain state-in-a-box (BSB) model[61] (would you like your brain state to be in a box?) sets of neurons are coupled to themselves rather than to each other. These feedback connections can be defined as a matrix of synaptic weights in which every neuron is connected to every other by way of modifiable synapses. Confronted with a stimulus, the BSB scheme works interactively in that feedback interacts with the continuing activity of the system. To ensure that activity in the system doesn't blow up, a nonlinearity is introduced, based on the fact that the individual neurons are limited in their range. In essence, the outcome of such limits is that the state vector representing a pattern of neural activity is confined to a hypercube, hence the term BSB.

If all this sounds a bit abstract, that's because it is. Apart from the use of neuralese, the whole approach corresponds mathematically to extracting the largest eigenvalues and eigenvectors from a matrix. Thus BSB is a general classifier algorithm[62] that maps all starting vectors (initial conditions) in a given region of the box into one corner, which corresponds to the pattern or feature classified. Intuitively (and formally as well), stable corners resemble the minima of an energy landscape akin somewhat to the fixed point attractors of our potential model.

By modifying the synaptic weights in BSB Kawamoto and Anderson were able to make their system shift between corners in a way that simulates multistable perception. Hysteresis, stimulus bias, and adaptation effects were also modeled in a similar fashion. Presumably our observations of *enhanced contrast* can be modeled, too, but Kawamoto and Anderson did not do this.

The Synergetic Computer   Whereas neural network approaches like BSB are based on low-level modifications at the synaptic level, Ditzinger and Haken begin from the top-down approach of synergetics to pattern recognition and decision making.[63] Theirs is a machine approach with significant consequences, I think, for modeling perception and cognition. It starts along conventional pattern-recognition procedures in which the pattern constitutes a two-dimensional array of colors or grey values (features) that are decomposed into cells (neurons) of a square grid. By labeling these cells with an index, j, and the gray-level in cell j by v_j, a set of vectors, $v = (v_1, v_2 \ldots, v_N)$, corresponding to different prototype patterns (faces, cars, pubs) is defined. In the low-level feature space, these prototype patterns constitute attractors. And, because there are potentially many prototypes, the entire system is multistable. When a pattern is presented to the system, its state vector is pulled to the attractor to which it comes closest. Once the specific attractor is reached, the pattern is recognized.

The equations that govern pattern recognition in the Ditzinger and Haken model are of exactly the same kind as govern pattern formation in other self-organizing, synergetic systems such as lasers and convective fluids. This in itself is a very interesting consequence of the synergetic approach, because it suggests that perceptual (and cognitive) organization is itself a pattern-forming process. Of significance in the present context, the dynamics of the synergetic computer (so called because the interactions among its many parts are occurring in a highly parallel fashion) are governed by just a few order parameters or collective variables whose growth rates are manipulable by control parameters. In Ditzinger and Haken, the latter are psychologically significant.

For example, the saturation of time-dependent attention parameters allows them to model oscillations in perception, which can occur when people look at reversible figures. In addition, by subjecting the attention parameters to fluctuating (stochastic) forces, they reproduce the distribution of switching times obtained in psychophysical studies of Necker cubes, as well as a whole variety of other effects. In fact, the Ditzinger-Haken model is one of the most comprehensive on the market: it is psychologically relevant in that it accounts for a large number of well known phenomena. Although no claims are made for its biological relevance, I find it interesting that synergetics is able to reconstruct the underlying wiring network, including coupling strengths between the neurons, that realize these effects. That approach is purely formal and is articulated in Haken's book *Information and Self-Organization* (1988).

Let me make a few final comments on this section. First, all the models described here stress the dynamical nature of information processing, from signal detection in the lowly crayfish to the rather cognitive processes involved in Fisher's man-girl figures. Although the models and algorithms are realized on digital computers, symbolic computation is not the primary focus. Second, whereas bottom-up and top-down approaches use, respectively, synaptic modifications or attention parameters to model intrinsic determinants of perception and cognition, the main goal of our approach thus far is to establish dynamics, especially stability, *as an essential property of perceiving.* This point tends to be lost in the morass of detail.

Finally, I want to leave the reader with what I hope is a helpful image that is created by the amalgamation of these dynamical models. It is an image, naturally enough, of an evolving attractor landscape shaped by a complicated web of interconnected nervous tissue.[64] The tension in the web and the location of its individual fibers are subject to patterns of activation and inhibition that shift and sculpt the attractor layout, which, as we will see, represents large-scale, metastable patterns of neuronal activity in the brain. This is an image that I'll return to when we consider the brain itself. For now, let's ask where the perceptual system likes to live on this landscape. I want to remind you of an old friend: intermittency.

METASTABILITY OF MIND

Ambiguous stimuli are those in which the stimulus input constrains the percept to one of several alternatives but does not determine which one. Once a particular configuration is perceived, the assumption is that perception remains stable for a period of time before spontaneously switching. Ambiguous stimuli thus provide a rare glimpse of a powerful disambiguation mechanism in perception and cognition.[65] For example, the Necker cube, which is a wire frame cube drawn in two dimensions, can be seen in one of two orientations, front face up and front face down. A well-known fact is that during continuous observation of the same stimulus, perception switches back and forth between the two alternatives. Using computer methods it's easy to record a sequence of switching times by having subjects press a button every time the percept switches.

Figure 7.12 shows a set of experimental conditions that Tom Holroyd and I studied in the laboratory.[66] The idea was to use the orientation of the Necker cube as a control parameter and to study the resulting pattern of switching times. One can see that at some orientations (bottom left) the figure is almost, but not quite, a flat, symmetrical, 2-D hexagon. Similarly, the stimulus shown above the near-hexagon is almost a flat square.

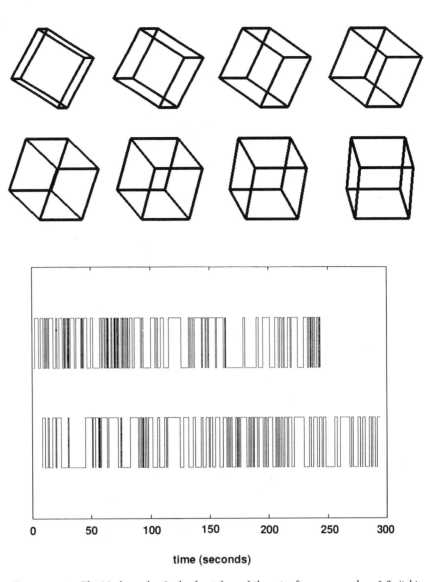

time (seconds)

Figure 7.12 The Necker cube. Is the front face of the wire figure up or down? Switching depends on the cube's orientation. An example of switching behavior is shown underneath for one of the viewing conditions. The time series of switching starts on the lower left of the box and continues to the upper right for a total of approximately 500 seconds.

Also shown in figure 7.12 is a time series for a single Necker cube orientation illustrating the pattern of switching time, and conversely, the duration of a given perceived form. The time series begins on the lower left of the box and continues to the upper right. One can see that bursts of switching are interspersed with prolonged periods during which no perceptual change takes place. No consistent pattern appears in the time series data. For example, switching does not appear to oscillate with viewing time; a sustained perception is not necessarily followed by briefly persisting percepts, and so forth.

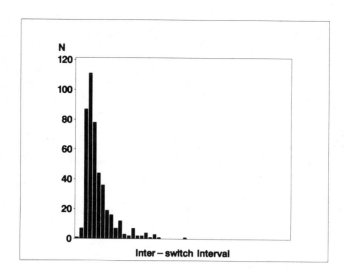

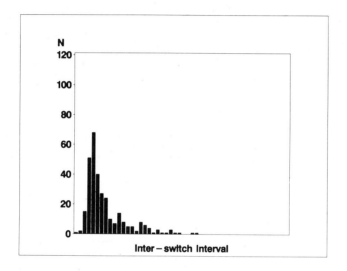

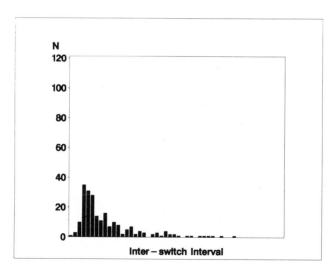

We were interested in seeing if and how the distribution of switching times changes as the orientation of the cube approached a flat figure, which shouldn't be ambiguous at all. We found that the distribution had an extended tail as stimulus configurations approached the 2-D form. This means that occasionally a given orientation is perceived for a long time without switching, an aspect that can also be seen in the time series. Figure 7.13 displays histograms of one subject's switching times when exposed to different Necker cube orientations. All three distributions have a single hump of varying height, but it's easy to see how they spread out as the stimulus parameter nears that of a flat figure (bottom).

Some years ago, Italian biophysicist Borsellino and colleagues[67] performed extensive switching time studies of ambiguous figures. They noted that it is "well known that subjects get blocked" (meaning they stick to one interpretation for a long time). Because of this they discarded switching times (percept durations) greater than about three standard deviations around the mean. Although the present distributions are similar to the gamma distribution fit by Borsellino et al., the parameter (orientation) dependence we observed may be, in part, because we did not discard outliers.

It seems to me that the technique of discarding data points that lie outside an observed distribution is dangerous. Such outliers might be telling us more than we think. Come to think about it, they might be telling us something important about *how* we think. Who among us hasn't struggled with a problem for a long time and then quit, only for the solution to pop up completely out of the blue?

What kind of dynamics might generate the parameter-dependent distributions shown in figure 7.13? Spontaneous switching in perception is usually assumed to necessitate a stochastic description based on random processes. But I will show that a simpler deterministic mechanism based on intermittency works just as well. The idea is to use our phase attractive map (introduced in chapter 4), which the reader will remember is a discrete description of coupled nonlinear oscillators. One can think of this map as a simplified description of neural coordination dynamics in which the neurons in this case are modeled as nonlinear oscillators coupled together by a phase relation, ϕ. This is by no means unrealistic: one of the most hotly discussed topics in the field of neuroscience is the discovery of phase-locked oscillations in the visual cortex of monkeys. Such phase and frequency synchronization has been interpreted as the way the brain links or binds stimulus features (e.g., form, color, motion) into a single unitary percept.[68]

I'll say more about this in the next chapter, but for the moment let me remind the reader of earlier evidence (chapter 4) backed by theory that suggests that only under very restricted conditions does the central nervous

Figure 7.13 Histograms of dwell time or interswitch interval when viewing the Necker cube in different orientations. Note how the distribution spreads out (bottom box) when the cube is perceived almost as a flat figure.

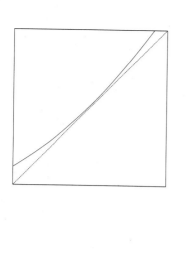
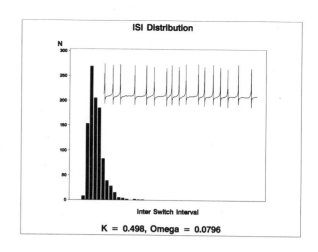

K = 0.498, Omega = 0.0796

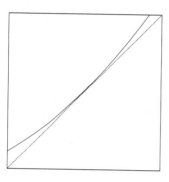
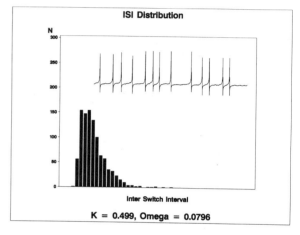

K = 0.499, Omega = 0.0796

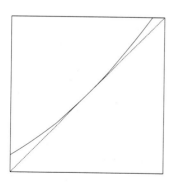
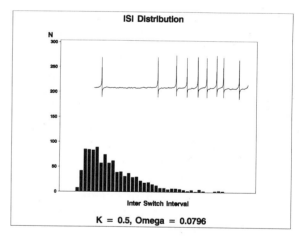

K = 0.5, Omega = 0.0796

Figure 7.14 Neural dynamics of switching behavior in the Necker cube experiment and corresponding distribution of dwell time or interswitch intervals. Flat parts of time series (inset) portray a nearly synchronized state between hypothesized neural modules where the percept does not switch. Viewing the figure from top to bottom, the system gets closer to the fixed point of the relative phase, ϕ (a phase and frequency locked state), rendering it less likely to switch (see text).

system exhibit pure frequency and phase locking. A system trapped in mode-locked oscillations, in my view, is too stable. I propose instead that to retain flexibility the nervous system should live near, but not in, mode-locked behavior. Using our map as a model, the essential theoretical idea is that the distance from a fixed point (which corresponds to a frequency- and phase-locked state of the coupled neural dynamics) varies directly with stimulus parameters (here the orientation of the Necker cube). Figure 7.14 shows the time-dependent behavior of ϕ, which captures the essential relationship between hypothesized neural modules.

The left part of figure 7.14 plots the function, ϕ_{n+1}, versus ϕ as the coupling parameter, K, is changed by a small amount. Note that this function never quite crosses the 45-degree line; that is, the local slope near the fixed point is never less than 1. Iterates of the function run through the corridor between the curve and the line; how fast they move through the gap depends on the width of this corridor.

The right part of figure 7.14 (top to bottom) shows the distribution of dwell times computed for the model as it approaches a 1 : 1 mode-locked state. The dwell time refers to how long the system spends in the narrow channel before exiting. Experimentally, it is equivalent to the persistence of a given perception before switching occurs. On the top right of figure 7.14 the dwell time distribution has a sharp peak and a rapid fall-off (compare with figure 7.13). The time series in the inset shows the evolution of the phase variable. Note the rather short pauses, seen as plateaus in the time series, where the variable is trapped in the channel, and the rapid escape, corresponding to a switch, followed by reinjection.

As the parameter moves the system closer (middle) and closer (bottom) to the tangent line, the system resides longer and longer near the fixed point. On the bottom right we see that the system stays in the neighborhood of such a point for a variable amount of time exhibiting occasional long dwell times and more frequent short dwell times. The comparison with the experimental data (figure 7.13) is quite compelling. Obviously, the closer the system is to a nearly mode-locked state, here corresponding to a single, unambiguously perceived form, the less likely it is to switch.

The dynamics shown in figure 7.14 are intermittent[69], near tangent or saddle node bifurcations. *The main mechanism of intermittency is the coalescence—near tangency—of stable (attracting) and unstable (repelling) directions in the neural coordination dynamics.* This convergence of data and theoretical model supports the hypothesis that the observed switching time behavior is generated by a coupled, nonlinear, neural dynamics residing in the intermittency regime. Intermittency means that the perceptual system (and the brain itself?) is intrinsically metastable, living at the edge of instability where it can switch spontaneously among collective states. Rather than requiring active processes to destabilize and switch from one stable state to another (e.g., through changes in parameter(s), increases in fluctuations), here intermittency appears to be an inherent built-in feature of the neural machinery that supports perception.

PRINCIPLES OF PERCEIVING: CALCULATING, SETTLING, RESONATING, AND TWINKLING

In a recent review of perceptual theory, the eminent psychologist William Epstein contrasts several main options: Gestalt theory, Constructivism, and the theory of Direct Perception (capital letters for effect only).[70] His thesis is that it may be time to rehabilitate Gestalt theory (in fairness to Epstein, he raises it as a question). *Constructivism* is a generic name given to information-processing approaches to perception. The perceptual process is one of arriving at the best solution to an ill-posed problem by inference or hypothesis testing. Nowadays, constructivism is viewed in computational-representational terms. Hypotheses entertained take the form of mental simulations executed in a representational medium. The late David Marr[71] was by far the most explicit in formulating computational strategies of visual information processing. For him, proof by computer implementation was adequate testament to the power of constructivism. Given sensory data and stored knowledge in a human-made or human machine, a solution can be *calculated*. In this chapter I've only fleetingly referred to constructivist approaches. One of the reasons I haven't elaborated them is because of the ghost in the machine; hypotheses are based on interpretations of raw sensations and have to be interpreted themselves ad infinitum.

A self-organized dynamical account of perceiving does not involve or invoke computations over representations. Certainly, perception can be described in terms of computational rules and strategies, but this does not necessarily mean it follows these rules. Gestalt and ecological approaches offer radically different alternatives to constructivist accounts of perception. They are usually viewed as diametrically opposed to each other as well. The Gestalt conception is highly consistent with computational approaches that claim to be free of homunculus-like agents. The perceptual process is one of unmotivated computation among many interconnected units. Excitatory and inhibitory activities promote the formation of stable coalitions within a network. Solutions to perception thus take the form of *Gestalten*: equilibrium states of minimum energy in large-scale neural networks.[72] The process is one of settling into one of these states, not *calculating* a solution. One of Köhler's chief desiderata when stressing intrinsic organizing factors in perception was equifinality: the tendency for biological systems to reach the same final state in different ways and from different initial conditions. Köhler and Koffka saw the inherent organization of perception as a result of natural forces, with no special agent necessary to produce it. In this chapter I have attempted to go beyond metaphor by establishing the existence of even richer dynamics—multistability, intermittency, and so forth—underlying the perceptual process.

How fares Gibson's theory of direct perception? Affordances, as we've seen, look to be self-organized and dynamical, although they have yet to be established as such. In Gibson's theory, the perceptual system *resonates* to invariants in the optic array, just as a properly tuned radio resonates to particular bandwidths of electromagnetic radiation. Cheekily one might ask who tunes the radio? A strange coincidence, although it bears absolutely no relation to Gibson's resonance metaphor, is that one of the proposed mechanisms by which the brain is thought to form percepts takes the form of phase and frequency synchronization among neural signals in the brain. Cells within single columns of the visual cortex respond to similar features. To represent the whole image of an object, so the story goes, calculations in many different columns must be somehow combined. One way this might be done is by a *neural resonance* mechanism, synchronized firing of cells that originates from stimulus features belonging to the same object.

So here we are. Calculating involves symbolic computation, but settling and resonating are dynamical processes typical of self-organizing systems. The distinction is not trivial. All computers are dynamical systems, but the opposite is not true. Settling and resonating are formally equivalent to fixed point and limit cycle attractors in dynamical systems. A new aspect introduced here is that the perceptual dynamics can also be metastable. Switching time distributions for ambiguous stimuli appear to be generated by an underlying neural dynamics that is intermittent. Such a generic mechanism, that I call twinkling, is essential for flexibly entering and exiting coherent neural patterns and avoiding resonant mode-locked states. Although settling, resonating, and twinkling are all properties of the same neurobehavioral dynamics, perception and action systems seem to reside mostly in the twinkling, metastable regime. There attraction (stabilizing) and repulsion (destabilizing) influences coexist in a finely balanced way. There is, strictly speaking, no asymptotic stability in the dynamics at all. This is necessary if the mind is not to get stuck or, worse still, fly apart.

8 Self-Organizing Dynamics of the Nervous System

All things by immortal power
Near and far
Hiddenly
To each other linked are,
That thou canst not stir a flower
Without troubling of a star
—Francis Thompson (1859–1907)

So now we come to the nervous system and the brain itself. Neuroscience and psychology are said to have a common goal: the scientific understanding of mind and behavior. That is a tall order. How are we to go about it? Let's first summarize the received views on the problem. There are similarities and differences between these, and how we'll proceed in this chapter and the next one.

One cannot fail to be impressed by the field of neuroscience. As a multidisciplinary enterprise, it has grown enormously in the last decade, both in terms of the number of people working in it (22,000 at the last meeting of the Society of Neuroscience) and the range of topics it investigates. With so much microscopic structure to explore and so much technology with which to explore it, one can lose sight of the big picture. What's it all for? Most neuroscientists are quite content to plug along happily in the experimental paradigm particular to their subfields using accepted tools and methods. One shouldn't knock them for this. Specialization is usually crucial to making original contributions in any branch of science. Yet specialization should not (I would say must not) deter one from thinking about relationships and asking how it all fits together.

Most neuroscientists are reductionists. They follow the time-honored thesis of classical physics, namely, that macroscopic states can be explained through microscopic analysis. Ultimately, the mind will boil down to molecular biology, which can be reduced to chemistry, which can be reduced to physics...where the ultimate goal for some is to find the "God particle." Reductionism explains, of course, but not all explanation is reductionistic.[1] That is not to say that the laws of physics (and thus of chemistry) do not hold also for (neuro) biology. They do. But studying the elementary components of the system is not enough. At each level of complexity, novel properties appear whose behavior cannot be predicted from knowledge of component processes alone. To reduce a person's behavior to a set of molecular configurations is, as English neurobiologist Steven Rose once said, to mistake the singer for the song.[2] As shown here, in complex systems new concepts must be found to transcend the purely microscopic description of systems. I will show that self-organization occurs on several levels from the single cell up, and shares many of the same dynamical characteristics across levels.

Even when we know what single neurons do (and keep in mind the neuron itself is a complex system), we still have to understand how they work together to support behavioral function. This brings me to one of the main alternatives to reductionism: systems or network modeling. This category contains very many approaches. The best of them try to incorporate anatomical, physiological, and chemical properties of cells and synapses, and propose mechanisms by which these are combined into networks. As a theoretical bridge between neuroscience and psychology, between the micro and macro levels of brain activity, network models possess some attractive features. For example, over ten years ago, in the first *Handbook of Cognitive Neuroscience*, Tuller and I drew some quite precise parallels between Edelman's theory of neuronal group selection[3] and concepts that we and others had proposed for the control and coordination of action.[4] Yet one has the queasy feeling, amplified somewhat when proud reference is made to computer simulations involving 129 networks composed of 220,000 cells and 8.5 million connections,[5] that neurobiology has been invaded by Carl Sagan. The only limitation to such modeling seems to rest on the amount of (super) computing capacity available.

When one probes a bit more deeply one realizes that such simulations do not necessarily capture the style of operation of the brain. Of course, a multi-level approach seems essential that "incorporates detailed models of synaptic function and cellular properties into large scale simulations of interconnected networks in a simulated animal functioning in an environment."[6] (I said, "Say what?") But the entire issue of what the *relevant* variables and their dynamics are on a given level of description is suppressed, if not avoided. In systems models, unfortunately, it is the modeler who defines those properties of the nervous system that seem important for the behavior he or she desires to model. Assumption after assumption has to be made about models at each level, never mind how levels are connected.[7] Hypotheses and speculations are made about how the system *might* do something, not how it actually does it.

My intent is not to run down computer models of large-scale neural networks. Far from it. In complex systems where you can't know all the details, some informed, even inspired, guesses must be made. But I am dubious when such modeling is touted as the only way to go about a theory of whole brain function. Computer simulations can be illuminating, but they do not take precedence over insight and understanding.

What then are the barriers to understanding brain and behavioral complexity? Reductionism *alone* is bankrupt, and systems models—at least of the entire nervous system—are so full of holes that the emperor might as well be naked. Here I reiterate a central premise of the present approach: no single level of description has ontological priority over any other. How then might levels be related? One problem is that each level has its own jargon that is familiar only to the specialists who speak it. The virtually insurmountable task seems to be one of translating back and forth among the many languages that separate molecular biology and behavior. Maybe some multilingual genius will show up. Maybe one is on the scene already, but the fragmentation of neuroscience prevents his or her voice from being heard.

Here I will adopt a rather more modest attitude that does not seek a systems modeling approach to the analysis of whole brain complexity. Rather, by examining specific processes and models at different levels of description, the aim is to look for common principles acting on each level that may, I claim, be the source of their unity. I admit to a single-minded goal: to show that on several levels the nervous system is an active, dynamic, self-organizing system. And I propose just one language to cut across biophysicese, biochemese, neurophysiologese, and psychologese: the language of pattern-forming dynamical systems. Putting my metaphysics out front, the linkage between coherent events at different scales of observation from the cell membrane to the cerebral cortex is *by virtue of shared dynamics*, not because any single level is more or less fundamental than any other.[8]

MICROSCALE EVENTS

Elementary Neurons and Synapses

For the biologist, understanding the relationship between structure and function is a top priority. When it comes to the nervous system this relation is by no means established. The anatomical unit of the nervous system is the neuron.[9] Neurons are cells whose primary function is thought to be one of passing signals through specific contact points called synapses from one part of the brain to another. Morphologically speaking, a synapse consists of a presynaptic-release site containing vesicles, a postsynaptic element, and a synaptic cleft between them. The number of release sites determines the properties of synaptic transmission between neurons. But neurons and synapses play many other roles beyond passing "messages" and modulating excitatory or inhibitory activities between cells. Some of the most important ones concern trophic or regulatory functions that are crucial for understanding injury and disease of the nervous system. And increasing evidence suggests that neurons can communicate without making intimate contact at synapses. Rather than information flowing along structured pathways like electricity flowing along wires in a circuit, such communication, called *volume transmission*, is more like a radio broadcast where signals traveling through the air

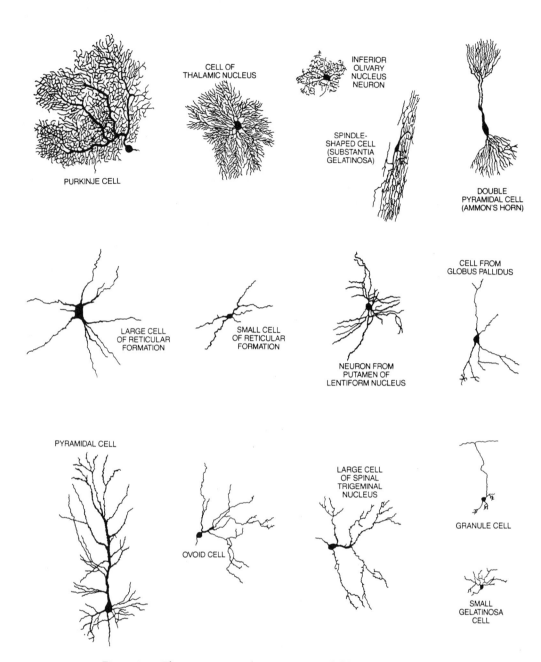

Figure 8.1 The great variety of neurons as revealed by staining techniques. (Adapted with permission from Fischbach G. D. (1992) Mind and brain. *Scientific American, 267,* 48–57.)

can be detected by a suitably attuned receiver.[10] This is our old resonance metaphor again, but in this case the medium is the fluid-filled space between neurons in the brain, the signals are chemical or electrical, and the receiver is a cell endowed with the appropriate receptor.

Neurons come in all shapes and sizes. Ever since the pioneering work of the great Spanish histologist Ramón y Cajal,[11] the beautiful arborization of neurons has been available for all to see. Most neurons have two types of processes that connect one part of the brain with another. *Axons* are specialized for transmitting electrical signals called *action potentials* from the cell body or *soma*. *Dendrites* are specialized to receive signals, mostly from impinging axons. They typically conduct passively rather than by action potentials. Although neurons usually have only one axon, they may have many branching dendrites(figure 8.1).

Although there are exceptions, axons are presynaptic, dendrites postsynaptic. At the terminal bouton of the axon the electrical signal is converted into a chemical one. In response to the arrival of an action potential, a chemical or *neurotransmitter* is released into a narrow synaptic cleft where it diffuses across the gap to contact specialized receptor molecules in the membrane of the postsynaptic neuron. There the chemical signal is transduced into an electrical one. Channels are opened, enabling ions to flow through them and thereby changing the electrical potential of the cell membrane. Intercellular communication by way of synapses depends on the transmission of action potentials, the frequency and duration of which are thought to be essential for information processing in the brain. The generation and propagation of action potentials depends on the *dynamics of ion flow*, the main topic of the next section.

Ion Channel Dynamics

Ion channel proteins are found in the fatty bilayer that forms the cell membrane.[12] These proteins are dynamic not static systems. Indeed, they were once described as "kicking and screaming!"[13] They can assume a large number of different shapes called *configurations* or *conformational states*. In some, ions can enter or exit the cell through holes or interior tunnels in the channel protein. In nerve cells, of course, it is the dynamics of ion flow, determined by the opening and closing of channel proteins, that leads to neural transmission. Incidentally, these channels are *function specific*. In nerve cells or neurons they control the flow of ions such as sodium and potassium. In muscle cells, channels control calcium levels that are involved in muscle contraction.

An important method called the *patch clamp technique*, which earned a Nobel prize in physiology or medicine in 1991 for its developers Bert Sakmann and Erwin Neher, allows one to observe the time sequence of open and closed states of a *single* protein channel. Roughly, a "patch" of cell membrane is "clamped" at constant voltage, and the current through it amplified and recorded. The channel protein acts like a gate. When it is open, ions flow through the channel, resulting in a measurable electrical current. When it is closed, no ions flow and no current is detected. Spontaneous switching between different conformational states—opening and closing—occurs under a variety of conditions, such as temperature fluctuations in the environment, binding with a ligand, and voltage change across the cell membrane.

The basic theoretical picture is not unlike the HKB model of coordination dynamics described in chapter 2. Conformational states (protein shapes) correspond to local minima in an energy landscape. To change from one state to another, the channel protein must cross an energy barrier. In reality, however, the picture is more complicated. In proteins, the conformational space is high dimensional and the number of minima extremely large. Moreover, experiments and computer simulations suggest that a coarse hierarchical organization exists in which the various substates can be categorized into tiers.[14]

Like the HKB model, however, it is the statistics of the landscape and its variation in time that govern protein (hand) motion. There are two main types of motion: a relaxational process in which the system moves from a nonequilibrium state to an equilibrium state, and equilibrium fluctuations in which spontaneous hopping from one state to the other occurs. Both are analogous to the kinds of time scales relations we considered when dealing with behavioral transitions at more macroscopic levels of description.

But our story has a further twist. Channel activity often changes abruptly from periods of great activity in which switching between opening and closing states occurs frequently, to periods of quiescent activity in which switching is much less frequent. The usual explanation of this phenomenon is that the physical parameters of the ion channel protein change abruptly. Larry Liebovitch was the first to realize that such behavior could also be produced by a channel protein with fairly *constant* physical parameters that generated a hierarchy of bursts within bursts of openings and closings.[15] For example, patterns of very rapid bursts that occur on a short time scale are seen in the *grouping* of those bursts within a longer time scale. Such an object or process in which patterns occurring on a small spatial or temporal scale are repeated at ever larger scales is called a *fractal*.[16] In the present situation, durations of open and closed time scales are *self-similar* to those measured at other time scales. An example is shown in figure 8.2 in which openings and closings are measured over a slow time scale. When the temporal resolution is increased 100-fold, the same basic pattern is observed!

What does this mean? One of the key observables (as might be imagined from figure 8.2) is the distribution of dwell times, the time spent in an open or closed state. Liebovitch showed by theoretical calculation that when there is only one closed (or open) conformational state and hence one pathway out of it across a single barrier, the dwell time distribution has a single exponential form, e^{-at}. When there are many states and many pathways, the exponential form is elongated, the stretched exponential, e^{-at^b}.

Finally, the dwell time histogram can take on a power law form, t^{-a}. This means there are huge numbers of closed and open states, and the distribution

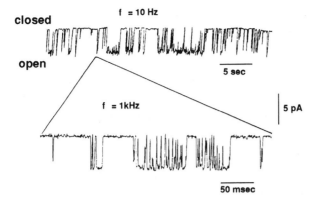

Figure 8.2 Opening and closing of ion channels can be fractal in time. Zooming in on a slowly sampled signal (*top*) by sampling it a hundred times faster (*bottom*) reveals a self-similar structure. A closing event in the upper trace actually consists of many openings and closings when the resolution is increased. (Adapted with permission from reference 12.)

of energy barriers is extremely broad. In short, the dwell time distributions reflect different kinds of energy landscapes. For some channels, such as the potassium channels in cells of the eye, Liebovitch found power law scaling that is characteristic of fractals. For other channels, combinations of scaling functions were observed.

Fractal scaling has some significant implications for neuroscience in particular and for my story in general. Arguably, the most famous mathematical equations in neurophysiology are those of Sir Alan Hodgkin and Sir Andrew Huxley, who formulated a model that accounts for sodium and potassium channels in the nerve membrane.[17] In the Hodgkin-Huxley theory, the probability of changing states depends only on the current state, and not on how long the system has already been in that state. That is, the history of previous states does not matter. Physically, this means that the ion channel protein has just a few conformational states (energy minima) that are independent of each other, separated by well-defined barriers.

Fractal scaling of dwell times suggests a very different physical picture. Power law functions indicate that the channel protein has many conformational states, a rugged energy landscape with many local minima. Processes at different time scales are not independent but are weakly correlated with each other. This relationship, as Liebovitch and Jon Koniarek note, is indicative of a global cooperative process among many different pieces in the channel. They use the words "working together." I say, *synergetic self-organization at the scale of the cell membrane.*

Modeling Ion Channel Switching

What's the minimum model that could produce fractal scaling of dwell times in the channel dynamics? Usually, the switching of a channel is assumed to be an inherently random process. Now we know, of course, that deterministic systems are capable of producing behavior so complex that it looks random. Using a nonlinear map in which the current flowing through the channel at time t is given by the value of the variable, x, and the current at the next time, $(t + \Delta t)$, is a function of the previous value of x, Liebovitch and Tóth[18] were able to produce stretched exponential and power law distributions of open and closed times similar to the data obtained from many channels. The generic mechanism in their map model is *intermittency* (cf. chapter 4).

Although the parameters of their map were not calculated from physical processes at the molecular level, it is possible to use the map's mathematical properties to suggest a physical interpretation. The motions of the channel protein in the model correspond to a nonlinear oscillator composed of masses (atoms) and springs (atomic bonds and electrostatic forces) subject to fluctuating forces. At first, this seems a very mechanical image of a biological process. Recall, however, what happened to the Tacoma Narrows bridge in Washington state.

Engineers used to think that what caused the bridge to wobble and go into wild oscillations and eventually disintegrate was *forced resonance*. In forced resonance, the amplitude of the response is greatest when the frequency of the periodic driving force (the wind) matches the natural frequency of the structure. But, as any sailor will tell you, the wind does not blow in a periodic fashion. Rather the bridge, as a nonlinear oscillator, "organized" the nonperiodic motion of the wind into a coherent pattern. The wind acting on the bridge distorted it in a way that altered the airflow to distort the bridge even more. Circular causality strikes again! The idea is that channel proteins do the same thing. Nonperiodic fluctuations may be organized into coherent motions that drive the channel protein from one conformational state to another. Sound familiar?

Ion Dynamics and Neural Geometry

When one looks at a single neuron, say a Purkinje cell in the cerebellum, one is struck by the anatomical beauty of dendritic trees (figure 8.1). This raises an intriguing question: what is the relationship between the *geometry* of the neuron and its *dynamics*? Might the fractal hierarchy of time scales observed in the switching behavior between states be related to a similar fractal hierarchy in space? Has nature chosen to couple nonlinear dynamic function with non-Euclidean geometrical structure? The answers to these questions are not known, although a promising start has been made by Andras Pellionisz and others.[19] Why would it matter? Well, for one thing, most current neural network models, real or artificial, completely ignore the morphology of the neuron. Furthermore, by treating the single neuron as an on-off device (the familiar flip-flop of digital computers), they also completely disregard ionic mechanisms, even those approximated by Hodgkin and Huxley.

Remember Hebb's rule, the foundation of associative learning in psychology and synaptic coupling in all neural network models? If neuron A repeatedly contributes to the firing of neuron B, then A's efficiency in firing B increases (Pavlovian conditioning in neuralese). Fractal geometry and dynamics imply that learning processes are distributed over many space and time scales and are not simplistically represented as a single scalar coupling strength quantity. The great achievement of the Hodgkin-Huxley approach was that they could mathematically *reconstruct* the action potential of real neurons. From there, a wide variety of behaviors is possible, from tonic spiking to periodic bursting.

Recent work by Larry Abbott and Gwendel LeMasson incorporates activity-dependent processes that occur over much longer time scales than modeled by the Hodgkin-Huxley equations.[20] They show that biochemical activities dynamically regulate membrane conductances indirectly by altering intracellular calcium concentration. So the picture continues to evolve. Moreover, it has been shown that ion channel fluctuations can cause *spontaneous* action potentials (the wink that causes the war again).[21] Such spontaneous action potentials have several implications for systems modeling or neural computation approaches that presently underestimate the role of intrinsic autonomous activity in the nervous system. Yet, as revealed throughout this book, without understanding such intrinsic processes it is difficult to explain how the nervous system is modified.

A Brief Reprise: Ions and the Mind

I've stressed all along how important it is to chose a level of description appropriate to the phenomenon one wants to understand. Since our ultimate goal is to understand mind, brain, and behavior in terms that reflect life itself (stabilities, transitions, crises, etc.), why are we talking about low-level ion channel kinetics at all? On the face of it, ion channels and single neurons are irrelevant to understanding brain and behavioral function. That's why biophysicists study ion channels and psychologists study mental abilities, right?

Of course, this is the very myth I want to debunk. Yes, the molecular mechanisms of ion flows in permeable membranes must be clarified in detail. Yes, it's useful to study cognitive functions in their own right. Here, however, the search is for *level-independent principles*. Self-organization—the spontaneous formation of spatiotemporal pattern that emerges under open conditions—is hypothesized to act on many scales of observation. Naturally enough, we don't expect the dynamics to be the same at all levels. It would be a miracle if they were.

Now it's quite obvious that we already have hints that this view has some merit. For instance, the patterns of switching that is seen between openings and closings of ion channels parallels the pattern of percept switching when people look at ambiguous figures (chapter 7). In fact, both can be understood in terms of intermittent processes in a nonlinear dynamical system. As I've

POWER LAW SCALING IN PERCEPTION AND LEARNING

Learning is usually thought to proceed *exponentially* as practice proceeds. That's in part due to the fact that most laboratory studies are time or practice limited. After several bouts of practice, a task is said to be learned if some performance criterion is met and sustained over some retention interval. Often the task itself has a ceiling in terms of how well it's performed. However, when very long periods of training are allowed on tasks that still allow room for measurable improvement, learning is better described by *power laws*. For example, in Crossman's study of cigar making,[24] the time for a human operator to make a single cigar continued to improve over seven years (and over ten million cigars!).

When a person listens to a syllable, a word, or a sentence that is repeated over and over, the person's perception is initially veridical. After some repetitions, perception begins to shift intermittently among a limited group of other syllables. These illusory changes have been referred to as the *verbal transformation effect*[25] and have received considerable attention in the speech perception literature. Tuller, Ding, and I[26] recently examined the time a given percept is maintained before a switch to another percept occurs. If the variable, l, denotes the length of such an event, power law scaling predicts that the logarithm of the number of occurrences, $N(l)$, of episodes of length l should scale as:

$$N(l) \sim l^{\alpha},$$

where α is the scaling exponent. We found a scaling exponent of $\alpha = 1.8$ over four orders of magnitude. The interesting implication of $\alpha < 2$ is that the average value of l is infinite.

What does power law scaling in perception and learning imply? Fundamentally, it means that there is no characteristic time scale for these behaviors; both processes proceed on a wide range of time scales. The lack of a characteristic time scale helps prevent the system from getting stuck in one state. Power law scaling is typically a sign of long range *anticorrelation*. That is, measured events tend to alternate between positive and negative values in time. There is a lot of evidence for this in studies of human timing behavior.[27] For example, if one response is long relative to some target behavior, the next will be short, and vice versa. Power law scaling, however, means that this tendency not only operates *locally* on an event-to-event basis, but *globally on all time scales*. It keeps the system away from settling on a steady state and staying there. Without such scale invariance, learning and perception would probably be nonadaptive and rigid. Viewed as a whole, these results suggest that the nervous system does not possess a characteristic time scale. Rather, it is flexibly self-organized on all scales.

tried to show, intermittency is a mechanism that is typical of *metastable* complex systems. Power law scaling, indicative of the fact that the underlying process cannot be characterized by a *single* scale of time, is not unique to ion channels. Recent examples of long-range power law scaling include nucleotide sequences in DNA,[22] different forms of which can be used to distinguish different kinds of genes, and successive increments in the beat-to-beat intervals of the heart.[23] More germane to our interests, power law scaling is a feature of both (long-term) learning and perception (see box).

MESOSCALE EVENTS

Let's take a tiny little step beyond the nerve membrane, but still with an eye for level-independent principles. Before proceeding, a few cautionary remarks are in order. First, the anatomical structures and physiological functions I'll describe here are enormously diverse. The patterns formed, however, are not. Second, to make a concept or principle clear, I'm going to be very selective of the examples I use. Even though I believe these examples to be representative, one should be wary of such selectivity. Third, orthodox empirical strategies quite rightly stress detailed descriptions. For the most part, I'm going to spare the reader these experimental details and try to extract the main points, even though these may not be the ones emphasized by the experimenters themselves.

Single Neuron Dynamics: Lamprey, Lobster, and Calamari

The lamprey was introduced in chapter 4. Neuroscientists like it because it's an experimentally amenable vertebrate whose nervous system is "simple" and can be maintained in vitro for a few days. Sten Grillner is one of the world's experts on lampreys. His goal is to understand the cellular basis of what lampreys do, which is mostly swim (at least when scientists study them). This means garnering detailed knowledge about the properties of nerve cells and how they interact synaptically.

For example, adding excitatory substances (amino acids) to the fluid perfusing the lamprey's isolated spinal cord (*a nonspecific control parameter?*) is sufficient to initiate and sustain a complex motor pattern.[28] Under certain conditions a single lamprey neuron exhibits pacemaker-like oscillations in membrane potential. These oscillations tend to be stable over long periods of time even when they are briefly perturbed. When Grillner now adds a very brief series of repetitive stimulus pulses to the continuing membrane oscillation,[29] guess what happens? The membrane becomes synchronized in 1 : 1 entrainment. When the pulses are delivered at a slightly higher rate than the intrinsic membrane oscillation, the two components remain phase and frequency locked. Grillner then notes that a "further frequency increase gave incomplete entrainment." But, as can be seen in figure 8.3, now only every *second* membrane oscillation is locked with the brief depolarizing pulses. In other words, they synchronize 2 : 1, the next most stable mode of coordination of the coupled nonlinear dynamics! (see plate 3) Of course, the physiological mechanism underlying 2 : 1 modelocking is specific to the lamprey neuron, and the way it's experimentally prepared (e.g., constant *N*-methyl-D-aspartate activation but deprived of all synaptic input. The *generic* mechanism, I submit, is a phase transition or bifurcation between mode-locked states, a signature feature of self-organization. However, since the paradigm of studying parameter-dependent transitions was not followed, one cannot be absolutely sure.

Self-Organizing Dynamics of the Nervous System

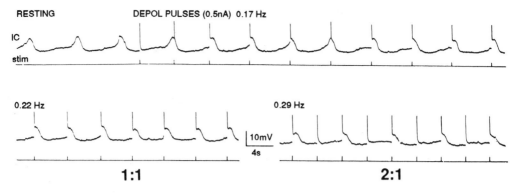

Figure 8.3 Membrane potential oscillations in the lamprey. Application of brief depolarizing pulses capture or entrain the resting oscillation in a 1:1 mode-locked coordination that is sustained when the pulse rate is increased. At higher rates, 2:1 mode locking occurs. (Reprinted with permission from reference 29.)

When we move on to *the* preparation for understanding the relationship between the nervous system and behavior—the stomatogastric ganglion of the lobster *Homarus*—we see exactly the same kind of behavioral dynamics. In lobsters, the movements of the posterior part of the stomach are coordinated by a small neural circuit called the *pyloric* pattern generator that consists of only fourteen neurons.[30] Cells located outside the pyloric pattern generator control its frequency of output. These neurons are intrinsically oscillatory and entrain the pyloric circuit with different coupling ratios. Beautiful experiments by French neurobiologists Maurice Moulins and Frédéric Nagy show the inherent flexibility of this coupling.[31] *Depolarizing* one of the cells in the pyloric circuit switches coordination from 1:2 to 1:1. *Hyperpolarization* results in shift from 1:2 to 1:3. Here again, these patterns are *elicited*, not formed and changed by parametric manipulation. Transition mechanisms per se are simply not elaborated. Nor is the connection between these neural patterns and behavior actually made. Nevertheless, strong hints of self-organization are again apparent.

Moving on to calamari, one way the squid escapes is by activating its giant axon. From its origin in the cell body of the neuron, an action potential travels down the axon conveying impulses that effect muscle contraction. The result is a powerful ejection of water, and the squid jets off. Squid can be caught, however, and brought into the experimental laboratory where it's possible to get their axons to produce spontaneous firing of action potentials, for example, by using the right concentration of salt in sea water.

The giant squid axon allowed us to understand the basic nature of the nerve impulse. Alan Hodgkin[32] remarks that squid were in poor supply in June 1949 around the seaport Plymouth, England. But by July, they were plentiful, and in August he and Huxley made all the experimental measurements that led to their Nobel prize-winning work. Progress was rapid because the two had spent so much time thinking about how to produce an action potential. As Hodgkin says, they knew what they had to measure in order to reconstruct it.

It turns out that the generation and propagation of action potentials in squid giant axons are also governed by nonlinear neural dynamics. When periodic stimulation is applied to a nerve that is already in a state of self-sustained oscillation, synchronization of various kinds occurs, including phase and frequency locking at different ratios (the Farey tree); quasiperiodicity, in which the two rhythms are incommensurate; and irregular, chaotic behavior, as revealed by analysis of Poincaré maps. Similar phenomena have been identified in a number of biological membranes, including chick heart cells, mollusc neurons, and *Onchidium* pacemaker neurons. Of significance, different bifurcation routes to chaos have been found, including the period-doubling cascade and intermittency. All of these effects have been modeled successfully using modifications of the Hodgkin-Huxley equations.[33] An essential physiological mechanism appears to be the interaction between fast spiking and slow membrane oscillations.[34]

These results indicate that individual neurons respond to input signals in a huge variety of ways. Their behavioral complexity far exceeds that of a simple threshold element. George Mpitsos et al have argued forcibly that chaos in individual neuronal spike trains is an essential source of variability, allowing for flexible and adaptive switching among different behaviors.[35] In this, they parallel the present view of the importance of fluctuations in general, whether of (deterministically) chaotic or stochastic origin.

My main point here is not the role of chaos per se, but that synchronization and its cognates (quasi-periodicity, intermittency, etc.) are collective, self-organized states that emerge spontaneously under nonequilibrium conditions. Regardless of whether a single neuron is induced to exhibit complex behavior or whether it does so spontaneously, the basic self-organizing mechanism is dynamic instability. This holds true for single neurons. But what about networks of neurons? Are similar principles in evidence?

Small-Scale Neural Dynamics

For a long time it has been recognized that the nervous system can generate complex behavioral patterns even when it is completely deafferented (bereft of all sensory input). This does not mean, of course, that sensory information is not used to sculpt behavior to the demands of the environment or the needs of the animal in the real world (cf. chapters 5 and 6). Nevertheless, intrinsic patterns may arise centrally in specific neural circuits or networks called *central pattern generators* (CPGs). Figure 8.4 presents a sampling of some of the best-known ones. I have long thought that CPGs offer the best and clearest chance to observe the principled relation of nervous system to behavior. The patterns of activity in these networks are often so well defined that they are given a name such as flight CPG, locomotor CPG, respiratory CPG, swimming CPG, and so forth. Notice in figure 8.4 that these behaviors arise as a result of the connections linking neurons together. It was once thought that single neurons called *command neurons* might be responsible for generating patterns, a kind of

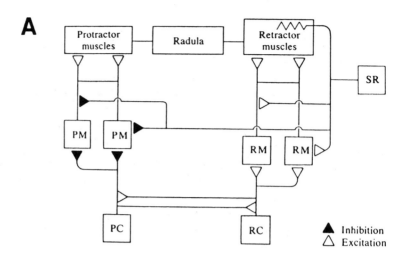

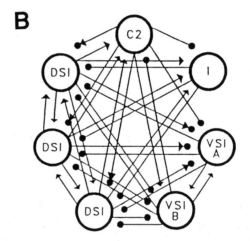

Figure 8.4 Examples of invertebrate neural circuits, showing some of the best worked-out cases. (A) Feeding circuit in the snail *Helisoma*. From *Cellular Basis of Behavior* by E. R. Kandel. Copyright © 1976 by W. H. Freeman & Co with permission.) (B) *Tritonia* escape swimming network. (From Getting, A. P. Reproduced with permission from *Annual Review of Neuroscience, 12* © 1989.) (C) Swimming circuit in the leech. (From Stent, G. L. et al. (1978) Reproduced with permission from *Science, 200,* 1348–1357. Copyright 1978 by AAAS.) (D) Stomatogastric ganglion of the lobster. (From A. I. Selverston et al. (1977). *Progress in Neurobiology 7,* 215–290. With kind permission of Elsevier Science.)

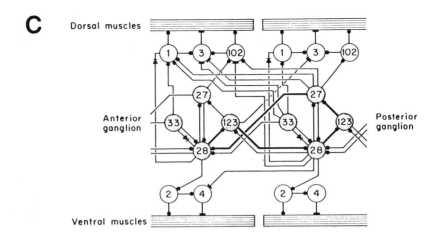

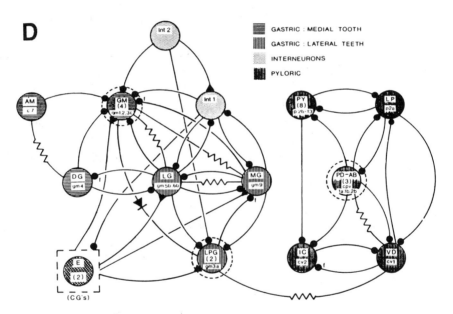

Figure 8.4 (continued)

superordinate motor program, but this is not generally so. Complex behavioral patterns in neuronal networks are the result of *cooperative effects* (or as neuroscientists now say, *emergent properties*) not typically found at the single neuron level.

How do we understand CPGs? The mechanistic answer is to identify the neurons in the network, determine the synaptic relations between neurons, and characterize the intrinsic properties of the neurons and the synapses.[36] This is all right as far as it goes. No one questions the importance of such

Self-Organizing Dynamics of the Nervous System

detailed knowledge. The problem is that even when we know the neurons and how they are connected, we still do not understand how the patterns are produced. Obviously, we have to know not just the particulars, but the principles of operation as well. In a complex system, even one as simple, say, as the gastric mill of the lobster, very many variables must be considered, and as technology advances they are bound to increase in number. Principles beyond those of neurophysiology per se are necessary to guide the selection of what is relevant and what is not. Detailed wiring diagrams are not enough.

Deeper knowledge of the nuts and bolts of CPGs indicates that the physiological mechanisms underlying complex behavioral patterns are *local* to the particular species member under investigation. This comes somewhat as a shock to those hoping for a nice one-to-one relation between neural mechanisms and overt behavior. *None of the well-described invertebrate CPGs uses the same mechanisms, despite the similarity in the patterns they generate.*[37] They may all use the same elements, neurons, but the way the patterns are assembled is flexible and unique (just like a person's golf swing, or his gait if you look closely, the way she speaks, etc.). This fact, that many physical mechanisms may instantiate the same basic pattern, hints strongly that context-dependent, self-organizing processes are active at the level of CPGs.

What form might such self-organization take? A crucial point is that of all the possible neuronal patterns that a circuit, in principle, could generate, only a few temporal orderings are actually produced. Such temporal constraints, for examples, in terms of synchronization and desynchronization, relative coordination, frequency and phase locking, relative timing among interacting components, and so on, are indicative of low-dimensional *cooperative effects*. Here again, we can't really be sure because, although much is known about the parameters and conditions that influence pattern-generating circuitry, relevant control parameters are seldom identified or experimentally manipulated. Usually, pattern change occurs spontaneously or is merely elicited, instead of being brought about by parametric control. This is important, because if central pattern generators are really self-organized, it should be possible to demonstrate loss of stability. Specifically, predictions regarding phase transitions (e.g., critical slowing, fluctuation enhancement) should be observable in putative pattern generators.

Along these lines, it has become quite clear that CPGs are multifunctional. For example, when low concentrations of the neuropeptide proctolin are injected into the lobster's circulation, its teeth operate in the squeeze mode; higher concentrations result in the cut and grind mode. Because proctolin is a circulating hormone in lobsters and affects their neurons, similar effects are observed when it is applied to the isolated lobster nervous system in vitro. This produces fictive feeding. Also, neuromodulatory substances released by identified neurons in the circuit itself can reconfigure the network, causing switching among different patterns and even creating new patterns.[38]

Not many years ago it was almost heresy to suggest, even when based on experimental evidence, that hard-wired neural circuits are the exception rather than the rule.[39] Now the "CPGers" talk about loosely organized systems under constantly changing chemical and/or sensory control that are easily sculpted into new functional circuits.[40] Others, comparing the results from many small-scale neuronal networks, have come to recognize one of the central principles of biological self-organization: similar tasks can be performed by different neuronal networks, and different tasks can be produced by the same neuronal network.[41] Still others[42] are ready to give up, or at least expand the (no longer recognizable) CPG concept to that of a motor pattern network (MPN).

The stated aim of the CPG field has always been to identify the principles of neural pattern generation amidst a mass of detail. How would we know a principle if we saw it? From my point of view, some of them are likely at hand. The fact that neural patterns in small-scale networks are reproducible attests to the central concept of stability (even if it is not directly tested). The fact that different patterns coexist in the same network (multifunctionality) is a natural consequence of multistability in a complex system possessing several attractive states. The fact that a neural network can switch flexibly among functional states and can reconfigure itself according to current conditions is likely a result of dynamic instabilities in a system whose functioning depends on interactions among many nonlinear processes at cellular, synaptic, and network levels. Neuromodulatory substances provide a kind of global arousal, leading the system into and maintaining it near transition regions. At such transitions, self-organization becomes apparent: new or different patterns arise as cooperative states of the coupling among neural elements (see box).

MACROSCALE EVENTS

Ion channels, single neurons, and collections of neurons are spontaneously active over a broad range of time scales. Regardless of level, however, nonlinear oscillatory processes, whether autonomous or driven, constitute the archetype of time-dependent behavior. There may be very many mechanisms at molecular, cellular, and network levels for constructing an oscillator. What is important to appreciate here is that when two or more nonlinear oscillators couple nonlinearly, the process of self-organization renders a wide variety of behaviors possible.

I have argued that behavioral function can serve as an excellent guide in the quest for collective variables and control parameters on the neural pattern level. If the behavior is fundamentally rhythmic, as it is in many CPGs, the underlying neural dynamics should exhibit strong hints of self-organization such as phase and frequency synchronization among the neural elements involved.

Oscillations and Resonance

Whereas the connection between functional description and neuronal dynamics is accessible in invertebrate and some vertebrate CPG preparations,

PHASE TRANSITIONS IN COUPLED NEURAL OSCILLATOR MODELS

If individual neurons do behave as nonlinear oscillators, and we have seen that under certain conditions they *are* nonlinear oscillators, then, when they are mutually coupled in circuits, the system may be characterized as an oscillatory neural network. Hirofumi Nagashino and I showed some years ago that phase transitions occur in such pattern-generating networks, thus providing a neural basis for behavioral transitions in bimanual coordination.[43] Our model is composed of excitatory (E) and inhibitory (I) neurons (figure 8.5, top), each with associated membrane time constants. These two elements are coupled to each other forming a negative feedback loop. Excitatory neurons have a self-excited connection forming a positive feedback loop. Without both loops, there is no oscillation in the network. Introducing only an interconnection from the inhibitory neuron in one oscillator to the excitatory neuron of the other turns out to be the simplest way to obtain coexistence of in-phase and antiphase modes and the transition from antiphase to in-phase. In our model, transitions occur through a change of coupling coefficients in the network that effects the change in frequency of oscillation. What makes the coupling coefficients change? We hypothesized that a tonic external input is capable of modulating the role of synapses. Neuromodulatory substances are likely involved.

Recently, Stephen Grossberg and co-workers produced a pattern generator model whose architecture is similar to ours (turn figure 8.5 bottom on its side to see the architectural similarity). In their model also, fast excitatory neurons are coupled non-linearly to slower inhibitory interneurons.[44] Notice that, in addition, they include external timing inputs to each unit. Their model also reproduces the stability of patterned modes and the transition between them. As I have stressed again and again, many mechanisms, both physiological and mathematical, can give rise to the same patterns and pattern change. But neither of these neural models presently accommodates the generic features of self-organized coordination dynamics, such as fluctuation enhancement and critical slowing down in collective states as the transition is approached, and the pattern of switching times at the transition itself. In my opinion, these pattern-generator models represent possible physiological instantiations of a more general physical principle, namely, synergetic self-organization. They do not replace it.

when we move up to higher levels of the nervous system the situation is not so clear, at least thus far. I think there are two main reasons for this state of affairs. First, behaviors such as perceiving, acting, remembering, learning, and developing have not been subjected to dynamic pattern analysis. Of course, much of the research described in the first part of this book is a step in that direction. Second, although dynamic phenomena, especially oscillations and resonance in the central nervous system (CNS), are hot stuff in the field of neuroscience these days, no one really knows what they are for. Between you and me, I suspect the emphasis on these particular phenomena is a bit misplaced. Here's why.

When people talk about oscillations in the nervous system they stress the *frequency* at which cells or cell groups oscillate. Forty oscillations per second (40 Hz) is the latest rage, and it can be found all over the place (if the

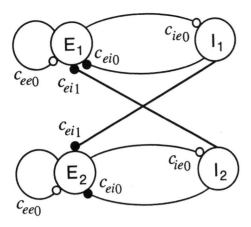

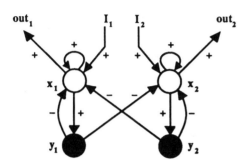

Figure 8.5 Neural oscillator models for phase transitions. (*Top*) Nagashino and Kelso's model. (*Bottom*) Cohen, Grossberg, and Pribe's model. (Reprinted with permission from reference 44.)

filters are set appropriately and one knows where to look for it). One day, 40 Hz in the mammalian brain is a cortical information carrier, the next it's the signature of consciousness. It's important to realize, however, that oscillation is itself a collective phenomenon, although it is seldom studied or described as such. It is nontrivial, for example, to establish how a particular frequency in the CNS arises from more microscopic scales.[45]

Similarly, resonance refers to an existing oscillation in an anatomical structure such as the cerebellum that is preferentially excited. Yet, ion channels and many neurons do not have a single characteristic frequency that can be excited. Resonance, as we have seen, is probably the least flexible and adaptable way to couple things together. In fact, I suspect that statements such as, "If the cell is slightly depolarized it may oscillate at 10 Hz. If the cell is hyperpolarized, it tends to oscillate at 6 Hz,"[46] are convenient fictions. I'd take the words "may" and "tends" seriously, because they hint at unexplained neuronal variability. And the numbers should be taken with a grain of salt.

It behooves me to give a few examples of oscillations and resonance in the nervous system and where possible, address their significance. One of the best-known cases is when you fall asleep. Quite beautiful work at several

Self-Organizing Dynamics of the Nervous System

levels from intracellular recordings in vivo and in vitro to computer simulations has been done by Mircea Steriade and colleagues.[47] They focus on the thalamus, one of the main gateways to the cerebral cortex and the first place that is blocked by synaptic inhibition when you go to sleep. The early stage of quiescent sleep is characterized by spindle waves in the thalamus, waxing and waning oscillations between 7 and 14 Hz that last a few seconds and recur once every three to ten seconds or so. Spindles mark the transition to sleep that is accompanied by loss of awareness. Nobody knows how they start (control parameters?) or why they wax and wane (transients, bifurcations, intermittency?). The particular frequencies are biophysically determined according to interactions among neurons with different cellular properties.

Several other oscillations are associated with sleep. The thalamus is also involved in generating these, although a recently discovered slow oscillation (<1 Hz) appears to originate in the neocortex. Thus, when you lesion the thalamus, this slow wave is still there. What about dreams? Despite a great deal of speculation, there is no generally accepted function for dreams or, indeed, sleep itself. And so far no real theory exists for the oscillations, beyond clues that they are cellularly based. How could it be otherwise?

Another famous oscillation in the brain is theta rhythm (4–10 Hz), which is considered the fingerprint of all the structures in the limbic system, especially the hippocampus. If you lesion the hippocampus, either by accident or design, memories that already exist are unimpaired, but new ones are very difficult to establish. Since the hippocampus receives inputs from many sensory modalities, it appears well suited to forming new associations with previously stored information.[48] But it is not possible to tie hippocampal rhythms to a single function. Theta is present when a rat or a cat explores, orients to relevant stimuli, and forages for food. It is extremely prominent in REM sleep, the phase of the sleep cycle associated with rapid eye movements and dreaming. And dreams, as we know, are the window into the unconscious. But theta itself is multifunctional. At present we understand neither its mechanisms nor the functions it subserves.

The same goes for alpha, the first oscillation found in the brain by Hans Berger in the 1930s, and the rest of the oscillatory abecedary. There are, as Grey Walter, the English electrophysiologist said, complex rhythms in everybody. Assigning them Greek names is no substitute for understanding what they mean, if anything. The significance of these oscillations, I will argue, does not really begin to become apparent until they are coupled or temporally correlated with other events in the nervous system.

Correlated Activity in the Nervous System

Oscillations and frequency-related measures refer essentially to events in component structures (e.g., neurons, neuronal groups), and have to be accounted for in terms of mechanisms at the cellular level and below. If behavioral functions such as perceiving, acting, learning, and developing are really due to

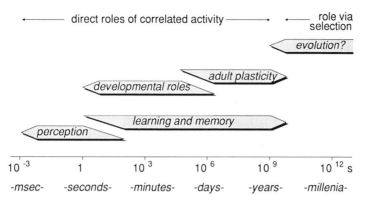

Figure 8.6 Time scales of the various functions that correlated activity might play a role in within the CNS. (Drawing courtesy of J. E. Cook. See reference 49.)

self-organizing processes in the nervous system, it seems more likely that relevant information is specified by coherent *relations* between neuronal events rather than by the oscillatory components per se. Indeed, in a complex system such as the brain we might expect the *timing* of interactions between neuronal activities or neuronal groups to be of major importance. Do we have any evidence that this is, in fact, the case?

The list of situations in which correlated activity is known to play, or suspected of playing, a crucial role embraces nearly all of neuroscience[49] (figure 8.6). For example, in the developing visual system (how each eye gets connected to its own territory in the brain), spatiotemporal relations among action potentials determine which synapses are strengthened and which are weakened and eliminated. Cells that fire together, it seems, wire together.[50] For now I want to focus on just two examples: learning and memory, and the so-called binding problem in perception.

Function of CNS-Correlated Activity. I. Learning and Memory

I have already mentioned the theta rhythm of the hippocampus. In a preparation called the hippocampal slice (exactly what it sounds like) it is possible to excite theta using cholinergic drugs such as carbachol. When neurons called CA1 are stimulated in the hippocampus with small shocks, they produce an elevated, sustained synaptic response called *long-term potentiation*, or LTP, which is thought to be one of the main cellular mechanisms underlying learning and memory. Without theta, however (or an elicited theta-like rhythm, see below), there is no synaptic enhancement. More intriguing is that LTP occurs only when applied stimuli are in phase with the theta rhythm. If conditioning stimuli are alternated antiphase with theta, the magnitude of the postsynaptic response is reduced. In other words, the *phase relation* between incoming stimuli and the intrinsic theta rhythm strongly affects synaptic efficacy.[51]

In the language of coordination dynamics developed in chapters 5 and 6, behavioral information (e.g., a task to be learned) *cooperates* with the intrinsic

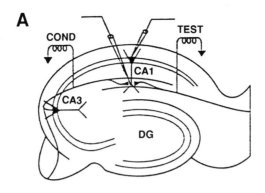

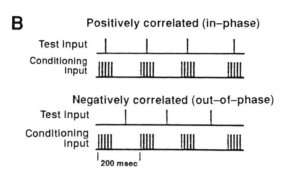

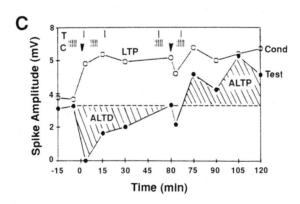

Figure 8.7 Hippocampal slice preparation and stimulus paradigms. (A) Diagram of the hippocampal slice showing recording and stimulus sites. (B) Conditioning paradigms. Test inputs are either in phase or antiphase with conditioning inputs. (C) Time course and magnitude of associative LTP or LTD depending on conditioning paradigm. Inset shows the stimulus patterns for the test (T) and conditioning (C) inputs; arrows indicate the time of stimulation. (Adapted with permission from reference 52.)

dynamics. Cooperation takes two essential, activity-dependent forms at this level: synaptic strengthening and synaptic weakening. Both are crucial aspects of plasticity. So if you want to keep something in memory, make sure it's coordinated with theta. Timing, it seems, is everything. Of interest, cholinergic activation, whether naturally occurring or induced by drugs, seems to enhance "learning" and "memory" in the hippocampus. This might explain why these functions are so disrupted in people with Alzheimer's disease, which is known to be accompanied by loss of cholinergic neurons.

There's more to it, of course. Even when theta rhythm is not present in the hippocampus, timing or temporal ordering is a major determinant of associative learning. Figure 8.7A and B show the hippocampal slice preparation and an experimental paradigm in which conditioning stimuli were input at 5 Hz (i.e., simulating theta frequency) and test stimuli (single shocks) inserted in phase or antiphase. Neither stimulus presented alone produced long-lasting changes in hippocampal cells. However, when test and conditioning inputs were synchronized in phase, synapse-specific LTP was observed. In contrast, when the two inputs were antiphase with each other, *long-term depression* (LTD) was observed. This is truly a cooperative dynamical effect: test and conditioning stimuli are identical in both cases, only their relative phase is altered. In-phase stimuli lower the threshold and increase the amplitude of the action potential in CA1 neurons. Antiphase stimuli arrive when the latter are hyperpolarized, increasing the firing threshold and eliciting a long-lasting reduction in amplitude.[52]

What happens when the phasing is not simply in-phase and antiphase? How does synaptic strength change? Viewed dynamically, if LTP and LTD correspond to two collective, self-organized states, our theory (see chapter 6) predicts a *phase transition* from LTP to LTD and vice versa as relative phase is altered, much like what we have seen in human learning. Moreover, if the phase relation between test and conditioning stimuli is systematically varied in one direction and then the other, hysteresis and hence multistability are expected. How context dependent are LTP and LTD? Which is the most attractive and long-lasting state? How much does the particular (theta) frequency matter? Thus far, no one knows. One can appreciate that the answers are important, not just for learning and memory, but for recovering from neuronal damage due to injury or seizure as well.

The same kinds of dynamics appear to operate at more macroscopic levels. Robert Vertes and Bernat Kocsis recently demonstrated that single unit activity in the dorsal raphe nucleus of freely behaving rats discharges rhythmically with hippocampal theta.[53] The dorsal raphe nucleus is a small structure in the brainstem that packs a big punch on two fronts. First, it is the reservoir of a huge source of serotonin in the brain, low levels of which are associated with depression. Second, dorsal raphe neurons project throughout widespread regions of the forebrain. Given first, that hippocampal LTP and LTD serve a vital role in learning and memory, and second, that they are best elicited at or around theta frequency, it seems possible that the dorsal raphe nucleus might be a key energizer of these functions. It's damned hard to learn or remember anything if you are depressed. Kocsis and Vertes find phase and frequency locking only when the animal is awake and actively exploring its environment.

Finally, one is reminded in this context of the classic work of Ross Adey and associates in the early 1960s.[54] As cats learned a task, they showed that irregular 4- to 7-Hz activity in hippocampal EEG evolved to a much more regular 5-Hz rhythm. Even more striking were systematic shifts in phase relations between hippocampal and cortical structures as the task was acquired. Recall that in chapter 6, environmental information was shown to modify existing coordination tendencies. Due to competition between "extrinsic" and "intrinsic" constraints, learning took the form of a phase transition, one of the chief signatures of spontaneous self-organization. It is intriguing to speculate that the neocortex, due to its direct contact with the environment, and the hippocampus, or hippocampal-brainstem relations representing intrinsic dynamics, also compete early in learning, but that during the learning process cooperation wins out, seen now as a stabilization of specific phase relations between cortical and hippocampal (or hippocampal-linked) structures.

Function of CNS-Correlated Activity. II. The Binding Problem

To separate figure from ground—to find the paperclip lying on my desk in the midst of a mass of papers—is easy for me, but at our current level of understanding is an insurmountable task for the nervous system. Cells responding to particular features of the paper clip, lines, edges, angles, and so forth, are scattered all over the visual cortex. To create a complete image of the object, it seems that cell activities must be integrated. Moreover, different attributes of an object in the field of view—its form, color, and motion—are thought to be analyzed by separate processing pathways. These too have to be coordinated or bound together somewhere or somehow so that they are seen as belonging to one object and not another. Of course, a binding problem exists only if this particular theory of perception is adhered to. Alternative viewpoints do exist, as we saw in the previous chapter (see also below).

Traditional neurophysiological approaches to the binding problem assumed that cells that respond to perceptual primitives converge on cardinal cells in higher cortical areas. According to this view, you know your grandmother because there are "grandmother" neurons that respond selectively to the configuration of features that denote your granny. What happens if this configuration changes? Well, then you have to activate another grandmother neuron. One problem with this concept is combinatorial complexity. The brain probably does not have enough cells to cope with the complexity of the world, at least when the world is described this way, as a collection of objects.

Recent evidence, which I've mentioned on a couple of previous occasions, indicates that neurons in the visual cortex activated by the same object fire in unison. Again, many investigators, including the discoverers of the phenomenon, Charles Gray and Wolf Singer, stressed the oscillation itself, which is in the region of 30 to 70 Hz.[55] Even when announced in the journal *Science*, the phenomenon was presented in terms of "*Oscillation* revealed" (italics mine). Yet, what is striking in figure 8.8 is the coherence or stable spatiotemporal relation between multiunit and local field activity in different parts of the cortex (here, area 17 in the adult cat). It is this coordination among distributed neuronal events that appears to be significant for binding, not the oscillation per se.

Notice in figure 8.8 that periodic spiking in the multiunit activity is synchronized with the local field potential, but the *intrinsic temporal structure of the two signals is very different indeed*. One is spiky, and the other goes up and

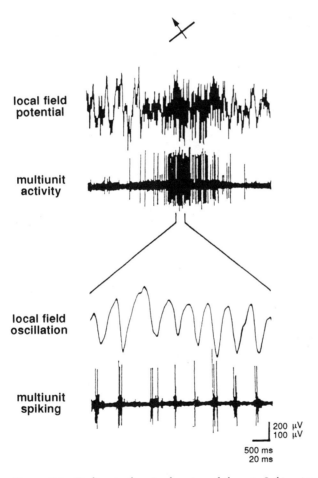

Figure 8.8 Binding in the visual cortex of the cat. Spiking in response to an optimally oriented light bar is synchronized with the local field potential oscillation. (Adapted with permission from reference 55.)

Self-Organizing Dynamics of the Nervous System

down in a quasi-regular fashion. We say the time series are synchronized or phase locked by virtue of discrete events, onsets and troughs in the respective signals.

The key idea behind all this, first articulated by theoretical physicist Christoph von der Malsberg, is that spatially separated cell groups should synchronize their responses when activated by a single object. Such appears to be the case both within and across cortical areas, including the two cerebral hemispheres.[56] Evidence suggests that Gestalt criteria, such as continuity, proximity, and so forth (see chapter 7) are important for the establishment of synchrony. However, neuroscientists have to be aware of a conceptual difference when discussing these results in terms of Gestalt theory. The Gestaltists emphasized the whole as *something other* than the parts. Here it is the coherence among the parts that gives rise to the whole.

Numerous models for generating oscillations and synchrony in the visual cortex are now available. Issues such as the origins of the oscillations, the kind of architecture that is required, how the coupling is achieved, and so forth are all being hotly pursued. Many of the models, naturally enough, are coupled oscillator networks that, as we now know ad nauseam, are remarkably robust and exhibit rapid self-organization. As just one example, H. G. Schuster used the phases, $\phi_{1,2}$, of limit cycle oscillators to describe two cortical columns in the active state[57]:

$$\dot{\phi}_1 = \omega_1 - K_{12} \sin(\phi_1 - \phi_2),$$
$$\dot{\phi}_2 = \omega_2 - K_{21} \sin(\phi_2 - \phi_1).$$

These equations are quite familiar and somewhat simpler than the HKB model: $\omega_{1,2}$ are the frequencies of the oscillators, and the K terms represent their mutual coupling. Schuster's model is shown in figure 8.9 and is similar to, although not of course the same as, models for phase locking and phase

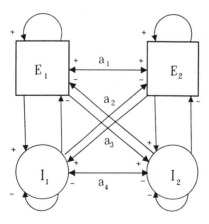

Figure 8.9 Schuster's model of binding. Two coupled cortical columns (oscillators) in the active state. + and − stand for excitatory and inhibitory coupling, respectively. (Reprinted with permission from reference 57.)

transitions in bimanual coordination (cf. figure 8.5). Schuster is able to trace the emergence of oscillations to a *Hopf bifurcation* (cf. chapter 3) in a single cortical column consisting of subpopulations of neurons coupled by excitatory and inhibitory synapses. The dynamics of coupled columns is described by the phases of oscillators, whose coupling strength depends on the external stimulus. Global features of the stimulus, its unity as a single coherent object, lead to the synchronization that is observed experimentally.

Let me make a few final points that suggest two different kinds of generalizations. First, although not everyone agrees that the foregoing theoretical scheme for binding is realistic, the consensus is that, compared with our knowledge of cellular and molecular levels of description, our understanding of integrative functions of the nervous system is poorly developed. The potential for coherent or correlated activity as a central unifying theme is great, but it needs elaboration. For example, similar mechanisms can be hypothesized not just for perceptual binding, but for concept formation and action as well. Yet, as we have demonstrated here, phase and frequency synchronization are only one fingerprint of self-organizing processes in the nervous system and behavior. Arguably, they represent the simplest kind of dynamical behavior (absolute coordination, in von Holst's words).

Second, and related, there is a huge difference between cells in the mammalian visual cortex and the central pattern generators that we described above. To take an extreme case, no normal, self-respecting neuroscientist would think of comparing the neural networks that coordinate the lobster's stomach or the lamprey's swimming with the neural network underlying visual perception. The neural substrates are enormously different in terms of their detailed electrochemical properties. But the cooperative, synergetic effects at the level of coordination itself are not. This, of course, is my central thesis: self-organizing principles lie at the level of patterns themselves, supported by a multitude of different kinds of material substrates and mechanisms.

EXTENDING THE BASIC PICTURE...AGAIN

Evidence for phase-locked oscillations in visual neurophsiology is very recent and has led to much speculation. In other domains, such as the olfactory system, similar phenomena have been studied for many years. Over the course of an outstanding research career, Walter Freeman was able to link quite conclusively gamma oscillations (ranging between 20 and 100 cycles per second) in the olfactory cortex to the sniffing and exploring behavior of cats and rabbits. Particular odors are associated with specific spatial patterns of neural activity in the olfactory bulb (see below). For this reason, phase-locked oscillations in visual cortex have been dubbed the visual sniff. As in the case of smell, the unity of visual perception appears to lie in a coordinated pattern among cell populations in the brain, not in single cells or cell groups.

As mentioned more than once, phase locking has been interpreted as a possible way the brain links features of an object. On the other hand, I have argued—with evidence to boot—that it is the *tendency* toward phase and frequency synchronization that is fundamental to coordination in complex living systems in which information must be communicated over large distances among different kinds of component subsystems, for example, cellular populations that exhibit a broad range of different frequencies. I stress the word *tendency* in contrast to pure mode locking, which if not wrong, is overly restrictive (see chapter 4).

Have we any evidence that the brain is nearly, not absolutely phase locked, exhibiting coherence on several time scales? Figure 8.10 shows two examples.

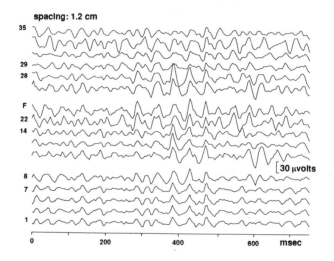

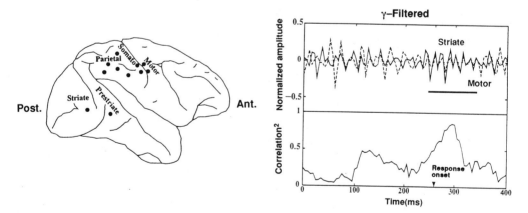

Figure 8.10 (*Top*) Spatial coherence in simultaneous recording of fast EEG activity of monkey visual cortex. (Reprinted with permission from reference 58.) (*Bottom left*) Location of recording sites in one of the monkeys used in Bressler et al's experiment. (*Bottom right*) Single waveforms from visual (striate) and motor areas show varying degrees of coherence despite wide spatial separation. Correlation achieves a maximum shortly after response onset. (Reprinted with permission from reference 59.)

One is from a study of visual discrimination in a monkey.[58] Shown are periodic waveforms recorded from subdural electrodes in the monkey's visual cortex over an interval of one and a half seconds. Widespread spatial coherence is evident characterized by transitory episodes of near synchronous activity at different frequencies. The other example is from recent work by Steven Bressler.[59] Shown are *single* waveforms of local field potentials obtained from visual (striate) and motor cortices in the monkey, again during a visual task. On recognizing, say, a diamond-shaped pattern, the monkey was required to lift his hand from a key to receive a juice reward. It is easy to see that the two signals wax and wane with each other to varying degrees, with the maximum correlation occurring shortly after response onset. The functional significance of these patterns of brain activity awaits clarification. The tasks employed did not involve any control parameter(s). The results do show, however, that broad-band coherence is widespread among cortical sites. Forms of coordination are not simply dependent on sensory input but on particular kinds of stimuli that the animal has learned to discriminate, and on an appropriate level of motivation.

The view of perception and cognition presented in chapter 7 is very different from one in which perception is constructed by binding together primitive features of stimuli. Rather, incoming information is categorized by entering a basin of attraction within which a range of patterns of brain activity is possible. Perceiving, the reader will recall, is not viewed as a phase-locked attractor in the brain but rather as an intermittent, metastable spatiotemporal pattern. In my view, a nervous system trapped even temporarily in resonant mode-locked states is too stable, and hence too rigid and inflexible to qualify as a model of brain and behavioral function. Such a picture fails to incorporate two key properties: one, it contains no mechanism for entering and exiting coherent spatiotemporal patterns of neural activity; and two, it contains no mechanism for flexibly engaging and disengaging participating subsystems. I will address these issues in more detail in the next chapter, the main focus of which is the human brain.

POSTSCRIPT ON ETYMOLOGY

Self-organization has become a kind of buzz word in systems neuroscience these days. Any time an investigator finds correlational activity between brain structures, the words "dynamic self-organization" are sure to trip glibly off the tongue. I find this dilution of a well-defined physical concept sad, to say the least. It reflects for the most part the sorry state of theoretical neuroscience (which is not the same as computational neuroscience), and perhaps also the failure of physics (synergetics excepting) to communicate essential concepts outside its own discipline.

A correlational pattern might be suggestive of a self-organizing process but it by no means proves it. Merely because some change in an animal's state (seldom, if ever well-defined) is associated with corresponding changes in the activated cortical sites, is not enough to invoke the principle of self-organization. For one thing, specific inputs (commands, instructions) from other cortical or subcortical sites could invoke correlations of the kind observed. Second, and more important, to pin down the chief mechanism of self-organization it is necessary to demonstrate either that macroscopic patterned states are spontaneously created or that they change drastically at critical points. Self-organization is not merely the mapping of some (usually) underspecified behavioral state, such as a monkey waiting for a stimulus, with a transient cortical pattern.

As I have underscored time and again in this book, the basic claim of self-organization rests on demonstrating pattern formation and change under nonspecific parametric influences. Most of cognitive and systems neuroscience has yet to incorporate this paradigm shift. Much work remains to be done to identify relevant control parameters and collective global variables that characterize brain activity. Once found, however, we will be on the way to understanding how the brain is modified by development, learning, and disease. The reason is that we will know what the relevant information is. Without that, buried in the mass of undigested detail, it is going to be hard to tell the forest from the trees.

9 Self-Organization of the Human Brain

To the theoretical question, can you design a machine to do whatever a brain can do? the answer is this: If you will specify in a finite and unambiguous way what you think a brain does ... then we can design a machine to do it.... But can you say what you think brains do?
—W. S. McCulloch

PROLEGOMENON

What kind of thing *is* the brain? What does it do? Stephen Hawking of *Brief History of Time* fame has said that because we know the basic laws that govern physics and chemistry, we should in principle be able to understand how the brain works.[1] Although it would be foolhardy to claim that the fundamental laws of physics and chemistry do not hold in biology, my position throughout this book is that the conceptual frame of known laws is too narrow. New concepts that transcend purely microscopic descriptions, especially of complex systems like brain and behavior, must be found. This may not be Hawking's business, but it is ours.

The central thesis in this chapter is simple and easy to express: the brain is fundamentally a pattern forming self-organized system governed by potentially discoverable, nonlinear dynamical laws. More specifically, behaviors such as perceiving, intending, acting, learning, and remembering arise as metastable spatiotemporal patterns of brain activity that are themselves produced by cooperative interactions among neural clusters. Self-organization is the key principle.

This may not seem like a major claim, but I submit it is, and I submit it is far from proven. How do we evaluate this thesis? What are the rules or laws by which self-organization is established in a complex system such as the human brain? How do we find these laws, and what forms do the equations take? Are they chaotic, in that small changes in initial conditions lead to very different outcomes? How are the putative laws of brain behavior related to what people do? These are the questions I will try to answer in this chapter. A key idea is to look beyond the single cell level, to try to find relevant variables that characterize the large-scale cooperative activity of the human brain. After all, we knew the important parameters of gases—pressure, volume, temperature, and their lawful relation—long before we knew the properties of individual gas molecules or how to derive the gas laws from statistical mechanics. Just as theoretical modeling without knowing particulars is an empty exercise—theorizing, I call it—so is collecting facts in the absence of understanding the basic operating modes of the human brain.

OBSTACLES TO UNDERSTANDING

How is it possible to capture the immense patterned complexity in space and time of Sherrington's "enchanted loom, where millions of flashing shuttles weave a dissolving pattern, always a meaningful pattern though never an abiding one?"[2] One of Sherrington's proteges, Sir John Eccles (also a Nobel laureate), is pessimistic that it is possible at all, either now or in the future. Echoing the master's voice, Eccles in his book *The Understanding of the Brain* (note the definitive article) argues:

The immense complexity of the patterns written in space and time ... and the emergent properties of the brain are beyond any levels of investigations by physics or physiology at the present time—and perhaps for a long time to come.[3]

Are emergent properties of the human brain beyond physics and physiology? Although I agree there are obstacles to realizing Sherrington's beautiful image of the brain as a transient, pattern-forming system, I do not believe that the situation is as hopeless as it appears. Progress, in my opinion, will rest on taking at least three mutually dependent steps. First, neuroscience must incorporate theoretical concepts of pattern formation and cooperative phenomena with appropriate methods for analyzing the spatiotemporal dynamics of the brain. This will involve a *mindshift* away from the single neuron doctrine that the *only* relevant level for understanding how nervous systems work is the single cell, toward the level of integrative cortical activity crucial for understanding higher brain functioning. Look at the picture of Einstein, for example (figure 9.1). Can you see the three girls bathing? Such perceptual ability is surely a result of the global dynamics of the brain, not a single neuron.

Ironically, over twenty years ago a meeting held under the auspices of the Neuroscience Research Program was entitled "Dynamic Patterns of Brain Cell Assemblies." The flavor of the research agenda that emanated from that workshop is captured in these prophetic words:

The *possibility* [emphasis mine] of waves, oscillations, macrostates emerging out of cooperative processes, sudden transitions, prepatterning etc. seem made to order to assist in the understanding of integrative processes of the nervous system—particularly in advancing questions of higher brain functions that remain unexplained in terms of contemporary neurophysiology.[4]

With a few notable exceptions such as Berkeley neurophysiologist, Walter Freeman,[5] that agenda died when its champion and inspiration biophysicist

Figure 9.1 This picture was drawn by Sandro Del-Prete. (Reproduced by permission from Del-Prete, S. (1984). *Illusorismen, Illusorismes, Illusorisms* 3rd ed. Bern: Benteli Verlag.)

Aharon Katchalsky was slain by terrorists in Tel Aviv Airport on May 30, 1972.

The second step toward realizing Sherrington's vision depends, in my opinion, on *technology*. New tools are necessary to observe and measure dynamic patterns of brain activity (preferably noninvasively) with sufficient spatial and temporal resolution. How sufficient is sufficient? This is a tricky issue, the pros and cons of which I'll describe later. Obviously, in addressing the putative role of dynamic self-organization, anatomical specificity and connectivity must be respected (see below). The third and perhaps most important step is to effect a *paradigm shift* that directs experimentation toward identifying control parameters and collective variables for dynamic patterns of cortical activity. I believe that only particular kinds of experimental probes are going to offer insight into self-organizing processes in the brain and how these relate to behavior. One is reminded of Otto Rössler's challenge to neuroscientists that in a complex system such as the brain (with more variables than the age of the universe in seconds), it is almost a miracle to find low-dimensional dynamics.[6] But that is, in fact, the miracle that we seek. In my view, it is the cooperative action of neurons functioning together to create dynamic patterns in the brain that permits this miracle to happen.

Later in this chapter I will show that when all three of these steps are implemented, specific testable models of brain and behavior are possible. More important, new insights into how the brain is coordinated during perception and sensorimotor function will emerge. A notable feature of the work I will describe is that pattern formation and change in the human brain take the form of a dynamic instability. This suggests a new mechanism—phase transitions—for the collective action of neurons in the human cerebral cortex.[7] But before getting into that, let's examine some suggestive evidence that supports the idea of a self-organizing brain.

THE BRAIN IS NOT A STATIC MACHINE

One of the main underlying themes in this book is that we need to break away from the classic dichotomy of structure versus function. The difference, I believe, may be one of appearance only. The framework I have presented here puts structure and function on an equal footing, namely, both are dynamical processes distinguished only by the time scales on which they live. Evidence from ion dynamics to perception indicates that this is at least possible, although this book is not the place for a detailed mathematical understanding of the structure-function relation.

In the following, I am going to focus on structural changes in the neocortex due, for example, to lesions in the brain or amputation of limbs. As we'll see, such changes have a dynamic basis corresponding to the way neurons are wired together and their correlated firing activities. The reason this work is important to society is that it allows us to understand better the processes underlying recovery of function after damage to the brain or the periphery. Of course, the reason I mention it here is that it supports my thesis that self-organizing principles are at work in the brain itself. Keep in mind, however, that self-organization, as used in this book is a well-defined (physical) concept that refers to the spontaneous formation of *pattern* or *pattern change* that arises due to nonlinear interactions among the components of a system. There is no mystery in this statement, difficult though it may be to establish theoretically and experimentally.

I object to the use of the word "self-organization" as a substitute for processes we do not yet understand, such as "how the brain coordinates activity to produce consciousness,"[8] or how "the brain produces thoughts without any input."[9] In fact, the present attitude to such issues would be to explain how consciousness and thought arise as a posteriori consequences of the dynamics of brain activity patterns. This viewpoint eliminates the thorny issue surrounding what philosophers call *self-actional* explanation: if the brain coordinates consciousness, who coordinates the brain?

Functional Instability in the Cortex

Were the title of this section in my own words, it would hardly be a surprise by now. But it isn't. I "stole" it from a paper published in 1917 by Leyton and Sherrington. The object of their research was to localize or map the functions of the brain by applying weak current to the motor cortex and determining

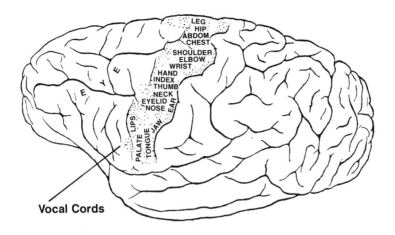

Vocal Cords

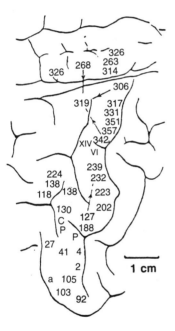

Figure 9.2 (*Top*) Perspective view of a gorilla's brain traced from a photograph. Body parts (leg, hip, etc.) refer to responses to stimulation. (*Bottom*) Mesial surface of chimpanzee brain. Numbers refer to responses obtained by electrical stimulation. (Adapted with permission from reference 10.)

which body part moved and what type of movement it made.[10] Figure 9.2 shows some of the anatomical parts that were activated in a gorilla (top) and a chimpanzee (bottom). Although the brain looks as if it behaves like a cortical keyboard à la "Simon Says" (stimulate here, the lips move; stimulate there, the hips move) nothing could be farther from the truth.

Leyton and Sherrington observed enormous variability. The forms this variability takes are worth a close look. For example, a phenomenon called *reversal of response* occurs: stimulating the same point in the cortex with the same stimulus produced diametrically opposite movements, flexion becoming extension and vice versa. Even more intriguing was a phenomenon they called *deviation of response*. This refers to a change in the entire character of the response so that instead of the originally stimulated movement appearing, some other movement or even some other body part was activated. Most fascinating of all were experiments in which stimulation proceeded systematically from point to point in one direction or another. Leyton and Sherrington describe one of these series as follows:

We turned to delimitation of the face area ... and thence proceeded point by point upward along the precentral gyrus not far in front of sulc. centralis. In due course the point yielding closure of opposite eye was again reached, and it was found that on proceeding farther upward to the point that had previously yielded elbow flexion as its primary movement, that point now yielded abduction of thumb as its primary movement and a little farther upward movement of index, chiefly extension was added to that of the thumb; and movements of thumb and index continued to be the primary movements right up through the region which previously had given elbow flexion as primary response, and thumb and index movements as primary responses trespassed actually into the area that had previously yielded shoulder movements as the primary response. Here, the "deviation of response" was seen to affect a whole series of points, influencing in its special direction a not inconsiderable fraction of the whole arm area.[11]

The poetry of the enchanted loom is substituted by the dry description of the laboratory scientist. But not quite. All these changes are viewed as "temporary expression of the functional instability of a cortical motor point." The motor cortex is clearly "a labile organ," instability serving as a basis on which is founded "the educability of the cortex." The conclusions are unavoidable, even obvious. Leyton and Sherrington demonstrated beautiful *context sensitivity* in the cortex. It's not a static keyboard playing out a "motor program" at all. Localization depends on a history of stimulation. That history takes the form of hysteresis and multistability, predominant features of self-organized, nonlinear, dynamical systems.

The Brain's Dynamic Way of Keeping in Touch

Nearly seventy years after Leyton and Sherrington, the words in the above heading appeared in the journal *Science*,[12] heralding results that are consistent with their work, or at least my interpretation of it. But whereas the former research stimulated the cortex and watched for the appearance of movements, this more recent work touches the animal's fingers and records from microelectrodes in the brain. The idea is to map the loci of peripheral sensations in the somatosensory cortex and study how these "brain maps" shift, say, after a finger is amputated or the nerve supply to the finger is cut off. The remarkable result is the fluidity and dynamism of the shifts that occur (figure 9.3). For example, after the middle finger is amputated, the regions corresponding

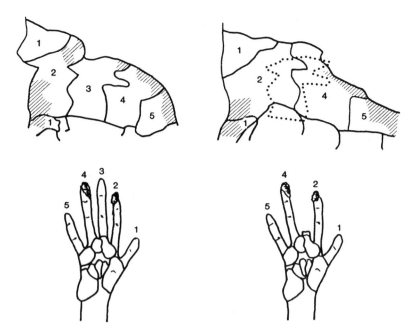

Figure 9.3 Brain maps of a monkey hand. The area representing the monkey's middle finger before it is amputated (left) is filled in by areas representing adjacent fingers a few weeks after the operation (right). (Adapted with permission from reference 12.)

to adjacent fingers expand into the region previously reserved only for the middle finger. In adult owl monkeys this expansion is gradual but steady over the course of a few weeks, and extends over a range of a couple of millimeters of cortical tissue. That in itself means involvement of many thousands of nerve cells.

What happens when a portion of the brain is injured; do nearby regions move into the damaged region? Or what happens to these maps if you practice a finger movement over and over again? In the first case, the cortical map moves into the region that is damaged. In the second, the brain region subserving the practiced movement grows bigger. As one of the main experimenters in this field, Michael Merzenich remarks:

The dominant view of the nervous system is of a machine with static properties. Our work shows this general view to be incorrect. The machine has embedded processes that make it self-organizing and that are driven by experience throughout life.[13]

Topographic maps of the brain are not hard-wired, anatomically frozen entities established just before or soon after birth. They are far more plastic and variable than previously thought. What's the mechanism? The evidence suggests (you guessed it) that *temporally correlated activity* among many neurons is the crucial force behind map (re)organization. Map movement is believed not to involve migrational movement or growth of neurons per se, but rather a spatial shift in their collective activity. The idea is one with which we are now quite familiar. Neurons that are active time after time tend to coalesce into cooperating groups, the collective behavior of which determines how they respond. Shades of cells that fire together wire together....

Gerald Edelman and colleagues have modeled some of these results.[14] In their model, which stresses the significance of neuronal variation at many different levels, a functional map represents the combined effect of what they call group confinement, selection, and competition processes. Movement of map boundaries is accounted for in terms of local fluctuations that occur as a result of trading of cells between cooperative groups. Large map movements, however, are viewed as the dissolution of a group due to major changes in peripheral inputs, such as nerve transection. Neuronal group selection provides a set of rules by which brain maps form and reorganize according to intrinsic or externally defined conditions. These rules involve, but do not invoke, synergetic concepts of self-organization such as fluctuations, instability, competition, and so forth. Again, appropriate experiments are still necessary to establish these concepts unequivocally (see box).

Such work provides a much more optimistic view of recovery of function after say, a stroke. In newborn creatures, of course, the possibility of massive functional reorganization in cortical maps is enormous, at least during the so called critical period. But just how restricted might it be in adults? How much "rewiring" goes on in the adult brain after injury to the nervous system? Earlier, I suggested that learning is a potentially lifelong (power law) process. It *is* possible to teach an old dog new tricks. But just how long-lived is the plasticity of the brain? Ironically, the answer lies with the animal rights movement and Edward Taub's confiscated Silver Spring monkeys. Briefly, Taub is a physiological psychologist who years ago showed that monkeys could eventually use a limb even when it was completely deafferented[16]; that is, after the nerves in the dorsal roots of the spinal cord carrying sensation to the brain had been completely severed. The end result is that the monkey can't *feel* a thing in its arm, but, after training can still *move* the limb purposefully. Taub wanted to understand how such improvements in goal-directed motor behavior were possible without normal sensory input. Perhaps, he hypothesized, this was due to the establishment of collateral nerve connections in the spinal cord. Unfortunately, he didn't get the opportunity to find out. His monkeys, housed at the Institute for Behavioral Research in Silver Spring, Maryland, were confiscated in 1981 after accusations that he was inflicting unnecessary pain and providing inadequate animal care.

Taub thought that reorganization might occur in the spinal cord. This was before Merzenich and colleagues revealed that such changes may occur in the brain as well.[17] This research showed only a small amount of cortical encroachment, about a couple of millimeters with short-term deafferentation. What does the somatosensory representation in the brain look like when the input from the entire limb has been cut off for ten years? Taub's incarcerated monkeys provided an unexpected opportunity to find out.

The first nonlinear effect I ever found was twenty years ago in experiments that functionally deafferented human limbs using blood pressure cuffs applied to the upper arm and wrist.[15] After a period of time the person loses sensation and movement in the fingers. Bill Wanamaker and I showed that the nerve fibers conducting impulses to and from the fingers are eventually blocked. Nerve conduction velocity gradually slows and then abruptly drops almost to zero. What happens to the brain maps as sensation is systematically lost in the fingers over a short interval of around twenty-five minutes? The answer in a monkey is that if sensation in the middle finger is temporarily blocked, adjacent finger representations encroach. When the block is removed and the finger is allowed to recover, the normal maps are reestablished. The plasticity of these maps and their dynamic capabilities are staggering. This would be a great experiment on humans in whom it is now possible to noninvasively localize the source of somatosensory evoked fields in the brain for each individual finger (see below).

Timothy Pons and colleagues received permission to study one of the last remaining Silver Spring monkeys, even as the custody battles raged on.[18] When they monitored the area of the brain typically dedicated to the deafferented arm and hand they found no response to sensory stimulation of the affected limb. When they stimulated the *face* by light touch, however, they reported "vigorous neuronal responses" distributed throughout the deafferented arm area! The normal face map had expanded some 10 to 14 mm into the deafferented zone, an order of magnitude greater than ever observed before.

No evidence to date exists for sprouting of new axonal projections across the deafferented zone of the neocortex in adult mammals after damage to the peripheral nervous system. Pons and company speculate that the reorganization is a reflection of changes that had taken place in subcortical areas such as the brain stem or thalamus, and then passed to the cortex. Neural maps in these areas are confined to a much smaller space, but because of extensive divergence in connecting pathways, they may be reflected in much larger changes in the cortex, a kind of neural spotlight effect. We don't know if this is true or not, and we don't know anything about the timing of these putative changes. If we did, the potential is certainly there for exploiting this kind of reorganizational capacity in rehabilitative therapy.

My point is that these results hint strongly of context-dependent self-organization in the brain. Notice that the form that cortical mappings and compensatory remappings take is far from haphazard. For instance, in the Pons et al study only a *nearby* region, the face, expands into the deafferented zone. On the other hand, the trunk area is also nearby, but does not expand. Why is that? Harking way back to our discussion of functional synergies in chapter 2, it is tempting (and admittedly a bit speculative) to propose that facial areas such as the chin and jaw are functionally connected with the hand and arm during many natural activities such as feeding and grooming. In my language, hand, arm, and mouth are functionally linked as a coordinative structure: a group of neurons, muscles, and joints whose activities covary as a result of shared afferent or efferent signals, or both. When normal input from the hand and arm is removed, one part of the linkage must compensate for other coupled parts. It is probably not the case that the face area grows into the deafferented zone. What is far more likely is that the face is *already* represented in nearby hand-arm areas, but under normal conditions responses to facial stimulation are suppressed or simply not seen. Only when hand-arm inputs are removed are they unmasked. Given the significance of hand-mouth coordination for survival and the extensive history of usage, it is perhaps not by chance alone that the oral area of the cortical map expands, and not the trunk area.

Neuroscience traditionally has assumed that cortical representations do not change throughout life, Leyton and Sherrington's "educability of the cortex" notwithstanding. Instead, they are fixed genetically early on. Remember— how could one forget—the thalidomide children with no arms, but whose feet were astonishingly prehensile? Evidence for massive cortical reorganization puts such phenomena in a new light. Rather than being some weird effect attributable to the plasticity of the infant brain, dynamic self-organization is probably the *normal* way the nervous system functions. The theoretical notions that I introduced in chapters 4, 5, and 6 of *intrinsic dynamics* (now referring to spatiotemporal patterns of brain activity) subject to *behavioral information* from both intrinsic motivational and extrinsic environmental sources do not seem far off the mark. These processes may cooperate or compete with each other in neocortical space, the stability of maps being a function of a dynamic balance between them.

Phantom Limbs

I mentioned that the timing of cortical changes—the dynamics of the dynamics—is not well understood. The way the experiments are done, my old saw, limits their interpretation. But some remarkable evidence shows that functional reorganization is far more rapid than previously thought. The evidence is indirect, but nonetheless compelling. Moreover, it goes some way beyond the notion of somatopic neural maps per se. The shaping and reshaping of maps turns out to be interpretable only in terms of a functional, task-relevant context.

The first intriguing tidbit comes from a study of phantom limbs by V. S. Ramachandran, an innovative perceptual neuropsychologist. In the Pons et al study, cells in the brain that originally received information from the arm came to respond to inputs from the face. Ramachandran and colleagues wondered whether a stimulus applied to the face is mislocalized to the arm in a person whose arm was amputated.[19] It is well known that people who have just had a limb or a part of the limb removed experience phantom limb sensations. Historically, such effects have been attributed to irritation of nerves close to the stump. Here we'll see that they arise in the brain.

Ramachandran et al made a number of fascinating observations on a 17-year-old patient whose hand had been amputated four weeks earlier. The one I like best was serendipitous. When a drop of warm water was placed on the young man's cheek he reported that his phantom hand felt warm. When water accidentally trickled down his face he described a vivid sensation of trickling warm water running down his phantom hand. These results suggest that highly organized, modality-specific rewiring of the brain can occur in as little as four weeks. The parallels with the Pons et al findings are unmistakable.

The Pinocchio Effect

An acquaintance of mine, James R. Lackner reported a strange and remarkable study in the British journal *Brain* in 1988.[20] Starting from the premise of topographically organized neural maps in the thalamus and cortex, Lackner wondered just how modifiable these maps might be by alterations of sensory input. For this purpose he used a vibrator to generate proprioceptive *misinformation* about limb position. Put in English, when you place a vibrator on a person's biceps and physically restrain the forearm from spontaneously flexing, after a few seconds the person feels that the arm is more *extended* than usual; vice versa for triceps vibration.

What Lackner did is vibrate the subject's biceps muscle while the subject was holding on to his own nose. What do you think the subject experienced? In a direct quote, provided by Lackner:

Oh my gosh, my nose is a foot long! I feel like Pinocchio.

Other subjects experience their fingers elongating but not their noses; still others feel both. What about the reciprocal experiment of triceps vibration? (Vibration, by the way is delivered at 120 pulses per second for 3 minutes). Many subjects feel that their nose has been pushed inside their head or that their fingers have passed through their nose and into their head!

One of my favorites is when Lackner vibrates the biceps of both arms when subjects are seated with their hands on their waist, arms akimbo. Many subjects feel their waist expand to keep pace with the apparent displacements of their hands. The reader can guess what happens when the triceps are vibrated bilaterally. Subjects report their arms and hands moving inward, compressing the waist into wasplike, svelte contours. Someone should patent this "instant dieting illusion."

Such bizarre sensations, stretching of the nose and shrinking of the waist, are not uncomfortable or painful. Instead, subjects display a sense of wonder as the dimensions of their bodies are perceived to change. All the reported effects represent a meaningful interpretation of the entire stimulation context. The perceptual representation is thus *relational* and highly *labile*. Only "good" solutions arise: the nose has to be perceived as a change in length, or the hand and fingers as elongated, because these are the only interpretations that are consistent with the hand and nose maintaining physical contact while the hand is moving (see box).

WHY DID THE BRAIN EVOLVE?

It is important to keep in mind when reading this material that the brain did not evolve merely to register representations of the world; rather, it evolved for adaptive action and behavior. Musculoskeletal structures coevolved with appropriate brain structures so that the entire unit functions together in an adaptive fashion. That was the main message of chapter 2, where it was demonstrated though the self-organizing mechanism of dynamic instability that the very many neurons, muscles, and joints act collectively as a single functional unit. Drawing on evolutionary arguments and morphogenetic evidence, Edelman arrived at a similar conclusion.[21] For him, like me, it is the entire system of muscles, joints, and proprioceptive and kinesthetic functions plus appropriate parts of the brain that evolves and functions together in a unitary way. Although Edelman's theory of neuronal group selection has been criticized recently for still failing to do justice to the brain as an active, dynamic, self-organizing system,[22] his attempt to remove the "homunculus" and his rejection of conventional information-processing models of brain function (who or what decides what is information?) are to be commended.

Summary

Anatomical structures, say, groups of cells in the cerebral cortex, are not frozen but remain fluid throughout adulthood. Neural representations or cortical maps are labile, alterable by life's experiences. They are dynamic and apparently self-organized on multiple space and time scales. Competitive and cooperative processes between intrinsic connectivity and external inputs, together with local fluctuations in cellular interactions, jointly sculpt these maps over time. Perceived body configuration involves more than just somatotopic neural maps. The entire task and stimulation context must be taken into account. The body schema, the specification of body parts in relation to each other and the layout of the environment, is subject to modification. Recall our work on learning described in chapter 6. There, not just one dynamic pattern was modified through learning, but the *entire coordination dynamics* or *attractor layout* was qualitatively and/or quantitatively changed. Once again, the nature of change is dictated by competition between the existing dynamical structure and extrinsic, task-related inputs.

Even the adult brain has a capacity for rapid functional reorganization. It is astonishing that established maps (e.g., for the length of one's finger or nose) appear to be subject to nearly immediate alterations by changed sensory conditions. This seems to override previous kinesthetic knowledge existing, in my case, for decades. Although receptive fields of particular neurons may remain fixed for a short time, any slight change in posture can upset the balance and shift or alter the cortical map. Sherrington and Leyton had it right when they described the cerebral cortex as functionally unstable. Ever since, it seems, we've been trying to turn it into a static, hard-wired machine.

THE BRAIN DYNAMICS APPROACH: FRACTAL DIMENSION

Complex systems such as the human brain have enormous numbers of interrelated dependent variables. It's not possible, even if it were useful, to measure all of them directly. I've stressed ad nauseum the need to identify the relevant ones at the level of description of interest. One popular approach in the brain business these days is to calculate the *dimension* of a time series from the nervous system (e.g., EEG) and study how it changes under different conditions (e.g., as cognitive demands vary; before, during, and after epileptic seizures; in various stages of sleep; during erotic thoughts, etc.). Recalling my discussion of dimension in chapter 2 in the context of simple hand movements, the idea is that this quantity tells you the number of relevant degrees of freedom in the system. Dimension is a quantitative measure of the geometric complexity of a set. Suppose an object traces a loop, such as a planetary orbit in three-dimensional space; the dimension of the set is 1, that of a line. In contrast, a finite fractional dimension suggests chaotic behavior. Calculation of a finite dimension may suggest the presence of deterministic chaos or simply reflect mixing of well-known frequency bands in the signals being monitored. Present analysis does not allow a distinction between noisy quasiperiodicity and deterministic chaos.

Just to get a flavor of this approach that emanates from the mathematical theory of dynamical systems, consider a set of voltage measurements obtained from a single electrode placed on the brain.[23] Suppose that M values of this voltage measure, V, are sampled, $V_1, V_2 \ldots V_M$. The idea is to reconstruct a trajectory in some N-dimensional embedding space as follows:

$X_1 = (V_1, V_2, \ldots V_N)$

$X_2 = (V_1, V_2, \ldots V_{N+1})$

$X_k = (V_K, V_{K+1}, \ldots V_M).$

Notice all that is done here is shift the values of the variable by a single sample. Or one could shift it by a few samples. This is called the method of *time delay*. K is the number of N-dimensional points that can be formed from M measured values. Thus, using the original time series, one constructs $K = M - N + 1$ *embedding vectors* (apologies for the jargon) and studies how these vectors evolve. The amazing mathematical result is that under certain conditions (e.g., unlimited amounts of noise-free data, stationarity, infinite measurement resolution) the dimension of the measured set, $\{X_K\}$, is the same as the dynamical system that generated the original time series.

Just consider what's at stake here. From any single time series generated by a complex system, in which it is impossible to measure everything, meaningful inferences about the dynamics of the entire system can be made. The reconstructed attractor, in other words, characterizes the *generator* of the time series—the brain or a part of it—not just the time series itself. Science for free.

There are drawbacks, of course, and I happen to think they are major. In part, they have to do with the experimental facts of life. Among these are that signals are *not* stationary (try not to think of elephants for 15 seconds; cf. chapter 7), measurement resolution does *not* go much beyond 12 bits (e.g., voltage measures are digitally scaled in the interval zero to 2^{12}, numbers ranging between 0 and 4096), and sampling frequencies tend to be determined by instrumentation capabilities, not by any intrinsic time scales established by the brain itself. As Al Albano, a physicist who knows what he's talking about, says, calculation of dimension from brain signals is a series of compromises.[24] This hasn't stopped people calculating them. You or your grandmother can easily get hold of the algorithm, called the *correlation integral*, apply it to your data, and come out with a number (and, dare I say, publish it?). Only in the hands of experts do these numbers mean anything and then only in a relative sense.[25] After providing an excellent review of the practical limitations and reliability of dimensions and other mathematical complexity measures, Albano and Rapp's evaluation of the situation is that "... the question to ask is ... whether these dimensions and related measures can help us better assess models of the brain's behavior, or better assess its state of health."[26]

Calculations of correlation dimension might be useful complexity measures in the diagnosis of diseases such as epilepsy and Parkinson's, or in evaluating states of consciousness, but they will never tell you what kind of thing the brain is. Nor will they tell you whether the measures that you observe are really relevant or not. Although complexity measures might indicate the presence of a deterministic dynamical law, sad to say they will never tell you what the law really is. Mathematically sophisticated methods only complement theory and experiment, they can't substitute for them.

SPATIOTEMPORAL PATTERNS OF THE BRAIN

The brain is a spatially extended, highly interconnected structure. Measures of correlation dimension from data obtained at one or a few locations (mathematical theorems notwithstanding) are thus limited. Ways have to be found to treat signals emanating from different brain sites and their complex time evolution. To obtain such a detailed understanding of the brain, especially with respect to its functional self-organization, we need, I have argued, three things: an appropriate set of theoretical concepts to motivate how to approach the business of the brain; a technology that affords analysis of the pattern-forming dynamics of the brain in space and time; and some clean experiments that prune away complications but retain the essence. The rest of this chapter tells the story of a research program in which theory, experiment, and advanced technology come together as a single synergistic unit.

The Technological Zoo: SQUID, PET, and so on

I have yet to say very much about how to obtain signals from the human brain without having to remove the top of the skull. I like my subjects too much to do that, and sometimes I need them to come back.[27] In the work that I'll describe, we employed a large array of SQUIDs in an attempt to discover self-organization in the human brain. When I talk about SQUIDs I don't mean the kind you eat. I refer, rather, to *superconducting quantum interference device*, the most sensitive detector of magnetic fields ever designed. At extremely low temperatures, near absolute zero, certain conductors undergo a transition to a *superconducting* state in which resistance to current flow vanishes. A crucial property of this state is that thermal noise is absent. The SQUIDs *have* to be supersensitive: the magnetic fields generated by neurons in the brain are about one billion times smaller than the earth's magnetic field.

In the nervous system, it is the *intracellular* dendritic currents that give rise to the magnetic field picked up by SQUID electronics. The more familiar method of electroencephalography (EEG) detects extracellular or volume currents (figure 9.4). It is estimated that it takes about 10,000 synchronously active neurons to generate a magnetic field that can be detected outside the head. A nice feature of pyramidal cells in the cerebral cortex is that they are arranged in columns contained within a sulcus or gyrus, and are so densely packed that a small volume of cortex is enough to produce a magnetic field. SQUID electronics detect these weak fields and transform them into voltages. Because the skull and scalp are transparent to magnetic fields and because newly developed sensor arrays cover a substantial portion of the head, it is

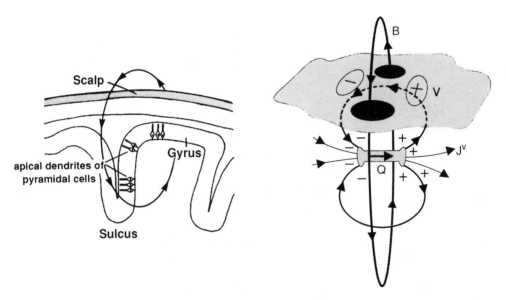

Figure 9.4 (*Left*) Apical dendrites of pyramidal cells in columns contained within a sulcus run perpendicular to the surface of the cortex and generate a magnetic field. (*Right*) A small piece of brain tissue produces a net intracellular current, Q, accompanied by a magnetic field, B, that emerges from the scalp and reenters nearby. It is the intracellular neuronal currents that are picked up by SQUIDs. Extracellular volume currents of density, J, create an electric field that diffuses to reach the scalp. The EEG electrodes detect this field as an electric potential, V.

now possible to record signals from most of the cortical mantle simultaneously.

Of course, SQUID technology is not the only means available to access the distributed functioning of ensembles of neurons in the human brain. Tantalizing glimpses are also provided by PET (positron emission tomography) from which measures of cerebral blood flow or metabolism may be obtained during different tasks or in various diseases and brain disorders. The PET technique uses a radioactive tracer added to the blood stream using a catheter (a wee bit invasive for some) to find out which parts of the brain work hardest when a person is doing something. Neurons require oxygen, and the more active they are, the more the local blood supply has to deliver. Positrons emitted by the tracer (a radioactive form of oxygen or fluorine) annihilate nearby electrons, producing detectable gamma rays.

An even newer technique, invented just a couple of years ago, functional magnetic resonance imaging (fMRI), is undergoing rapid evolution. This too assumes that increases in neural activity are accompanied by increases in blood flow and oxygen uptake. Because deoxygenated blood acts as a paramagnetic agent (a blood vessel containing deoxyhemoglobin placed in a magnetic field alters that field), the technique is entirely noninvasive. The magic of the method lies in the effect of distortions of the magnetic field on nearby water molecules. This has the consequence of amplifying otherwise difficult-to-measure signals 100,000-fold. A great advantage of fMRI is that it also permits much faster scanning times than the old MRI, fractions of a second compared with minutes.

The main disadvantage of both PET and fMRI, for me, at least, is the hemodynamic response time. In English, this refers to the unfortunately inherent time lag of more than two seconds between vascular response and neural activity. A lot of action can go on in the brains of people on time scales faster than a couple of seconds. A world-class sprinter takes less than a fifth of a second to move in response to the starter's pistol, and it takes much less than a second to recognize a familiar object. As I have mentioned a few times already, short bursts of cortical activity that last hundredths of a second might be very important in perception and other brain functions. The only existing techniques with such *millisecond* temporal resolution are EEG and neuromagnetic source imaging using SQUIDs. As we'll see, combining information obtained from multichannel electrodes and large SQUID arrays with other modalities such as MRI provides better spatial and temporal resolution than either one alone.

In what follows, it's important to keep in mind that although spatiotemporal resolution is paramount, during the course of normal or aberrant behavior brain activity sources can be anywhere and everywhere. To study and understand the patterned complexity of the human brain on different time scales, it is the *relative* magnitudes and interrelations among neural signals that are crucial, not localization or activation foci per se. Despite the present focus on nonmedical applications of imaging technologies, some of our analyses reveal advance warning signals that anticipate the onset of changes in brain patterns that may occur in functional brain disorders. Such advanced warning signs cannot be seen with slower techniques such as PET and MRI. As a basic research scientist, I'm quite excited that signatures of upcoming changes in the brain associated with dynamic instabilities—critical slowing, critical fluctuations—might actually be useful, providing an opportunity to halt or even reverse drastic changes before significant tissue damage occurs. Again, all this is predicated on identifying control parameters and the key collective variables that characterize self-organized neuronal networks in the brain, the problem to which I turn at last.

Paradigm Shift

The pretty PET scan pictures that show up in popular magazines such as *Time* (e.g., Fall, 1992) are based on highly dubious *subtractive* methods. You see one part of the brain "light up" for sad thoughts, another for seeing, another for learning, another for speaking. The idea is to have a subject perform tasks that involve only the cognitive or emotional components that you want to manipulate and isolate. For example, reading and speaking are compared with reading alone. Then, by subtracting the resulting images in each task, you isolate the area in the brain that's responsible for one of them, here, speaking. Seems simple and logical. The problem, however, is that it is not the way the brain works. Subtraction is a *linear* operation, fine for computers but not necessarily for brains. Neither the brain nor its individual neurons are linear. The brain is parallel, highly distributed, and nonlinear. When one examines brain images before they are subtracted from each other, one sees activity distributed all over the place. There are no centers for reading and speaking, even though each task may selectively involve *in time* certain areas more than others.

Compared with the numerous experiments that use subtractive methods, studies employing the nonlinear paradigm promoted in this book have hardly begun. Investigators have not manipulated control parameters systematically, in part because they are ignorant as to what the control parameters might be; they have not thought this way, although the language of brain states and behavioral states is rampant; and the data-processing requirements are enormous and require sophisticated computer methods. Instead of subtracting one image from another and showing what remains, special methods have to be employed to capture relevant patterns in the entire spatial array of, say, SQUIDs or electrodes, as it evolves millisecond by millisecond. When this is done, the enchanted loom—or better, the weaving process of an apparently seamless garment—begins to be revealed.

A Brain-Behavior Experiment

The key idea is to exploit experimental paradigms in which *qualitative* changes in behavior are known to occur (cf. chapters 2–7) as a way to determine if self-organization occurs in the brain.[28] This is not obvious to a lot of people. A few years ago, when I proposed such an experimental program to a federal funding agency, a group of peer reviewers was (rightly) skeptical:

The investigator assumes transitions or abrupt nonlinear shifts in motor or perceptual variables/states will be accompanied by detectable changes in evoked magnetic response (EMR). There is simply no evidence in the current literature to indicate that detectable changes in even single site EMRs can occur in relation to shifts in behavior or perceptual state.

I don't have to tell you what happened to this proposal. A lot of dead trees in vain. But here's some advice to prospective grantees. You'll always know the sound of the death knell when a sentence starts as follows:

The key problem with this program is that it is trying to do too many new things at once [my comment: so much for the parallel brain]: first to connect qualitative effects in the brain predicted by nonlinear systems theory to behavior [sic]; second to look physiologically at an intermediate scale of neural integration where no one has looked before; and third, to use experimental techniques where there is little experience with analysis [my comment: only twenty years of my life].

Undaunted by such setbacks, my colleagues and I continue to venture into the unknown, comforted by the knowledge that our research, although obviously misguided, is at least original.

Remember the Juilliard experiment described in chapter 4? Tones are presented to the subject starting at a frequency of once per second, increasing every ten tones in small steps. The subject's task is to syncopate with the stimulus. At a certain critical frequency, however, subjects are no longer able to syncopate and switch spontaneously to a coordination pattern that is now synchronized with the stimulus. This paradigm deals, therefore, not simply with auditory stimulus processing and manual response processes, but with *patterns* or *modes of anticipation and sensorimotor coordination*. Now imagine that during runs of this experiment, activity is recorded over the left side of the brain using an array of thirty-seven SQUIDs. Plate 5a shows a computer reconstruction of the subject's head and the location of the SQUID array. Underneath the latter is a re-construction of the cortex made using MRI (plate 5b). An example of the magnetic field activity detected by the SQUIDs is displayed on the cortical surface in plate 5c, and on a single slice in plate 5d.

The top of figure 9.5 shows the averaged data from sensors before and after the transition from syncopation to synchronization. Before the transition, the stimulus and the response are antiphase. After the transition, the subject's responses are nearly in phase with the stimulus. The brain's magnetic field shows a strong periodicity during this perception-action task, expecially in the pretransition region. After the transition the amplitude drops, even though the stimuli and responses are more rapid, and the signals look noisier. This result, although paradoxical, is extremely interesting. On the one hand, *behavioral* synchronization is more stable and feels "easier" than syncopation. On the other, brain activity is less coherent during synchronization than syncopa-

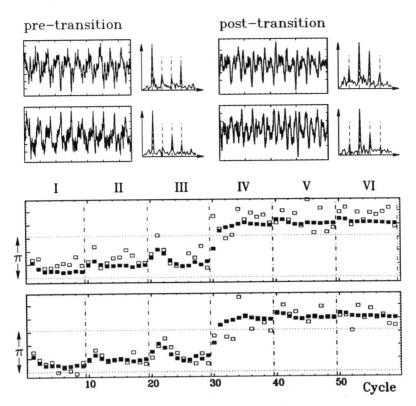

Figure 9.5 (*Top*) Time series from two single sensors before and after the transition, together with corresponding power spectra showing peaks at the stimulus (and movement) frequency (first dashed vertical line). Note that the frequency doubles and even triples after the transition. (*Bottom*) Superimposed relative phase (*y*-axis) calculated at the stimulus frequency for each cycle of the sensorimotor behavior over time (solid squares) and two of the SQUIDs (open squares). See text for details.

tion. Why is this? It's as if the more difficult syncopation task induces a more coherent state in the brain; the less difficult synchronization task puts the brain in a less coherent state, free, as it were, to do other things.

Analysis of Single SQUIDs: The Mesoscopic Level As I mentioned earlier, estimates are that at least 10,000 neurons must synchronize their activities to produce a coherent magnetic field in the brain as detected by a single SQUID. We might therefore consider this level as an intermediate *mesoscopic* scale of brain function, somewhere between the single-neuron microscopic level and the activity of major areas of the brain, the macroscopic level. Power spectra, which reveal the amplitude of the signals at particular frequencies, are shown in figure 9.5. Before the onset of the task, the correlated action among neurons is comparatively weak. A striking experimental result is the near instantaneous onset of spatial and temporal coherence in the brain when a meaningful behavioral event is introduced. Both before and after the transition, the brain signals exhibit dominant spectral peaks at the stimulus (and movement) frequency. After the transition, a number of SQUIDs exhibit spectral power at double or even triple the stimulus-response frequency. Such coherent states are obviously produced by the cooperative participation of very many neurons.

Such findings, by the way, appear to be consistent with other work on cats and rabbits.[29] For example, Reinhard Eckhorn and colleagues present signal correlations of different types from single cells, multiple units, and local slow waves that show shifts from "stochastic uncoupled states" to "stimulus-driven states," including coherent oscillations in the gamma frequency (20–100 Hz) range.[30] Figure 9.5 shows evidence of strong, task-related temporal correlations among neurons contributing to a single SQUID sensor. Moreover, such cooperativity is spatially pervasive *across* SQUID sensors, although spectral power varies from one region of the brain to another.

A most remarkable result is shown in the bottom of figure 9.5, which superimposes the relative phase between stimulus and response (solid squares) with the relative phase between stimulus and brain signals from two representative SQUIDs (open squares). The dotted vertical lines indicate points where the stimulus frequency changes in the brain-behavior experiment. The horizontal lines represent a phase difference of π radians or 180 degrees. As to be expected, the SQUID data are somewhat noisier than the behavioral data. Nevertheless, a transition in both brain and behavior is clearly evident in the relative phase, which typically drifts upward and fluctuates before switching, a sign of approaching instability. Such *critical fluctuations* reflect the increasing susceptibility of the brain toward changing to a new coherent state. *Critical slowing down* is indicated by the fact that both brain and behavior are perturbed more by the *same* magnitude of perturbation (a step change of 0.25 Hz) as the critical point approaches. It takes longer and longer to return to the preperturbation relative phase value as the transition draws near. Pattern formation and switching, in other words, take the form of a dynamic instability. Notably, the coherence of both brain and behavioral signals is captured by the same collective variable, relative phase. There is, as it were, an abstract order parameter isomorphism between brain and behavioral events that cuts across the fact that different things are being coordinated.

Modes of the Brain: The Macroscopic Level   The patterns of brain activity picked up by the SQUID array turn out to be very coherent in space. How then do we capture this macroscopic behavior? One way is to treat the signals from the thirty-seven SQUIDs as a *spatial pattern* that evolves in time. To do this, a decomposition was performed using a method called the Karhunen-Loève (KL) expansion.[31] The mathematical details won't concern us here: the magic of the method is to decompose the signals into a set of spatial patterns or modes and corresponding amplitudes that vary in time. The idea is to see whether an essentially infinite-dimensional system, the human brain with around 100 billion neurons and 60 trillion synapses, exhibits a small (or at least restricted) set of time-varying global modes.

Plate 6 shows the spatial modes on the different plateaus of our Julliard experiment. The numbers underneath each pattern represent that mode's contribution to the entire signal as calculated from the KL decomposition. These numbers add up to one, so that an (eigen)value, say, of $\lambda = 0.75$ means that a given mode contributes 75% to the whole signal. Notice the beautiful symmetry of the patterns and the fact that only a small set of modes is sufficient to characterize the signals obtained from the entire SQUID array. Indeed, the top mode contributes about 60% of the energy contained in the original signal and the top two modes nearly 80%. Notice also that the first two modes change before and after the transition, and that the spatial dependence of the higher modes is almost the same on all plateaus.

In plate 7 (top, middle) I show the spatial form of the functions obtained by the KL expansion performed on each plateau and their amplitudes for the two most dominant modes. For the top mode, a strong periodic component is evident over the entire time series. However, the spectra show that a qualitative change occurs between pretransition and posttransition regions. In the pretransition regime, brain behavior is dominated by the first KL mode oscillating at the frequency of the sensorimotor behavior. After the transition, the biggest peak (most energy) is located at twice and even thrice this frequency. The temporal behavior of the second mode (middle) is rather less periodic and consumes less of the total energy in the signal. Note, however, that after the transition, most of the power is again expressed at double the behavioral frequency.

As previously mentioned, the antiphase syncopation pattern is not stable beyond a certain critical frequency, and a spontaneous switch to an in-phase synchronization pattern is observed. As shown in plate 7 (*bottom*) the first KL mode (*solid squares*) exhibits a clear transition of π at the transition point. Notice that the phase of the brain activity and the sensorimotor behavior are almost identical in the pretransition region, whereas after the transition brain activity becomes more diffuse even as the sensorimotor behavior becomes more regular. Relaxational behavior, typical of critical slowing down in self-organizing systems, is once again evident.

A good way to visualize the spatiotemporal dynamics of the brain is to plot its behavior in *phase space*. Often, this involves taking the derivatives of a time series and plotting, say, position, velocity, and acceleration against each other in 3-D space. If one is lucky, it's possible to see the *geometry* of the system's dynamical behavior. Another way is to plot the time series versus a delayed version of itself. Using this method of delay coordinates, the top mode's trajectory is shown in phase space in figure 9.6. To obtain a better view of the geometry of the evolving trajectory, the same plots are shown in figure 9.7, but for smaller time windows. In each box, about two or three cycles are taken from the second (pretransition) and fifth (posttransition) plateaus (upper and lower rows, respectively). Obviously, this phase space image results in a very interesting structure whose geometry changes qualitatively before and after the transition. In the next section I'll describe our recent attempts to model these findings theoretically.

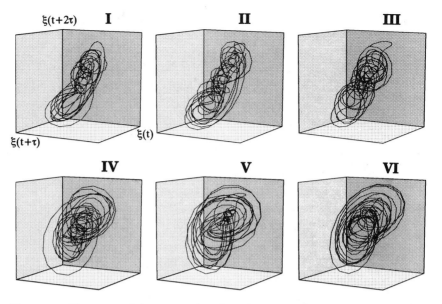

Figure 9.6 Trajectories of the top spatial mode in phase space plotted in time-delay coordinates for the first six frequency plateaus in the Juilliard experiment (upper left to lower right, I–VI).

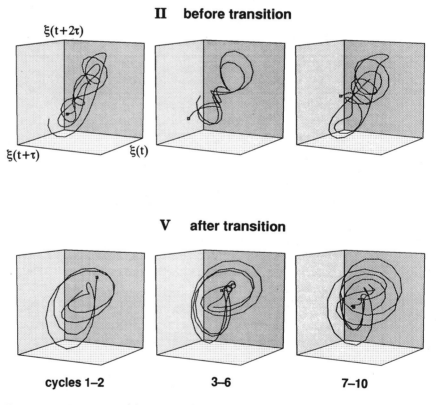

Figure 9.7 Trajectories of the top spatial mode on plateaus II (upper row) and V (lower row) for smaller time windows. Cycle numbers are shown for each plateau.

MODELS OF BRAIN BEHAVIOR: COUPLED MODES AND ŠIL'NIKOV CHAOS

What type of dynamical system can be used to model brain behavior? For the moment, let's just concentrate on some of the main observed facts. The first concerns the behavior of the spatial modes, the top one dominating before the transition and the second one after (see plate 7). The second fact concerns the behavior of the relative phase between stimulus and brain signals, a spontaneous transition by a factor of π occurring from one phase state to the other. The third concerns the intriguing geometry of the trajectory of the top mode before and after the transition (see figures 9.6 and 9.7).

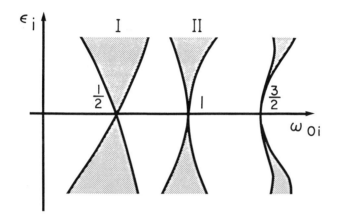

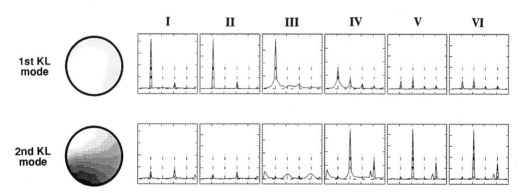

Figure 9.8 (*Top*) The stable and unstable regions of the Jirsa et al model of brain mode coupling. The strength of driving, ε_i, is plotted as a function of the mode frequency, ω_{oi}. Before the transition, the first spatial mode is in the instability region, I, where it operates at the stimulus (and movement) frequency. At a critical value, the second mode enters instability region II, where it operates at twice the stimulus frequency. (*Bottom*) Simulation of the brain dynamics. Fourier spectra of the first two KL mode coefficients on plateaus I to VI obtained by numerical simulation of the theoretical model. Vertical lines correspond to the stimulus (and movement) frequency, and multiples of it.

Victor Jirsa, Rudolf Friedrich, Hermann Haken, and I modeled the first two facts using nonlinearly coupled oscillators representing the first two spatial modes, driven, in this case, by an external force, periodically occurring tones.[32] This process is called *parametric excitation*. Sparing the mathematical details, the main conceptual idea is that the equations are structured such that before the transition the first KL mode operates in the instability region I in figure 9.8 where it oscillates with the stimulus frequency, while the second KL mode (due to its coupling with the top mode) is damped. When the stimulus frequency is increased to a critical value, the second mode enters the instability region II and starts oscillating with twice the stimulus frequency, and the first mode becomes damped. The increase of the amplitude of the second mode after the transition has a further consequence, namely, the phase Φ of the first mode oscillating at the stimulus frequency exhibits a transition of π. In the sense of synergetics, the first two KL modes can be identified with the order parameters of the system, since they contain all (or most) of the information about the observed macroscopic behavior of the brain, namely, the transition from one frequency state to the other and the transition from one phase relation to the other. These two coupled collective modes operate as a competition mechanism, with the first one winning before the transition and the second one afterward.

Notice that the mode coupling model has a number of distinctive features. In contrast to many studies that evaluate single time series, it treats the *entire spatial array and its evolution in time*. The patterns or modes of brain activity are spatially coherent, but their temporal evolution is complex. By modeling switching dynamics and its multistability, this theory connects brain events (internal behavior) to behavioral events (overt behavior). Finally, the waxing and waning of modes is a result of *nonlinear coupling* between the outside world and the internal spatial modes of the brain.

An amazing aspect of this coupling, revealed by detailed mathematical analysis, is that it takes exactly the same form as the nonlinear coupling between the oscillators in the original HKB model of bimanual coordination. My colleagues and I suspect that this nonlinear coupling is of a fundamental biophysical nature, providing the simplest means for a system to express basic functional properties such as multistability and switching. Obviously, the coupling is biophysical in a *coordinative* or *informational* sense, linking component processes of very different kinds (cf. chapters 3 and 4).

There is something terribly appealing about the notion—even though it may turn out to be wrong—that the human brain warbles among its major macroscopic states by the mechanism of mode-mode interaction among spatial oscillators. One can see in plate 6 that at least seven coherent spatial modes are present in the brain, the contributions of which may vary dynamically according to the kinds of tasks people (and brains) have to perform. It's tempting to speculate that each of us is born with a brain that operates globally in a relatively small set of basic modes whose contributions vary with life's trials and tribulations. The clinical implications of this picture, as in psychiatric disorders such as schizophrenia, bipolar disorders, and other brain diseases, are tantalizing. In some cases, the pathological brain may be restricted in its modal content and/or its dynamics. In others, too many modes may be excited and/or their dynamics never settle.

It is important to understand, moreover, how the brain exhibits modal behaviors that are not strictly periodic. A detailed look at the phase space trajectory of the top mode (figures 9.6 and 9.7) illustrates what I mean. The trajectories never repeat or settle down to a fixed point, even though a clearly defined geometry is apparent. Remember that this is just the top mode in a system containing several, so it is clearly possible that its behavior is of chaotic origin. We know from studies of a variety of nonequilibrium systems that chaotic behavior generated by mode interactions is a cooperative effect produced by the entire system. Isolated parts or single subsystems do not usually undergo chaotic behavior on their own. Also, a chief characteristic of chaotic behavior generated by mode interactions in a spatially extended system is the existence of homoclinic (self-connecting) and heteroclinic loops in the phase space of the collective modes. These loops or orbits exist for particular values of a control parameter and are composed of trajectories connecting saddle points.

An example due to the Russian mathematician L. P. Šil'nikov provides one possibility for creating chaotic attractors. In figure 9.9, x and y denote the coordinates in a two-dimensional manifold, and z the coordinate in the one-dimensional manifold of a *saddle focus*. Notice the two different types of homoclinic orbits: on the left, the two-dimensional manifold is stable and the trajectory is kicked out when it approaches the unstable fixed point (note arrows); on the right, the two-dimensional manifold is unstable and the trajectory leaves the fixed point on an unstable limit cycle, only to be reinjected later.

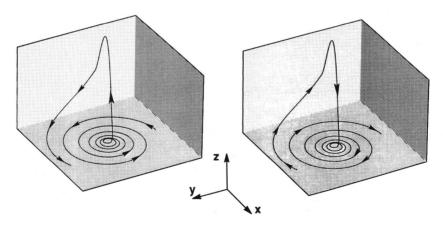

Figure 9.9 Homoclinic orbits involved in Šil'nikov attractors. (*Left*) The two dimensional x-y manifold is stable; z is unstable. (*Right*) x-y manifold is unstable; z is stable. Arrows denote the direction of flow (see text for details).

Self-Organization of the Human Brain

In such a three-dimensional dynamical system, Šil'nikov showed that if a certain condition holds, an infinite number of unstable limit cycles exist, thereby allowing the system to enter a chaotic state.[33] Chaotic systems are, of course, sensitive to initial conditions. This means there are directions in the phase space where trajectories that are very close to each other diverge exponentially. On the other hand, such nonlinear systems can also exhibit long and complicated *transients* in which all the fingerprints of chaos may be viewable, but the systems are not chaotic at all. For example, it is impossible to tell whether the phase space trajectories of the brain's top mode in the SQUID data before and after the transition (see figures 9.6 and 9.7) are chaotic or the result of a long transient toward a limit cycle.

This does not mean we cannot model such effects. Figure 9.10 (top) compares the phase space trajectories of the experimental data before (plateau II) and after (plateau V) the transition with a simulation of a theoretical model developed by Armin Fuchs and me.[34] By varying only a single parameter corresponding to stimulus frequency in the experiment, the main features of brain behavior before and after the transition are reproduced. Not only are the phase space portraits very similar, the *quantitative* match between theory and experiment is good. Figure 9.11 compares time series and spectra of the model and the SQUID data on plateaus II and V. Although the time series of the model looks less noisy than the data, all the main features are still present. Thus, it is quite obvious that this model generates the essential behavior of

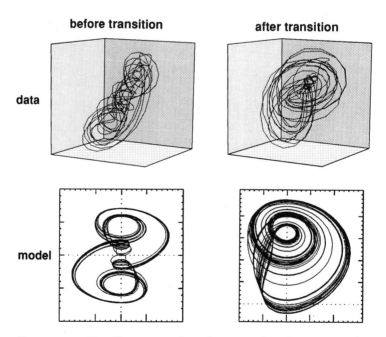

Figure 9.10 (*Top*) Phase space plots of top spatial mode using time-delay coordinates before (plateau II) and after (plateau V) the transition. (*Bottom*) Model simulation using a three-dimensional dynamical system typical of Šil'nikov chaos.

the brain's top mode. More intriguing, it underscores the fascinating connection between the geometry and spatiotemporal dynamics of the brain.

CODA: BRAIN BEHAVIOR

What is this mind of ours? Last week's potatoes! ... The atoms come into my brain, dance and dance and then go out—there are always new atoms but always doing the same dance, remembering what the dance was yesterday.
—R. P. Feynman

The brain possesses tremendous heterogeneity of structure, and its dynamics, in general, are nonstationary. Nevertheless, I've shown in this chapter that under well-defined experimental conditions and using a new tool—a SQUID

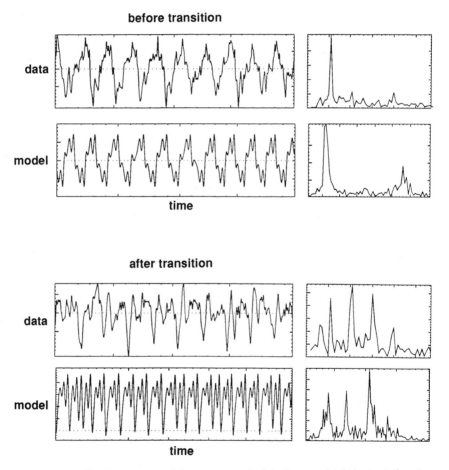

Figure 9.11 (*Top*) Time series and Fourier spectra (right) of top mode's behavior before the transition (plateau II) and corresponding model simulation. (*Bottom*) Time series, Fourier spectra, and model simulation after the transition is over (plateau V).

array—the brain exhibits low-dimensional dynamical behavior. From a relatively incoherent or rest state (with a $1/f$ distribution of component frequencies) the brain manifiests coherent spatiotemporal patterns immediately it is confronted with a meaningful task. This fact attests to both the *adaptability* and *information-compression* capability conferred on the brain by its inherently pattern-forming and nonlinear character.

It can hardly be overemphasized that the patterns formed by the nervous system depend on environmental or task requirements. The brain does not exist in a vacuum, detached from context. If the brain is intrinsically chaotic, possessing, by definition, an infinite number of unstable periodic orbits, it has the capacity to match an equally unpredictable environment. Being chaotic at rest allows the brain access to any of these unstable orbits to satisfy functional requirements. Thus, when a cognitive, emotional or environmental demand is made on the organism, an appropriate orbit or sequence of orbits is selected and then stabilized through a kind of chaotic synchronization mechanism.[35]

A crucial aspect of pattern-forming dynamics in both brain behavior and overt behavior pertains to *critical instabilities*. Like many complex, nonequilibrium systems in nature, at critical values of a control parameter, the brain undergoes spontaneous changes in spatiotemporal patterns, measured, for example, in terms of relative phases, spectral properties of spatial modes, and so forth. Remarkably, these quantities exhibit critical slowing down and fluctuation enhancement, predicted signatures of pattern-forming instabilities in self-organizing (synergetic) systems. Even more remarkable, coherent states and state transitions in both brain and behavior are often captured by the same collective variable, characterizing the spatiotemporal phase relations among obviously very different component processes. Similarly, the linkage between materially different entities appears to be governed by a fundamental biophysical coupling. The resulting collective dynamics bridges the language of neuronal ensembles and behavioral function.

The discovery of critical instabilities in the brain highlights the importance of fluctuations—whether of stochastic or deterministic origin—in probing the stability of coherent patterns and creating new patterns when the environmental, task, or internal conditions demand it. Not only do nonequilibrium phase transitions offer a new mechanism for the collective action of neurons, they provide the brain with a switching mechanism, essential for rapidly entering and exiting various coherent states. Thus, phase transitions confer on the brain the hallmark of flexibility.

An even more flexible and fluid view emerges from phase space reconstruction of the spatial modes in which a class of behavior known as Šil'nikov chaos is revealed. Rather than requiring an active process to destabilize and switch from one stable state to another, the existence of Šil'nikov chaos suggests that the brain possesses an inherently *intermittent* character.

How, then, do we reconcile the mode-mode interaction picture of self-organization based on coupled nonlinear oscillators with the apparently chaotic trajectory of the brain's top mode? Is this an either-or situation to be settled by experiment, or just an inherent ambiguity of theoretical modeling? I submit it is neither. Rather, *both* features are essential to the global operation of the brain, each capturing critical features, one at a more detailed grain of analysis than the other. A further consideration surrounds Šil'nikov's condition. If the condition is *not* fulfilled, the homoclinic orbit around an unstable focus is deformed into a *stable* limit cycle. A consequence of this scenario is that *any* slight change or fluctuation in a parameter may lead the system from a stable limit cycle to a chaotic attractor. So it is with the human brain. Both complex, chaotic-looking spatiotemporal behavior and mode-mode interactions are possible, and even necessary.

In a rigorous sense, the term "attractor" is defined only under *stationary* conditions as time goes to infinity. For practical purposes, this definition may be weakened and an attractor defined assuming no changes in the parameters of the system for a time much longer than the relaxation time (cf. chapter 2). The human brain (except, perhaps when it obsesses) does not have time to wander around in phase space before it finally finds an attractor, whether periodic or chaotic. Transient behavior for both dynamical systems is very similar and depends on the location and stability of the fixed points in phase space. I suspect, therefore, that it is the geometry of fixed points and the changes in this geometry that hold the key to understanding the coherent, self-organizing dynamics of the brain.[36]

Although this argument is general, perhaps too general for anatomically minded cognitive scientists who seek to locate mental processes in the brain, the facts presented here cannot be denied. The actual trajectory of the main spatial pattern of the brain displays the geometry, a saddle focus connecting stable and unstable manifolds by homoclinic orbits, characteristic of Šil'nikov chaos. Thus the brain, as I have said is inherently intermittent, never actually in a stable fixed point or mode-locked state, but continuously evolving in the vicinity of a saddle focus, poised on the brink. Although brain activity patterns are spatially coherent, their temporal evolution traces a complex geometry that, as stressed time and again, is context-dependent. In reply to Warren McCulloch's question at the beginning of this chapter, then, the brain ain't a computer. The brain is a self-organized, pattern-forming, dynamical system. And its coherent, but unpredictable spatiotemporal trajectories— *brain behavior*—is the mind.

Epilogue: Mind, Matter, and Meaning

The only laws of matter are those which our minds must fabricate and the only laws of mind are fabricated for it by matter.
—J. C. Maxwell

Self-organized matter, cracking itself into meaningfully coherent modes whose time-dependent behavior expresses the mind itself. For centuries, scientists have wondered how mind fits into the natural world described by physics. In his classic little book *Matter and Motion*, the great Scottish physicist James Clerk Maxwell restricted physics to a consideration of those phenomena "of the simplest but most abstract kind, excluding the consideration of the more complex phenomena such as those observed in living things." A beautifully profound contradiction.

But science, in my view, does not need some special, extraphysical approach to accommodate how human beings come to know their world. I have argued in this book that the concepts and tools we need to ground a science of mind and behavior (that strange, almost paradoxical dictionary definition of psychology) are only now emerging. This new science of self-organization in complex systems has its roots in theories of pattern formation in open, non-equilibrium systems, especially Haken's synergetics, a synthesis of enormous range and power. Instead of trying to *reduce* biology and psychology to chemistry and physics, the task now is to *extend* our physical understanding of the organization of living things.

Why is such understanding so important? People seem to be more captivated by remote events that are light years away (e.g., the Big Bang) than how their own brains work or how they can wiggle their toes. But this *is* the "decade of the brain," and some 40 odd million people in the United States alone suffer from functional brain disorders, causing a huge strain on health care resources and the very fabric of society. There is a fundamental need to understand the most complex system of all, ourselves. Even the most ardent reductionists now admit that the brain cannot be understood solely on the basis of the chemistry and biophysics of single cells. But there is a huge void in our knowledge of what single cells do versus what many of them do when they cooperate. That's why it is crucial to discover the laws and principles of *coordination* in living things. It is this coordination that lies at the root of understanding ourselves and the world we live in.

Detailed information about single neurons or, in general, the individual components of a system is of course important, even necessary (my old saw, a single component at one level may be a coordinated system with respect to levels below). But assembly instructions and a detailed wiring diagram don't explain how a radio works. A cathedral, it is said, is more than and different from bricks and mortar. But where is the science of coordination that presumably explains how things are put together? The answer is that, for the most part, the laws and principles of coordination that produce dynamic behavioral patterns are not known. This book aims to fill this breach.

Is the conceptual framework of dynamic patterns, with its emphasis on synergetic self-organizing processes on several scales of observation and its creative use of dynamical language, up to the task of understanding how brain relates to mind? Only time will tell. Nowadays it's quite commonplace for neuroscientists and others to assert that mind can be explained by the interactions of nerve cells. Seldom if ever is the nature of these interactions formulated in any rigorous fashion. In a parallel development, mind-body philosophers are advised to study *neurons*, everyday language being an obstacle to understanding.

I think this advice is misguided and that the real solution to the mind-brain-body problem rests in how *information* is to be conceived in living things, in general, and the brain in particular. In the first part of this book I showed that pattern variables such as ϕ capture the coherent relations among different kinds of parts. These patterns are governed by laws of dynamical systems that I refer to variously as coordination or pattern dynamics. The pattern dynamics contain couplings that are quite independent of the physical medium through which they are realized. They are, strictly speaking, informational couplings. So the self-organized pattern dynamics is, at core, an informational structure. (Though not, I hasten to add, in the purely syntactic sense of information used by physicists and engineers.)

In the second half of this book, I described how the pattern dynamics may be modified by the "will," the environment, by learning and so forth. Theoretically, "the will," or willing, has no meaning outside its influence on the order parameters or pattern variables. To be informationally meaningful, any conscious thought, will or intention must be expressed, in this theory, in terms of relevant pattern variables. Intending a behavior or learning a behavior or perceptually specifying a behavior means sculpting the coordination dynamics.

If thoughts, according to theory, must be expressed in terms of ϕ-like collective variables that characterize dynamic patterns of spatiotemporal activity in the brain, then the following conclusion appears logically inescapable: an order parameter isomorphism connects mind and body, will and brain, mental and neural events. Mind itself is a spatiotemporal pattern that molds the metastable dynamic patterns of the brain. Mind-body dualism is replaced by a single isomorphism, the heart of which is semantically meaningful pattern variables. The present analysis of dynamic patterns suggests we should take Sherrington's "enchanted loom" image of the brain very seriously indeed.

My aim here was to join together neural processes at one end of the scale and mental or cognitive processes at the other, in a common language. This is the language of dynamic patterns, not the neuron per se. But more than a language, shared principles of self-organization provide the linkage across levels of neural and cognitive function. These principles encompass spontaneity, attraction, repulsion, broken symmetry, intermittency, crises, instability, transitions, and synchronicity, phenomena that can be clothed mathematically and biologically, yet touch everyday life. I have tried to create an intellectual image of brain, mind, and behavior that feels closer to life, that gives life a certain coherence.

Notes and References

CHAPTER 1

1. Sherrington, C. S. (1951). *Man and His Nature*, 2nd ed., London: Cambridge University Press, p. 178.

2. This is also the view of the late theoretical physicist David Bohm. See Bohm, D. (1980). *Wholeness and the Implicate Order*. London: Ark.

3. From the point of view of principles of self-organization, both organisms and inanimate systems exhibit behavioral characteristics that may be said to be alive (*defn*: in a state of action; active, lively, vibrant).

4. *Webster's Encyclopedic Unabridged Dictionary*. (1989). New York: Portland House.

5. One is reminded of the words of the great French mathematician Henri Poincaré: "The aim of science is not things themselves, as the dogmatists in their simplicity imagine, but the relations among things; outside these relations there is no reality knowable." Poincaré, H. (1905, reprinted 1952). *Science and Hypothesis*. New York: Dover, p. xxiv.

6. Turing, A. M. (1952). The chemical basis of morphogenesis. *Philosophical Transactions of the Royal Society*, B237, 37–52.

7. Meinhardt, H. (1982). *Models of Biological Pattern Formation*. London: Academic Press; and Wolpert, L. (1991). *The Triumph of the Embryo*. Oxford: Oxford University Press.

8. Hermann Haken introduced the term "synergetics" in a lecture given at the University of Stuttgart in 1969. Synergetics is an interdisciplinary field concerned with cooperation of individual parts of a system that produces macroscopic spatial, temporal, or functional structures. It deals with deterministic as well as stochastic processes. There are now over sixty volumes in the Springer Series in Synergetics edited by Haken. For an excellent technical treatment see Haken, H. (1983). *Synergetics: An Introduction*, 3rd ed.; and (1984). *Advanced Synergetics*, 2nd ed. Berlin: Springer-Verlag. A very readable and accessible introduction is Haken, H. (1984). *The Science of Structure: Synergetics*. New York: Van Nostrand Reinhold.

9. For example, Babloyantz, A. (1986). *Molecule, Dynamics and Life*. New York: Wiley; Bak, P. (1993). Self-organized criticality and Gaia. In: W. B. Stein & F. J. Varela (Eds.). *Thinking About Biology*. Reading, MA: Addison-Wesley, pp. 255–268; Bergé, P., Pomeau, Y., & Vidal, C. (1984). *Order Within Chaos*. Paris: Hermann; Collet, P., & Eckmann, J. P. (1990). *Instabilities and Fronts in Extended Systems*. Princeton, NJ: Princeton University Press; Nicolis, G., & Prigogine, I. (1989). *Exploring Complexity: An Introduction*. San Francisco: Freeman; and Yates, F. E. (Ed.). (1987). *Self-Organizing Systems*. New York: Plenum Press.

10. Schrödinger, E. (1944). *What Is Life?* Cambridge: Cambridge University Press.

11. Haken's slaving principle states that in the neighborhood of critical points, the behavior of a complex system is completely governed by few collective modes, the order parameters, that slave all the other modes. See Haken, H. (1983) *Synergetics*. An early background reference and informal discussion is Burgers, J. M. (1963). On the emergence of order. *American Mathematical Society Bulletin, 69,* 1−25.

12. Ashby, W. R. (1956). *An Introduction to Cybernetics*. London: Methuen.

13. Velarde, M. G., & Normand, C. (1990). Convection. *Scientific American, 243,* 92.

14. Castets, V., Dulos, E., Boissonade, J., & de Kepper, P. (1990). Experimental evidence of a sustained standing Turing-type nonequilibrium chemical pattern. *Physical Review Letters, 64,* 2953, 2956; and Ouyang, Q., & Swinney, H. L. (1991). Transition from a uniform state to hexagonal and striped Turing patterns. *Nature, 352,* 610−612. For general discussions see Pool, R. (1991). Did Turing discover how the leopard got its spots? *Science, 251,* 627; and Amato, I. (1993). A chemical loom weaves new patterns. *Science, 261,* 165.

15. Turing, A. M. (1952). Chemical basis of morphogenesis.

16. Murray, J. D. (1989). *Mathematical Biology*. Berlin: Springer-Verlag. See especially chapter 15.

17. Lengyel, I., & Epstein, I. R. (1991). Modeling of Turing structures in the chlorite-iodide-malonic acid-starch reaction system. *Science, 251,* 650−652.

18. Haken, H. (1983). *Synergetics*.

19. Murray, J. D. (1989). *Mathematical Biology*.

20. Robbins, W. J. (1952). Patterns formed by motile *Euglena gracillis* var. *bacillaris. Bulletin of the Torrey Botanical Club, 79,* 107−109. See also Goldbeter, A., & Decroly, O. (1983). Temporal self-organization in biochemical systems: Periodic behavior versus chaos. *American Journal of Physiology, 245,* R478−R483.

21. Miller, K. D., Keller, J. B., & Stryker, M. P. (1989). Ocular dominance column development: Analysis and simulation. *Science, 245,* 605−615. See also Obermayer, K., Ritter, H., & Schulten, K. (1992). A model for the development of the spatial structure of retinotopic maps and orientation columns. In: J. Mittenthal & A. Baskin (Eds.). *Principles of Organization in Organisms*. Reading, MA: Addison-Wesley.

22. Kelso, J. A. S. (1988). Introductory remarks: Dynamic patterns. In: J. A. S. Kelso, A. J. Mandell, & M. F. Shlesinger (Eds.). *Dynamic Patterns in Complex Systems*. Singapore: World Scientific.

23. Hénon, M., & Pomeau, Y. (1976). Two strange attractors with a simple structure. In: *Turbulence and Navier-Stokes Equations*. Springer Lecture Notes in Mathematics, Vol. 565. New York: Springer-Verlag.

24. May, R. M. (1976). Simple mathematical models with very complicated dynamics. *Nature, 261,* 459−467.

25 Dawson, S. P., Grebogi, C., Yorke, J. A., Kan, I., & Kocak, H. (1992). Antimonotonicity: Inevitable reversals of period doubling cascades. *Physics Letters A, 162,* 249−254.

26. A critical point of a map is a local maximum or minimum. These points are useful for studying the behavior of maps because they are the points where trajectories are folded together. The forward orbits of these points can be seen in bifurcation diagrams and form a kind of skeleton supporting the entire structure. Notice the curved lines running through the dark portions of figures 1-6 and 1-8.

27. Schrödinger, E. (1944). *What Is Life?*

28. Anderson, P. W. (1972). More is different. *Science, 177,* 293–296.

29. Penrose, R. (1989). *The Emperor's New Mind.* Oxford: Oxford University Press.

30. Barinaga, M. (1990). The mind revealed? *Science, 249,* 856–858.

31. Gray, C. M., König, P., Engel, A. K., & Singer, W. (1989). Oscillatory responses in cat visual cortex exhibit inter-columnar synchronization which reflects global stimulus properties. *Nature, 338,* 334–337.

32. Crick, F. H. C., & Koch, C. (1990). Toward a neurobiological theory of consciousness. *Seminars on Neurosciences, 2,* 263–275; and (1992). The problem of consciousness. *Scientific American, 267,* 152–159.

33. Levin, R. (1992). *Complexity.* New York: Macmillan, pp. 164–165.

34. One is reminded of the words (reported by R. Rosen (1985) in his book *Anticipatory Systems,* Oxford: Pergamon Press, p. 3) of Robert M. Hutchins, President of the University of Chicago before he reached the age of 30: The gadgeteers and data collectors, masquerading as scientists have threatened to become the supreme chieftains of the scholarly world.... As the Renaissance could accuse the Middle Ages of being rich in principles and poor in facts, we are new entitled to enquire whether we are not rich in facts and poor in principles.

CHAPTER 2

1. For an excellent treatment of Skinner's work, including some of his classic articles, see Catania, C. A., & Harnad, S. (Eds.). (1988). *The Selection of Behavior.* Cambridge: Cambridge University Press.

2. Köhler, W. (1940). *Dynamics in Psychology.* New York: Liveright. See especially chapter 1 in which Köhler draws distinctions between psychological facts as such (experience) and facts of functional dependence that are seldom directly observed.

3. This story is told in a book (not yet translated into English) by Bernstein, N. A. (1991). *On Dexterity and Its Development.* Moscow: Physical Culture and Sport Press. It is repeated in a review of the book by Latash, L. P., & Latash, M. L., *Journal of Motor Behavior,* 1994, *26,* 56–62.

4. Tinbergen, N. (1951). *The Study of Instinct.* New York: Oxford University Press; and Lorenz, K. (1950). The comparative method in studying innate behavior patterns. *Symposium of the Society for Experimental Biology, 4,* 221–268. Together with Karl von Frisch, Tinbergen and Lorenz received the Nobel Prize in medicine and physiology.

5. Golani, I. (1976). Homeostatic motor processes in mammalian interactions: A choreography of display. In: P. G. Bateson & P. H. Klopfer (Eds.). *Perspectives in Ethology,* Vol. 2. New York: Plenum Press.

6. Fentress, J. C. (1984). The development of coordination. *Journal of Motor Behavior, 16,* 99–134.

7. Golani, I., Wolgin, D. L., & Teitelbaum, P. (1979). A proposed natural geometry of function from akinesia in the lateral hypothalamic rat. *Brain Research, 164,* 237–267. A good review of works that use the E-W system is Jacobs, W. J., et al. (1988). Observations. *Psychobiology, 16,* 3–19.

8. Chomsky, N. (1968). *Language and Mind.* New York: Harcourt Brace Jovanovich.

9. Searle, J. R. (1992). *The Rediscovery of the Mind.* Cambridge: Bradford Books/MIT Press.

10. In a paper delivered to the International Society for Attention and Performance at Jesus College, Cambridge, in 1980, I argued against the computer metaphor and for dynamical

self-organization. See Kelso, J. A. S. (1981). Contrasting perspectives on order and regulation in movement. In: J. Long & A. Baddeley (Eds.). *Attention and Performance IX.* Hillsdale, NJ: Erlbaum. References to earlier work from both camps can be found in this paper.

11. This notion stems from a famous work by Lashley, K. D. (1951). The problem of serial order in behavior. In: L. A. Jeffress (Ed.). *Cerebral Mechanisms in Behavior.* New York: Wiley.

12. Dennett, D. C. (1978). *Brainstorms.* Montgomery, VT: Bradford Books, p. 12.

13. See reference 2. Also Köhler, W. (1947). *Gestalt Psychology.* New York: Liveright. Chapter 4, "Dynamics as opposed to machine theory," is a classic.

14. Epstein, W., & Hatfield, G. (1994). Gestalt psychology and the philosophy of mind. *Philosophical Psychology 7,* 163–181; and Stadler, M., & Kruse, P. (1990). Cognitive systems as self-organizing systems. In: W. Krohn et al. (Eds.). *Selforganization: Portrait of a Scientific Revolution.* Norwell, Ma.: Kluwer academic, pp. 181–193.

15. Gibson, J. J. (1979). *The Ecological Approach to Visual Perception.* Boston: Houghton Mifflin. Gibson stimulated a remarkable body of research and even a journal (*Ecological Psychology*) devoted to the study of animal-environment systems. See, for example, Kugler, P. N., Kelso, J. A. S., & Turvey, M. T. (1982). On the control and coordination of naturally developing systems. In: J. A. S. Kelso & J. E. Clark (Eds.). *The Development of Movement Control and Coordination.* Chichester: Wiley; Carello, C., Turvey, M. T., Kugler, P. N., & Shaw, R. E. (1984). Inadequacies of the computer metaphor. In: M. S. Gazzaniga (Ed.). *Handbook of Cognitive Neuroscience.* New York: Plenum Press; Shaw, R. E., Turvey, M. T., & Mace, W. (1983). Ecological psychology: The consequences of a commitment to realism. In: W. Weimer & D. Palermo (Eds.). *Cognition and Symbolic Processes,* Vol. 2. Hillsdale, NJ: Erlbaum; and Reed, E. S., & Jones, R. (1982). *Reasons for Realism: Selected Essays of James J. Gibson.* Hillsdale, NJ: Erlbaum.

16. British psychologist David Lee was the first to exploit and analyze the optic flowfield. His 1981 paper with P. E. Reddish, Plummeting gannets: A paradigm of ecological optics. *Nature, 293,* 293–294, is a classic. See also Wagner, H. (1982). Flow-field variables trigger landing in flies. *Nature, 297,* 147–148.

17. Rössler, O. (1987). Chaos in coupled optimizers. In: S. H. Koslow, A. J. Mandell, & M. F. Shlesinger (Eds.). *Perspectives in Biological Dynamics.* New York: Academy of Sciences, pp. 229–240.

18. For a remarkable attempt to analyze such events, see Condon, W. S., & Ogston, W. D. (1967). A segmentation of behavior. *Journal of Psychiatric Research, 5,* 221–235.

19. See Latash, L. P., & Latash, M. L. Ref. 3. In Bernstein, N. A. (1967). *The Coordination and Regulation of Movements.* London: Pergamon Press.

20. Greene, P. H. (1972). Problems of organization of motor systems. In: R. Rosen & F. Snell (Eds.). *Progress in Theoretical Biology,* Vol. 2. New York: Academic Press, pp. 303–338; and Greene, P. H. (1982). Why is it easy to control your arms? *Journal of Motor Behavior, 14,* 260–286. Peter Greene was one of a handful of researchers who were keenly attuned to the special problems of coordination and control in complex systems. See also Boylls, C. C. (1975). A theory of cerebellar function with applications to locomotion. COINS Technical Report 76-1. Amberst, MA: Department of Computer and Information Science, University of Massachusetts; Easton, T. A. (1972). On the normal use of reflexes. *American Scientist, 60,* 591–599; Szentagothai, J., & Arbib, M. A. (1974). Eds. Conceptual models of neural organization. *Neuroscience Research Program Bulletin, 12(3);* Turvey, M. T., Shaw, R. E., & Mace, W. (1978). Issues in the theory of action: Degrees of freedom, coordinative structures and coalitions. In: J. Requin (Ed.). *Attention and Performance,* Vol. 7. Hillsdale, NJ: Erlbaum; and Kelso, J. A. S., Southard, D. L., & Goodman, D. (1979). On the nature of human interlimb coordination. *Science, 203,* 1029–1031.

21. Gelfand, I. M., Gurfinkel, V. S., Tsetlin, M. L., & Shik, M. L. (1971). Some problems in the analysis of movements. In: I. M. Gelfand, V. S. Gurfinkel, S. V. Fomin, & M. L. Tsetlin (Eds.). *Models of the Structural-Functional Organization of Certain Biological Systems.* Cambridge: MIT Press. See also Kots, Y. A. (1977). *The Organization of Voluntary Movement.* New York: Plenum Press; Fukson, O. I., Berkinblit, M. B., & Feldman, A. G. (1980). The spinal frog takes into account the scheme of its body during the wiping reflex. *Science, 209,* 1261–1263; and Feldman, A. G. (1966). Functional tuning of the nervous system with control of movement or maintenance of a steady posture. III. Mechanographic analysis of execution by man of the simplest motor tasks. *Biophysics, 11,* 766–775.

22. Sherrington, C. S. (1906). *The Integrative Action of the Nervous System.* London: Constable.

23. For a review and analysis, see Kelso, J. A. S., & Tuller, B. (1984). A dynamical basis for action systems. In: M. S. Gazzaniga (Ed.). *Handbook of Cognitive Neuroscience.* New York: Plenum Press, pp. 321–356.

24. Kelso, J. A. S., Southard, D. L., & Goodman, D. (1979). On the nature of human interlimb coordination; and Kelso, J. A. S., Goodman, D., & Putnam, C. A. (1983). On the space-time structure of human interlimb coordination. *Quarterly Journal of Experimental Psychology, 35A,* 347–375. See also Marteniuk, R. G., & MacKenzie, C. L. (1980). A preliminary theory of two handed coordinated control. In: G. E. Stelmach & J. Requin (Eds.). *Tutorials in Motor Behavior.* Amsterdam: North-Holland. There has been a good deal of further work using this paradigm in adults, children, and the neurologically impaired. See Sherwood, D. E. (1989). The coordination of simultaneous actions. In: Wallace, S. A. (Ed.). *Perspectives on the Coordination of Movement.* Amsterdam: North-Holland.

25. Kelso, J. A. S., Tuller, B., & Fowler, C. A. (1982). The functional specificity of articulatory control and coordination. *Journal of the Acoustical Society of America, 72,* S103; and Kelso, J. A. S., Tuller, B., Vatikiotis-Bateson, E., & Fowler, C. A. (1984). Functionally specific articulatory cooperation following jaw perturbations during speech: Evidence for coordinative structures. *Journal of Experimental Psychology: Human Perception and Performance, 10,* 812–832. For review, see Abbs, J. H., & Gracco, V. L. (1983). Sensorimotor actions in the control of multimovement speech gestures. *Trends in Neuroscience, 6,* 391–395.

26. Folkins, J. W., & Abbs, J. H. (1975). Lip and jaw motor control during speech: Responses to resistive loading of the jaw. *Journal of Speech and Hearing Research, 18,* 207–220; and Folkins, J. W., & Zimmermann, G. N. (1982). Lip and jaw interaction during speech: Response to perturbations of lower-lip movement during bilabial closure. *Journal of the Acoustical Society of America, 71,* 1225–1233.

27. Turvey, M. T. (1990). Coordinatior *American Psychologist, 45,* 938–953.

28. Kugler, P. N., Kelso, J. A. S., & Turvey, M. T. (1980). On the concept of coordinative structures as dissipative structures. I. Theoretical lines of convergence; and Kelso, J. A. S., Holt, K. G., Kugler, P. N., & Turvey, M. T. (1980). On the concept of coordinative structures as dissipative structures. II. Empirical lines of convergence. In: G. E. Stelmach & J. Requin (Eds.). *Tutorials in Motor Behavior.* Amsterdam: North-Holland, pp. 3–70.

29. For a highly informed analysis of the main themes in the work of Haken, Prigogine, and Thom, see Landauer, R. (1981). Nonlinearity, multistability and fluctuations: Reviewing the reviewers. *American Journal of Physiology, 241,* R107–R113.

30. Kugler, P. N., & Turvey, M. T. (1987). *Information, Natural Law and the Self-Assembly of Rhythmic Movement.* Hillsdale, NJ: Erlbaum.

31. Turvey, M. T. (1990). Coordination.

32. Pennycuick, C. J. (1992). *Newton Rules Biology.* Oxford: Oxford University Press, p. 1.

33. Bentley, E. C. (1936). *Trent's Own Case*. New York: Perennial Press.

34. Kelso, J. A. S. (1981). On the oscillatory basis of movement. *Bulletin of Psychonomic Society, 18,* 63; and (1984). Phase transitions and critical behavior in human bimanual coordination. *American Journal of Physiology: Regulatory, Integrative and Comparative Physiology, 15,* R1000–R1004.

35. Tuller, B., & Kelso, J. A. S. (1989). Environmentally specified patterns of movement coordination in normal and split-brain subjects. *Experimental Brain Research, 74,* 306–316.

36. Haken, H., Kelso, J. A. S., & Bunz, H. (1985). A theoretical model of phase transitions in human hand movements. *Biological Cybernetics, 51,* 347–356. For a recent discussion of theoretical models of coordination, see Fuchs, A., & Kelso, J. A. S. (1994). *Journal of Experimental Psychology: Human Perception and Performance, 20,* 1088–1097.

37. These remarks follow along the lines made by Beek, W. J. (1990). Synergetics and self-organization: A response. In: H. T. A. Whiting, O. G. Meijer, & P. C. W. van Wieringen (Eds.). *The Natural-Physical Approach to Movement Control.* Amsterdam: VU Press. See also chapter 3.

38. Kelso, J. A. S., & Scholz, J. P. (1985). Cooperative phenomena in biological motion. In: H. Haken (Ed.). *Complex Systems: Operational Approaches in Neurobiology, Physical Systems and Computers.* Berlin: Springer-Verlag; Kelso, J. A. S., Scholz, J. P., & Schöner, G. (1986). Non-equilibrium phase transitions in coordinated biological motion: Critical fluctuations. *Physics Letters A, 118,* 279–284; and Kelso, J. A. S., Schöner, G., Scholz, J. P., & Haken, H. (1987). Phase-locked modes, phase transitions and component oscillators in biological motion. *Physica Scripta, 35,* 79–87.

39. Scholz, J. P., Kelso, J. A. S., & Schöner, G. (1987). Non-equilibrium phase transitions in coordinated biological motion: Critical slowing down and switching time. *Physics Letters A, 123,* 390–394; and Scholz, J. P., & Kelso, J. A. S. (1989). A quantitative approach to understanding the formation and change of coordinated movement patterns. *Journal of Motor Behavior, 21(2),* 122–144.

40. Schöner, G., Haken, H., & Kelso, J. A. S. (1986). A stochastic theory of phase transitions in human hand movement. *Biological Cybernetics, 53,* 442–452.

41. Kay, B. A., Kelso, J. A. S., Saltzman, E. L., & Schöner, G. (1987). The space-time behavior of single and bimanual movements: Data and model. *Journal of Experimental Psychology: Human Perception and Performance, 13,* 178–192; and Kay, B. A., Saltzman, E. L., & Kelso, J. A. S. (1991). Steady-state and perturbed rhythmical movements: Dynamical modeling using a variety of analytical tools. *Journal of Experimental Psychology: Human Perception and Performance, 17,* 183–197.

42. For excellent expositions of phase resetting methodology and analysis, see Winfree, A. T. (1980). *The Geometry of Biological Time.* New York: Springer-Verlag; and Glass, L., & Mackey, M. C. (1988). *From Clocks to Chaos.* Princeton, NJ: Princeton University Press.

43. Many books on fractals and chaos exist. For a recent thorough example see Peitgen, H.-O., & Saupe, D. (Eds.). (1988). *The Science of Fractal Images.* Berlin: Springer-Verlag.

44. Grassberger, P., & Procaccia, I. (1983). Measuring the strangeness of strange attractors. *Physica D, 9,* 189.

45. Haken, H., Kelso, J. A. S., & Bunz, H. (1985). A theoretical model of phase transitions in human hand movements. See also Kelso J. A. S., & Scholz, J. P. (1985). Cooperative phenomena in biological motion.

46. For more on the distinction, see Rosen, R. (1991). *Life Itself.* New York: Columbia University Press; and Haken, H., Karlquist, A., & Svedin, U. (Eds.). (1993). *The Machine as Metaphor and Tool.* Heidelberg: Springer-Verlag.

47. Haken, H. (1988). *Information and Self-Organization*. Berlin: Springer-Verlag; and Stadler, M., Vogt, S., & Kruse, P. (1991). Synchronization of rhythm in motor actions. In: H. Haken & H. P. Köpchen (Eds.). *Rhythms in Physiological Systems*. Berlin: Springer-Verlag. See also Kruse, P., & Stadler, M. (1993). The significance of nonlinear phenomena for the investigation of cognitive systems. In: H. Haken & A. Mikhailov (Eds.). *Interdisciplinary Approaches to Nonlinear Complex Systems*. Berlin: Springer-Verlag.

48. Turvey, M. T. (1990). Coordination.

CHAPTER 3

1. Haken, H. (1988). Synergetics in pattern recognition and associative action. In: H. Haken (Ed.). *Neural and Synergetic Computers*. Berlin: Springer-Verlag, p. 14.

2. Beek, W. J. (1990). Synergetics and self-organization: A response. In: H. T. A. Whiting, O. G. Meijer, & P. C. S. van Wieringen (Eds.). *The Natural-Physical Approach to Movement Control*. Amsterdam: Free University Press.

3. Muybridge, E. (1955). *The Human Figure in Motion*. New York: Dover.

4. Hoyt, D. F., & Taylor, C. R. (1981). Gait and energetics of locomotion in horses. *Nature, 292*, 239–240.

5. Kelso, J. A. S., & Scholz, J. P. (1985). Cooperative phenomena in biological motion. In: H. Haken (Ed.). *Complex Systems: Operational Approaches in Neurobiology, Physical Systems and Computers*. Berlin: Springer-Verlag, 124–149.

6. Shik, M., Orlovskii, G. N., & Severin, F. V. (1966). Organization of locomotor synergism. *Biophysics, 13*, 127–135.

7. Haken, H., Kelso, J. A. S., & Bunz, H. (1985). A theoretical model of phase transitions in human hand movements. *Biological Cybernetics, 51*, 347–356; and Schöner, G., Jiang, W. Y., & Kelso, J. A. S. (1990). A synergetic theory of quadrupedal gaits and gait transitions. *Journal of Theoretical Biology, 142*, 359–393. See also Collins, J. J., & Stewart, I. N. (1992). Coupled nonlinear oscillators and the symmetries of animal gaits. *Journal of Nonlinear Science, 3*, 349–392.

8. Kelso, J. A. S., & Jeka, J. J. (1992). Symmetry breaking dynamics of human multilimb coordination. *Journal of Experimental Psychology: Human Perception and Performance, 18*, 645–668; Jeka, J. J., Kelso, J. A. S., & Kiemel, T. (1993). Pattern switching in human multilimb coordination dynamics. *Bulletin of Mathematical Biology, 55*, 829–845; and (1993). Spontaneous transitions and symmetry: Pattern dynamics in human four-limb coordination. *Human Movement Science, 12*, 627–651.

9. Swinnen, S., Massion J., & Heuer, H. (1994). Topics on interlimb coordination. In: S. P. Swinnen, H. Heuer, J. Massion, & P. Casaer (Eds.). *Interlimb Coordination: Neural, Dynamical, and Cognitive Constraints*. San Diego: Academic Press.

10. This case has been made forcefully by Peter Beek and ourselves. See Beek, P. J. (1989). *Juggling Dynamics*. Amsterdam: Free University Press; Kelso, J. A. S., & DeGuzman, G. C. (1988). Order in time: How cooperation between the hands informs the design of the brain. *Neural and Synergetic Computers*. Berlin: Springer-Verlag, pp. 180–196; and DeGuzman, G. C., & Kelso, J. A. S. (1991). Multifrequency behavioral patterns and the phase attractive circle map. *Biological Cybernetics, 64*, 485–495. See also my interpretation of gating in neurophysiological studies of locomotion reviewed in Kelso, J. A. S. (1981). Contrasting perspectives on order and regulation in movement. In: J. Long & A. Baddeley (Eds.). *Attention and Performance IX*. Hillsdale, NJ: Earlbaum, pp. 437–457.

11. Schöner, G., Jiang, W., & Kelso, J. A. S. (1990). A synergetic theory of quadrupedal gaits

and gait transitions; and Collins, J. J., & Stewart, I. N., (1992). Coupled nonlinear oscillators and the symmetries of animal gaits.

12. He also says, "The empty space is full of symmetry," See his beautiful book Caglioti, G. (1992). *The Dynamics of Ambiguity*. Berlin, Heidelberg: Springer-Verlag.

13. Glass, L., & Young, R. E. (1979). Structure and dynamics of neural network oscillators. *Brain Research, 179*, 207–218.

14. Stewart, I. (1991). Mathematical recreations. Why Tarzan and Jane can walk in step with the animals that roam the jungle. *Scientific American, 264*, 158–161; Collins, J., & Stewart, I. (1993). Coupled nonlinear oscillators . . . ; Schöner, G., Jiang, W., & Kelso, J. A. S. (1990). A synergetic theory . . . ; and Kelso, J. A. S., & DeGuzman, G. C. (1988). Order in time.

15. Kelso, J. A. S., Buchanan, J. J., & Wallace, S. A. (1991). Order parameters for the neural organization of single, multijoint limb movement patterns. *Experimental Brain Research, 85*, 432–444.

16. Baldissera, F., Cavallari, P., & Civaschi, P. (1982). Preferential coupling between voluntary movements of ipsilateral limbs. *Neuroscience Letters, 34*, 95–100; and Baldissera, F., Cavallari, P., Marini, G., & Tassone, G. (1991). Differential control of in-phase and anti-phase coupling of rhythmic movements of ipsilateral hand and foot. *Experimental Brain Research, 83*, 375–380.

17. Buchanan, J. J., & Kelso, J. A. S. (1993). Posturally induced transitions in rhythmic multijoint limb movements. *Experimental Brain Research, 94*, 131–142.

18. Quoted in Pattee, H. H. (1976). Physical theories of biological coordination. In: M. Grene & E. Mendelsohn (Eds.). *Topics in the Philosophy of Biology*. Boston: Reidel, pp. 153–173.

19. Kelso, J. A. S., & Scholz, J. P. (1985). Cooperative phenomena in biological motion. In: H. Haken (Ed). *Complex Systems: Operational Approaches in Neurobiology, Physics and Computer*. Berlin: Springer-Verlag.

20. Kelso, J. A. S., Buchanan, J. J., DeGuzman, G. C., & Ding, M. (1993). Spontaneous recruitment and annihilation of degrees of freedom in biological coordination. *Physics Letters A, 179*, 364–371.

21. Collins, J. J., & Stewart, I. (1993). Coupled nonlinear oscillators . . . ; and Stewart, I. N., & Golubitsky, M. (1992). *Fearful Symmetry*. Oxford: Blackwell.

22. See Marsden, J. E., & McCracken, M. (1976). *The Hopf Bifurcation and Its Applications*. New York: Springer-Verlag.

23. For review, see Huang, M. X., & Waldron, K. J. (1989). An efficient rate allocation algorithm in redundant kinematic chains. *Journal of Mechanisms, Transmissions, and Automation in Design, 111*, 545–554.

24. Seminal work in the field of biological motion perception has been done by Gunnar Johansson. See Johansson, G. (1975). Visual motion perception. *Scientific American, 232*, 76–88. For review, see Bertenthal, B. I., & Pinto, J. (1993). Complimentary processes in the perception and production of human movements. In: L. B. Smith & E. Thelen (Eds.). *A Dynamic Systems Approach to Development*. Cambridge: MIT Press.

25. Gibson's position (see chapter 2), to which I am sympathetic, would argue that the light to the eye is far from underdetermined. Rather, the optic array undergoes disturbances of structure which reveal properties of the environment; e.g., the existence and layout of surfaces. In chapter 7 I'll argue that the death and destruction/growth and creation of objects during the locomotion of an observer is a pattern-formation process.

26. Haken, H. (Ed.). (1979). *Pattern Formation by Dynamic Systems and Pattern Recognition*. Berlin: Springer-Verlag.

27. Haken, H., Kelso, J. A. S., Fuchs, A., & Pandya, A. (1990). Dynamic pattern recognition of coordinated biological motion. *Neural Networks, 3,* 395–401. Haas, R., Fuchs, A., Haken, H., Horvath, E., & Kelso, J. A. S. (in press). Recognition of dynamic patterns by a synergetic computer. *Progress in Neural Networks, 3.*

28. See reference 24.

29. Quoted by Herbert Muschamp in the New York Times, May 30, 1993.

30. Tuller, B., & Kelso, J. A. S. (1990). Phase transitions in speech production and their perceptual consequences. In: M. Jeannerod (Ed.). *Attention and Performance XIII.* Hillsdale, NJ: Erlbaum, pp. 429–452.

31. For biographical details of this remarkable man see Kelso, J. A. S., & Munhall, K. G. (1988). *R. H. Stetson's Motor Phonetics: A Retrospective Edition.* San Diego: College Hill.

32. Tuller, B., Shao, S., & Kelso, J. A. S. (1990). An evaluation of an alternating magnetic field device for monitoring articulatory movements. *Journal of the Acoustical Society of America, 88*(2), 674–679.

33. Schmidt, R. C., Carello, C., & Turvey, M. T. (1990). Phase transitions and critical fluctuations in the visual coordination of rhythmic movements between people. *Journal of Experimental Psychology: Human Perception and Performance, 16,* 227–247.

CHAPTER 4

1. See note 5, chapter 1.

2. von Holst, E. (1939/1973). The behavioral physiology of man and animals. In: R. Martin (Ed.). *The Collected Papers of Erich von Holst.* Coral Gables, FL: University of Miami Press, p. 29.

3. Cook, J. E. (1991). Correlated activity in the CNS: A role on every timescale? *Trends in Neurosciences, 14,* 397–401.

4. For a good review, see Haken, H., & Köpchen, H. P. (1991). *Rhythms in Physiological Systems.* Berlin: Springer-Verlag, pp. 3–20.

5. Glazier, J.A., & Libchaber, A. (1988). Quasiperiodicity and dynamical systems: An experimentalist's view. *IEEE Transactions on Circuits and Systems, 35,* 790–809.

6. Of course, other dynamical phenomena including deterministic chaos are also possible in many of these systems.

7. For reviews, see Kelso, J. A. S., & DeGuzman, G. C. (1988). Order in time: How cooperation between the hands informs the design of the brain. In: H. Haken (Ed.). *Neural and Synergetic Computers,* Berlin: Springer-Verlag, pp. 180–196; Kelso, J. A. S. (1991). Behavioral and neural pattern generation: The concept of neurobehavioral dynamical systems (NBDS). In: H. P. Köpchen & T. Huopaniemi (Eds.). *Cardiorespiratory and Motor Coordination,* Berlin: Springer-Verlag, pp. 224–238; Haken, H., & Köpchen, H. P. (1991). *Rhythms in Physiological Systems;* Mackey, M. C. & Glass, L. (1988). *From clocks to chaos.* Princeton: Princeton University Press, and Rensing, L., an der Heiden, U., & Mackey, M. C. (Eds.). (1987). *Temporal Disorder in Human Oscillatory Systems.* Berlin: Springer-Verlag.

8. Kelso, J. A. S., DelColle J., & Schöner, G. (1990). Action-perception as a pattern formation process. In: M. Jeannerod (Ed.). *Attention and Performance XIII.* Hillsdale, NJ: Erlbaum, pp. 139–169.

9. Wimmers, R. H., Beek, P. J., & van Wieringen, P. C. W. (1992). Phase transitions in rhythmic tracking movements: A case of unilateral coupling. *Human Movement Science, 11,* 217–226.

10. For examples see Kelso, J. A. S., & Jeka, J. J. (1992). Symmetry breaking dynamics of human multilimb coordination. *Journal of Experimental Psychology: Human Perception and Performance, 18(3),* 645–668; Schmidt, R. C., Shaw, B. K., & Turvey, M. T. (1993). Coupling dynamics in interlimb coordination. *Journal of Experimental Psychology: Human Perception and Performance, 19,* 397–415; and Treffner, P. J. (1993). Unpublished doctoral thesis, University of Connecticut.

11. Kelso, J. A. S., DelColle, J., & Schöner, G. (1990). Action-perception as a pattern formation process.

12. Of course, a noise term $F(t)$ must also be included on the right-hand side because all real systems contain fluctuations. As described in chapters 1 and 2, fluctuations are conceptually and practically important. For simplicity of exposition, a full treatment is avoided here.

13. Jeka, J. J. (1992). Unpublished doctoral thesis, Florida Atlantic University; Jeka, J. J., & Kelso, J. A. S. (in press). Manipulating symmetry in the coordination dynamics of human movement. *Journal of Experimental Psychology: Human Perception and Performance;* and Kelso, J. A. S., & Jeka, J.J. (1992). Symmetry breaking dynamics of human multilimb coordination. Sternad, D., Turvey, M. T. & Schmidt, R. C. (1992) Average phase difference theory and 1:1 phase entrainment in interlimb coordination. *Biological Cybernetics, 67,* 223–231. Earlier data can be interpreted in this light, e.g., Rosenblum, L. D., & Turvey, M. T. (1988). Maintenance tendency in coordinated rhythmic movements: Relative fluctuation and phase. *Neuroscience, 27,* 298–300.

14. Kelso, J. A. S., & Ding, M. (1993). Fluctuations, intermittency and controllable chaos in biological coordination. In: K. M. Newell & D. M. Corcos (Eds.). *Variability and Motor Control.* Champaign, IL: Human Kinetics.

15. Pomeau, Y., & Manneville, P. (1980). Intermittent transition to turbulence in dissipative dynamical systems. *Communications in Mathematical Physics, 74,* 189–197.

16. See Kauffman's contribution in Mittenthal, J. E., & Baskin, A. B. (Eds.). (1992). *Principles of Organization in Organisms.* Reading, MA: Addison-Wesley, and comments by the editors, e.g., p. 39.

17. Ermentrout, G. B., & Rinzel, J. (1984). Beyond a pacemaker's entrainment limit: Phase walk-through. *American Journal of Physiology, 246,* R102–R106.

18. Mirollo, R. E., & Strogatz, S. H. (1990). Synchronization of pulse coupled oscillators. *SIAM Journal of Applied Mathematics, 50,* 1645–1662.

19. For excellent reviews, see contributions in Cohen, A. H., Rossignol, S., & Grillner, S. (Eds.). (1988). *Neural Control of Rhythmic Movements in Vertebrates.* New York: Wiley.

20. Rand, R. H., Cohen, A. H., & Holmes, P. J. (1988). Systems of coupled oscillators as models of central pattern generators (pp. 333–367); and Kopell, N. (1988). Toward a theory of modeling central pattern generators (pp. 369–413). In: Cohen et al. (Eds.). *Neural Control of Rhythrmic Movements in Vertebrates.*

21. Iberall, A. S., & McCulloch, W. S, (1969). The organizing principle of complex living systems. *Transactions of the American Society of Mechanical Engineers* (June), 290–294.

22. For mathematical details, see DeGuzman, G. C., & Kelso, J. A. S. (1991). Multifrequency behavioral patterns and the phase attractive circle map. *Biological Cybernetics, 64,* 485–495.

23. See note 7.

24. For discussion, see Bergé, P. Pomeau, Y., & Vidal, C. (1984). *Order Within Chaos.* New York: Wiley.

25. Direct evidence that this is in fact the case comes from Kelso, J. A. S., & DeGuzman, G., (1988). Order in time; Peper, C. E., Beek, P. J., & Van Wieringen, P. C. W. (1991). Bifurcations in polyrhythmic tapping; in search of Farey principles. In: J. Requin & G. E. Stelmach (Eds.).

Tutorials in Motor Neuroscience, Norwell, MA: Kluwer; and Treffner, P. J., & Turvey, M. T. (1993). Resonance constraints on rhythmic movement. *Journal of Experimental Psychology: Human Perception and Performance, 19,* 1221–1237.

26. Kelso, J. A. S., & DeGuzman, G. C. (1988). Order in time; Kelso, J. A. S., DeGuzman, G. C., & Holroyd, T. (1991). The self-organized phase attractive dynamics of coordination. In: A. Babloyantz (Ed.). *Self-Organization, Emerging Properties and Learning, Series B, Vol, 260,* New York: Plenum Press, pp. 41–62. In a recent study, Treffner, P. J., & Turvey, M. T. (1993). Resonance constraints on rhythmic movement, provided evidence that the Farey structure is a major constraint on rhythmic movement (unimodular shifts predominate, simple rhythms are most stable, Fibonacci ratios are least stable). They did not, however, evaluate the phase relation inside the frequency ratios studied.

27. Libchaber, A. (1987). From chaos to turbulence in Bénard convection. *Proceedings of the Royal Society of London, A 413,* 63–69.

28. See also Bak, P. (1986). The devil's staircase. *Physics Today,* 38–45; and Shenker, S. J. (1982). Scaling behavior in a map of a circle onto itself: Empirical results. *Physica,* 5D, 405–411.

29. Kelso, J. A. S., & DeGuzman, G. C. (1988). Order in time; and DeGuzman, G. C., & Kelso, J. A. S. (1991) Multifrequency behavioral patterns and the phase attractive circle map.

30. Peper, C. E., Beek, P. J., & Van Wieringen, P. C. W. (1991). Bifurcations in polyrhythmic tapping; and Treffner, P. J., & Turvey, M. T. (1993). Resonance constraints . . .

31. Pomeau, Y., & Manneville, P. (1980). Intermittent transition to turbulence in dissipative dynamical systems.

32. Kelso, J. A. S., DeGuzman, G. C., & Holroyd, T. (1991). The self-organized phase attractive dynamics of coordination. In: A. Babloyantz (Ed.) *Self-Organization, Emerging Properties* and *Learning* New York: Plenum.

33. See also Freeman, W. J. (1991). The physiology of perception. *Scientific American, 264,* 79–85.

34. Kelso, J. A. S., & Holt, K. G. (1980). Exploring a vibratory systems account of human movement production. *Journal of Neurophysiology, 43,* 1183–1196.

35. Kelso, J. A. S. (1991). Anticipatory dynamical systems, intrinsic pattern dynamics and skill learning. *Human Movement Science, 10,* 93–111.

36. Pais, A. (1982). *Subtle Is the Lord.* Oxford: Oxford University Press, p. 455.

CHAPTER 5

1. Edelman, G. M. (1992). *Bright Air, Brilliant Fire.* New York: Basic Books.

2. Quoted in Needham, J. (1968). *Order and Life.* Cambridge: MIT Press, p. 9.

3. Even F. H. C. Crick acknowledges open systems as one of the two minimum requirements for life. He quickly proceeds however to the second, namely, the need for a copying device. See Crick, F. (1966). *Of Molecules and Men.* Seattle: University of Washington Press, p. 9.

4. See, for example, Wolpert, L. (1991). *The Triumph of the Embryo.* Oxford: Oxford University Press. But see also Meinhardt, H. (1982). *Models of Biological Pattern Formation.* London: Academic Press.

5. Berg, P., & Singer, M. (1992). *Dealing with Genes.* Mill Valley, CA: University Science Books.

6. Granit, R. (1977). *The Purposive Brain.* Cambridge: MIT Press.

7. Crick, F. H. C. (1966). *Of Molecules and Men.*

8. Mayr, E. (1988). *Toward a New Philosophy of Biology*. Cambridge: Harvard University Press, p. 45.

9. Berg, P., & Singer, M. (1992). *Dealing with Genes*.

10. Lewontin, R. C. (1992). The dream of the human genome. *New York Review*, May 28, pp. 31–40.

11. Fox Keller, E. (1983). *A Feeling for the Organism*. New York: Freeman.

12. von Bertalanffy, L. (1968). *General System Theory*. New York: Braziller.

13. In an excellent article, Gene Yates raises the question, in what sense genes can be said to "specify" an organism. See Yates, F. E. (1993). Self-organizing systems. In: C. A. R. Boyd & D. Noble (Eds.). *The Logic of Life*. Oxford: Oxford University Press.

14. Wolpert, L. (1991). *The Triumph of the Embryo*.

15. Blaedel, N. (1988). *Harmony and Unity. The Life of Niels Bohr*. Berlin: Spring-Verlag.

16. I named this the Bohr effect in Kelso, J. A. S. (1991). Anticipatory dynamical systems, intrinsic pattern dynamics and skill learning. *Human Movement Science, 10*, 93–111.

17. Entire books have been written on these topics. See for example, Kots, Y. A. (1977). *The Organization of Voluntary Movement*. New York: Plenum Press; and Wise, S. P. (Ed.). (1987). *Higher Brain Functions*. New York: Academic Press.

18. Kelso, J. A. S. (1991). Anticipatory dynamical systems ...

19. For review, see Kelso, J. A, S., & Wallace, S. A. (1978). Conscious mechanisms in movement. In: G. E. Stelmach (Ed.). *Information Processing and Motor Control*. New York: Academic Press, pp. 79–116.

20. Deecke, L. Scheid, P., & Kornhuber, H. H. (1969) Distribution of readiness potential, premotion positivity and motor potential of the human cerebral cortex preceding voluntary finger movements. *Experimental Brain Research, 7*, 158–168. For a good review, see Deecke, L. (1990). Electrophysiological correlates of movement initiation. *Reviews of Neurology, 146*, 612–619.

21. Allen, G., & Tsukahara, N. (1974). Cerebrocerebellar communication systems. *Physiological Reviews, 54*, 957–1006.

22. Wiesendanger, M. (1981). Organization of secondary motor areas of cerebral cortex. In: V. B. Brooks (Ed.). *Handbook of Physiology, The Nervous System, Motor Control*. Washington, DC: American Physiology Society, pp. 1121–1147.

23. Brinkman, C., & Porter, R. (1983). Supplementary motor area and premotor area of monkey cerebral cortex: Functional organization and activities of single neurons during performance of a learned movement. In: J. E. Desmedt (Ed.). *Motor Control in Health and Disease*, New York: Raven Press.

24. Roland, P. E., Larsen, B., Lassen, N. A., & Shihøj, E. (1980). Supplementary and other cortical areas in organization of voluntary movements in man. *Journal of Neurophysiology, 43*, 118–136.

25. Goldenberg, G., Wimmer, A., Holzner, F., & Wessely, P. (1985). Apraxia of the left limbs in a case of callosal disconnection: The contribution of medial frontal lobe damage. *Cortex, 21*, 135–148.

26. Lang, W., Obrig, H., Lindinger, G., Cheyne, D., & Deecke, L. (1990). Supplementary motor area activation while tapping bimanually different rhythms in musicians. *Experimental Brain Research, 79*, 504–514; and Deecke, L. (1990). Personal communication.

27. Deecke, L (1990). See ref. note 20.

28. Pattee, H. H. (1972). Laws and constraints, symbols and language. In: C. H. Waddington (Ed.). *Towards a Theoretical Biology*. Chicago: Aldine; and (1977). Dynamic and linguistic modes of complex systems. *International Journal of General Systems, 3,* 259–266.

29. Kugler, P. N., Kelso, J. A. S., & Turvey, M. T. (1982). On the control and coordination of naturally developing systems. In: J. A. S. Kelso & J. E. Clark (Eds.). *The Development of Movement Control and Coordination.* New York: Wiley, pp. 5–78. See also Kugler, P. N., & Turvey, M. T. (1987). *Information, Natural Law and the Self-Assembly of Rhythmic Movement* Chapter 13.

30. Kelso, J. A. S., Scholz, J. P., & Schöner, G. (1988). Dynamics governs switching among patterns of coordination in biological movement. *Physics Letters, A, 134,* 8–12; Schöner, G., & Kelso, J. A. S. (1988). A dynamic pattern theory of behavioral change. *Journal of Theoretical Biology, 135,* 501–524; and Scholz, J. P., & Kelso, J. A. S. (1990). Intentional switching between patterns of bimanual coordination is dependent on the intrinsic dynamics of the patterns. *Journal of Motor Behavior, 22,* 198–124.

31. Dretske, F. (1988). *Explaining Behavior.* Cambridge: MIT Press.

32. Deneubourg, J. L. (1977). Application de l'ordre par fluctuations à la description de certain ètapes de la construction du nid chez les termites. *Social Insects, 24,* 117–130. See also Deneubourg, J. L, & Goss, S. (1990). Collective patterns and decision making. *Ecology, Ethology and Evolution, 1,* 295–311. A nice summary is also presented in Kugler, P. N., & Turvey, M. T. (1987). *Information, Natural Law and the Self-Assembly of Rhythmic Movement.*

33. Kugler, P. N., Shaw, R. E., Vincente, K. J., & Kinsella-Shaw, J. (1990). Inquiry into intentional systems. I. Issues in ecological physics. *Psychological Research, 52,* 98–121.

34. Thom, R. (1990). *Semiophysics: A Sketch.* Redwood City, CA: Addison-Wesley.

35. This statement is not to deny the laws of electromagnetism, only to stress that the couplings in intentional systems are informationally meaningful. See also Kelso, J. A. S. (1994). The informational character of self-organized coordination dynamics. *Human Movement Science, 13,* 393–413.

36. Beck, F., & Eccles, J. C. (1992). Quantum aspects of brain activity and the role of consciousness. *Proceedings of the National Academy of Sciences, 89,* 11357–11361.

37. Stapp, H. P. (1993). *Mind, Matter and Quantum Mechanics.* Berlin: Springer-Verlag.

38. Sherrington, C. S. (1906). *The Integrative Action of the Nervous System.* New Haven, CT: Yale University Press.

39. Granit, R. (1977). *The Purposive Brain.*

CHAPTER 6

1. Kandel, E. R., & Hawkins, R. D. (1992). The biological basis of learning and individuality. *Scientific American, 267,* 78–86.

2. Kandel, E. R., & Hawkins, R. D. (1992). The biological basis of learning and individuality. See also Thompson, R. F. (1988). The neural basis of basic associative learning of discrete behavioral responses. *Trends in Neurosciences, 11,* 152–155.

3. Abbott, L. F. (1990). Learning in neural network memories. *Network, 1,* 105–122.

4. Schmidt, R. A. (1975). A schema theory of discrete motor skill learning. *Psychological Review, 82,* 225–260.

5. Adams, J. A. (1971). A closed-loop theory of motor learning. *Journal of Motor Behavior, 3,* 111–149.

6. Bower, B. (1992). Brain clues to energy-efficient learning. *Science News, 141,* 215.

7. Fitts, P. M. (1964). Perceptual-motor skill learning. In: A. W. Melton (Ed.). *Categories of Human Learning*. New York: Academic Press, pp. 243–255.

8. Tinbergen, N. (1951). *The Study of Instinct*. New York: Oxford University Press, p. 6. Key experimental results were first reported in Kelso J. A. S. (1990). Phase transitions: Foundations of behavior. In: H. Haken, & M. Stadler (Eds.). *Synergetics of Cognition*. Berlin: Springer-Verlag; and Zanone P. G., & Kelso, J. A. S. (1992). The evolution of behavioral attractors with learning: Nonequilibrium phase transitions. *Journal of Experimental Psychology: Human Perception and Performance, 18*(2), 403–421. Recent reviews including discussion of new experimental work are Schöner, G., Zanone, P. G., & Kelso, J. A. S. (1992). Learning as a change of coordination dynamics: Theory and experiment. *Journal of Motor Behavior, 24,* 29–48; and Zanone, P. G., & Kelso, J. A. S. (1994). The coordination dynamics of learning: Theoretical structure and experimental agenda. In: S. Swinnen, H. Heuer, J. Massion, & P. Casaer (Eds.). *Interlimb Coordination: Neural, Dynamical, and Cognitive Constraints*. New York: Academic Press, pp. 461–590.

9. These theoretical ideas were first expressed in a pair of papers. Schöner, G., & Kelso, J. A. S. (1988). A synergetic theory of environmentally specified and learned patterns of movement coordination. I. Relative phase dynamics. *Biological Cybernetics, 58,* 71–80; and II. Component oscillator dynamics. *Biological Cybernetics, 58,* 81–89. See also Schöner, G. (1989). Learning and recall in a dynamic theory of coordination patterns. *Biological Cybernetics, 62,* 39–54.

10. Tuller, B, & Kelso, J. A. S. (1985). *Coordination in Normal and Split-Brain Patients*. Boston: Psychonomic Society; and Tuller, B., & Kelso, J. A. S. (1989). Environmentally-specified patterns of movement coordination in normal and split-brain subjects. *Experimental Brain Research, 75,* 306–316.

11. Yamanishi, J., Kawato, M., & Suzuki, R. (1980). Two coupled oscillators as a model for the coordinated finger tapping by both hands. *Biological Cybernetics, 37,* 219–225.

12. Schöner, G., Zanone, P. G., & Kelso, J. A. S. (1992). Learning as a change of coordination dynamics.

13. Katz, D. (1951). *Gestalt Psychology*. London: Methuen.

14. Vihman, M. M. (1991). Ontogeny of phonetic gestures: Speech production. In: I. G. Mattingly & M. Studdert-Kennedy (Eds.). *Modularity and the Motor Theory of Speech Production*. Hillsdale, NJ: Erlbaum, pp. 69–104.

15. van Geert, P. (1991). A dynamic systems model of cognitive and language growth. *Psychological Review, 98,* 3–53.

16. van der Maas, H. L. J., & Molenaar, P. C. M. (1992). Stagewise cognitive development: An application of catastrophe theory. *Psychological Review, 99,* 395–417.

17. van der Maas, H. L. J., & Molenaar, P. C. M. (1992). Stagewise cognitive development, p. 398.

18. van der Maas, H. L. J., & Molenaar, P. C. M. (1992). Stagewise cognitive development, p. 405.

19. For example, Thelen, E. (1985). Developmental origins of motor coordination: Leg movements in human infants. *Developmental Psychobiology, 18,* 1–22; Thelen, E. (1988). Dynamical approaches to the development of behavior. In: J. A. S. Kelso, A. J. Mandell, & M. F. Shlesinger (Eds.). *Dynamic Patterns in Complex Systems*. Singapore: World Scientific, pp. 368–369; Clark, J. E., Whitall, J., & Phillips, S. J. (1988). Human interlimb coordination: The first six months of independent walking. *Developmental Psychobiology, 21,* 445–456; and Roberton, M. A. (1993). New ways to think about old questions. In: L. B. Smith & E. Thelen (Eds.). *A Dynamic Systems Approach to Development*. Cambridge: MIT Press, pp. 95–117.

20. For an excellent review, see Thelen, E., & Ulrich, B. D. (1991). Hidden skills. *Monographs of the Society for Research in Child Development, 56,* 1–98.

21. Thelen, E. (1988) Dynamical approaches to the development of behavior. In: Kelso, J. A. S. et al. (Eds.).

22. Wolff, P. H. (1991). How are new behavioral forms and functions introduced during ontogenesis? Commentary on Thelen and Ulrich, *Monographs of the Society for Research in Child Development, 56*, R103.

23. Barlow, N. (1969). *The Autobiography of Charles Darwin*. New York: Norton, p. 89.

24. Newman, S. A. (1992). Generic physical mechanisms of morphogenesis and pattern formation as determinants in the evolution of multicellular organization. In: J. Mittenthal & A. Baskin (Eds.). *Principles of Organization in Organisms*. Reading, MA: Addison-Wesley.

25. Thompson, D. (1942). *On Growth and Form*, 2nd ed. Cambridge: Cambridge University Press.

26. See Kaufman, S. A. (1993). *Origins of Order: Self-Organization and Selection in Evolution*. Oxford: Oxford University Press.

27. Kaufman, S. A. (1992). *Origins of Order*, p. 538.

28. Mayr, E. (1982). Questions concerning speciation. *Nature, 296*, 309.

29. Williamson, P. G. (1981). Paleontological documentation of speciation in cenozoic molluscs from Turkana basin. *Nature, 294*, 214.

30. For example, Bolles, R. C. (1970). Species-specific defense reactions and avoidance learning. *Psychological Review, 77*, 32–48; Garcia, J. & R. Garcia (1985). Evolution of learning mechanisms. In B. L. Hammond (Ed.) *Psychology and Learning: The master lecture series* (pp. 187–243). Washington, DC: American Psychological Association. For recent review, Timberlake, W. (1993). Behavior systems and reinforcement: An integrative approach. *Journal of the Experimental Analysis of Behavior, 60*, 105–128.

31. Barnes, D. M. (1986). Lessons from snails and other models. *Science, 231*, 1246–1249.

CHAPTER 7

1. Gibson, J. J. (1950). *The Perception of the Visual World*. Boston: Houghton Mifflin; (1979). *The Ecological Approach to Visual Perception*. Boston: Houghton Mifflin; Turvey, M. T., & Kugler, P. N. (1984). An ecological approach to perception and action. In: H. T. A. Whiting (Ed.). *Human Motor Actions: Bernstein Reassessed*. Amsterdam: North-Holland; Turvey, M. T. (1977). Contrasting orientations to the theory of visual information processing. *Psychological Review, 84*, 67–88; Turvey, M. T., Shaw, R. E., Reed, E. S., & Mace, W. M. (1981). Ecological laws of perceiving and acting. In reply to Fodor and Pylyshyn (1981). *Cognition, 9*, 237–304; Lee, D. N. (1976). A theory of visual control of braking based on information about time to contact. *Perception, 5*, 437–459; and Warren, W. H., & Shaw, R. E. (1985) (Eds.). *Persistence and Change*. Hillsdale, NJ: Erlbaum.

2. Köhler. W. (1940). *Dynamics in Psychology*. New York: Liveright; (1947). *Gestalt Psychology*. New York: Liveright; Koffka, K. (1935). *Principles of Gestalt Psychology*. London: Routledge & Kegan Paul; Stadler, M., & Kruse, P. (1986). Gestalttheorie und Theorie der Selbstorganisation. *Gestalt Theory, 8*, 75–98; Kanizsa, G., & Luccio, R. (1990). The phenomenology of autonomous order formation in perception. In: H. Haken & M. Stadler (Eds.). *Synergetics of Cognition*. Berlin: Springer-Verlag; Kruse, P., & Stadler, M. (Eds.). (1994), *Multistability in Cognition*. Berlin: Springer-Verlag. For recent reviews and analyses, see Rock, I., & Palmer, S. (1990). The legacy of Gestalt psychology. *Scientific American, 267*, 84–90; and Epstein, W. (1988). Has the time come to rehabilitate Gestalt theory? *Psychological Research, 50*, 2–6.

3. Epstein, W. (1988). Has the time come to rehabilitate Gestalt theory?

4. Epstein, W. (1988). Has the time come...; and Rock, I., & Palmer, S. (1990). The legacy of Gestalt psychology.

5. Rota, G.-C. (1986). In memoriam of Stan Ulam: The barrier of meaning. *Physica, 22D,* 1–3. See also Ulam, S. M. (1991) *Adventures of a Mathematician.* Berkeley: University of California Press.

6. A very readable account of Gibson's theory of information is Michaels, C. F., & Carello, C. (1981). *Direct Perception.* Englewood Cliffs, NJ: Prentice-Hall.

7. For a good overview of the possibilities, consult Warren, W., & Shaw, R. E. (1985). *Persistence and Change.*

8. Lee, D. N., & Reddish, P. E. (1981). Plummeting gannets: A paradigm of ecological optics. *Nature, 293,* 293–294.

9. Raviv, D., Orser, D., & Albus, J. S. (1992). On logarithmic retinae. National Institute of Standards and Technology Report (NISTIR #4807).

10. Wagner, H. (1982). Flow-field variables trigger landing in flies. *Nature, 297,* 147–148.

11. Excellent reviews of time-to-contact applications in skilled activities may be found in Lee, D. N., & Young, D. S. (1985). Visual timing of interoceptive action. In: D. Ingle, M. Jeannerod, & D. N. Lee (Eds.). *Brain Mechanisms and Spatial Vision.* Dordrecht, The Netherlands: Martinus Nijhoff; and Bootsma, R. (1988). *The Timing of Rapid Interoceptive Actions: Perception-Action Coupling in the Control and Acquisition of Skill.* Amsterdam: Free University Press.

12. In fact, Wagner found that the relative retinal expansion velocity (RREV) must exceed a threshold of $\sim 13s^{-1}$ to trigger landing (the onset of deceleration) in the household fly.

13. Gregor Schöner and I implemented this model in a research proposal submitted to the U.S. Office of Naval Research (1988) but did not publish it. For a similar and well-developed approach to the moving room paradigm, see Schöner, G. (1991). Dynamic theory of action-perception patterns: The "moving room" paradigm. *Biological Cybernetics, 64,* 455–462.

14. Raviv, D. (1991). *Invariants in Visual Motion.* National Institute of Standards and Technology Report (NISTIR #4722).

15. Kaiser, M. K., & Phatak, A. V. (1993). Things that go bump in the light: On the optical specification of contact severity. *Journal of Experimental Psychology: Human Perception and Performance, 19,* 194–202.

16. See also Turvey, M. T., & Kugler, P. N. (1984). An ecological approach to perception and action.

17. Kelso, J. A. S., & Kay, B. A. (1987). Information and control: A macroscopic analysis of perception action coupling. In: H. Heuer & A. F. Sanders (Eds.). *Perspectives on Perception and Action.* Hillsdale, N.J.: Erlbaum, pp. 3–32.

18. Lishman, J. R., & Lee, D. N. (1973). The autonomy of visual kinesthesis. *Perception, 2,* 287–294.

19. See page 91 of Lee, D. N., & Lishman, J. R. (1975). Visual proprioceptive control of stance, *Journal of Human Movement Studies, 1,* 87–95.

20. See especially Schöner, G. (1991). Dynamic theory of action-perception patterns.

21. Kelso J. A. S., DelColle, J, D., & Schöner, G. (1990). Action-perception as a pattern formation process...

22. This view is spelled out in Turvey, M. T., & Kugler, P. N. (1984). An ecological approach..., and other writings of the Connecticut school, e.g., Michaels, C. F., & Carello, C. (1981). *Direct Perception.*

23. Wang, Y., & Frost, B. J. (1992). Time to collision is signalled by neurons in the nucleus rotundus of pigeons. *Nature, 356,* 236–237.

24. Tanaka, K., & Saito, H. (1989). Analysis of motion of the visual field by direction, expansion/contraction and rotation cells clustered in the dorsal part of the medial superior temporal area of the macaque monkey. *Journal of Neurophysiology, 62,* 626–641; Orban, G. A., Lagae, L., Verri, A., Raiguel, S., Xiao, D., Maes, H., & Torre, V. (1992). First-order analysis of optic flow in monkey brain. *Proceedings of the National Academy of Sciences, 89,* 2595–2599; and Duffy, C. J., & Wurtz, R. H. (1990). Organization of optic flow sensitive receptive fields in cortical area MST. *Society for Neuroscience Abstracts, 16,* 6.

25. Gibson, J. J. (1959). Perception as a function of stimulation. In: Koch, S. (Ed.). *Psychology: A Study of Science.* New York: McGraw-Hill, p. 465.

26. For an excellent analysis, see Epstein, W. (1993). The representational framework in perceptual theory. *Perception and Psychophysics, 53,* 704–709.

27. Warren, W. H., Jr. (1984). Perceiving affordances: Visual guidance of stairclimbing. *Journal of Experimental Psychology: Human Perception and Performance, 10,* 683–703.

28. But actually this hypothesized bifurcation was not studied as such.

29. For example, Mark, L. S., Balliett, J. A., Craver, K. D., Douglas, S. D., & Fox, T. (1990). What an actor must do in order to perceive the affordance for sitting. *Ecological Psychology, 2,* 325–366.

30. von Holst, E. (1939/1973). Relative coordination as a phenomenon and as a method of analysis of central nervous function. In: R. Martin (Ed.). *The Collected Papers of Erich von Holst.* Miami: University of Miami Press, pp. 33–135.

31. Gibson, E. J. (1988). Exploratory behavior in the development of perceiving, acting and the acquiring of knowledge. *Annual Review of Psychology, 39,* 1–41.

32. Kruse, P., & Stadler, M. (1990). Stability and instability in cognitive systems: Multistability, suggestion and psychosomatic interaction. In: H. Haken & M. Stadler (Eds.). *Synergetics of Cognition.* Berlin: Springer-Verlag, pp. 201–215.

33. Köhler, W. (1947). *Gestalt Psychology.*

34. Harnad, S. (1987). *Categorical Perception: The Groundwork of Cognition.* Cambridge: Cambridge University Press.

35. Haken, H., Kelso, J. A. S., Fuchs, A., & Pandya, A. (1990). Dynamic pattern recognition of coordinated biological motion. *Neural Networks, 3,* 395–401.

36. Glass, L. (1960). Moiré effect from random dots. *Nature, 223,* 578–580; and Glass, L., & Perez, R. (1973). Perception of random dot interference patterns. *Nature, 246,* 3603–362.

37. Gallant, J. L., Braun, J., & Van Essen, D. C. (1993). Selectivity for polar, hyperbolic and cartesian gratings in macaque visual cortex. *Science, 259,* 100–103.

38. Motter, B. C., Steinmetz, M. A., Duffy, C. J., & Mountcastle, V. B. (1990). Functional properties of parietal visual neurons: Mechanisms of directionality along a single axis. *Journal of Neuroscience, 7,* 154–176.

39. For a review, see Kelso, J. A. S., Case, P., Holroyd, T., Horvath, E., Rączaszek, J., Tuller, B., & Ding, M. A. (1994). Multistability and metastability in perceptual and brain dynamics. In: M. Stadler & P. Kruse (Eds.). *Multistability in Cognition.* Springer Series in Synergetics. Berlin: Springer-Verlag.

40. von Schiller, P. (1933). Stroboskopische alternativbewegungen. *Psychologische Forschung, 17,* 179–214; and Ramachandran, V. S., & Anstis, S. M. (1985). Perceptual organization in multistable apparent motion. *Perception 14,* 135–143.

41. Attneave, F. (1971). Multistability in perception. *Scientific American, 225*, 62–71.

42. Hock, H. S., Kelso, J. A. S., & Schöner, G. (1993). Bistability, hysteresis and loss of temporal stability in the perceptual organization of apparent motion. *Journal of Experimental Psychology: Human Perception and Performance, 19* , 1, 63–80.

43. Held, R., & Richards, W. (Eds.) (1976). The organization of perceptual systems. In: *Perception: Mechanisms and Models*. San Francisco: Freeman.

44. Grosof, D. H., Shapley, R. M., & Hawken, M. J. (1992). Macaque V1 neurons can signal "illusory" contours. *Nature, 365*, 550–552.

45. Tuller, B., Case, P., Ding, M., & Kelso (1994). The nonlinear dynamics of categorical perception. *Journal of Experimental Psychology: Human Perception & Performance, 20(1)*, 3–16.

46. Tuller, B. et al. (1994). The nonlinear dynamics of categorical perception.

47. Ditzinger, T., & Haken, H. (1989). Oscillations in the perception of ambiguous patterns: A model based on synergetics. *Biological Cybernetics, 61*, 279–287; (1990). The impact of fluctuations on the recognition of ambiguous patterns. *Biological Cybernetics, 63*, 453–456; and Haken, H. (1990). Synergetics as a tool for the conceptualization and mathematization of cognition and behavior—How far can we go? In: H. Haken & M. Stadler (Eds.), *Synergetics of Cognition*. Berlin: Springer-Verlag, pp. 2–31.

48. Helson, H. (1964). *Adaptation Level Theory: An Experimental and Systematic Approach to Behavior*. New York: Harper & Row.

49. Tuller, B. et al. (1994). The nonlinear dynamics ...

50. Kelso, J. A. S., et al. (1994). Multistability and metastability in perceptual and brain dynamics ... Tuller, B., & Kelso, J. A. S. (1994). Speech dynamics. In: F. Bell-Berti & L. J. Raphael (Eds.). *Studies in Speech Production: A Festschrift for Katherine Safford Harris*. New York: American Institute of Physics.

51. Kawamoto, A. H., & Anderson, J. A. (1985). A neural network model of multistable perception. *Acta Psychologica, 59*, 35–65.

52. Williams, D., Phillips, G., & Sekuler, R. (1986). Hysteresis in the perception of motion direction as evidence for neural cooperativity. *Nature, 324*, 253–255.

53. Carpenter, G. A., & Grossberg, S. (Eds.). (1992). *Pattern Recognition by Self-Organizing Neural Networks*. Cambridge: MIT Press.

54. Haken, H. (1979). Pattern formation and pattern recognition—An attempt at a synthesis. In: H. Haken (Ed.). *Pattern Formation by Dynamic Systems and Pattern Recognition*. Berlin: Springer-Verlag.

55. Ditzinger, T., & Haken, H. (1989). Oscillations in the perception of ambiguous patterns; and (1990). The impact of fluctuations on the recognition of ambiguous patterns.

56. For an excellent review, see Moss, F. (1992). *Stochastic Resonance: From the Ice Ages to the Monkey's Ear*. St. Louis: University of Missouri at St. Louis.

57. Douglass, J. K., Wilkens, L., Pantazelou, E., & Moss, F. (1993). Noise enhancement of information transfer in crayfish mechanoreceptors by stochastic resonance. *Nature, 365*, 337–340; and Longtin, A., Bulsara, A., & Moss, F. (1991). Time-interval sequences in bistable systems and the noise induced transmission of information by sensory neurons. *Physical Review Letters, 67*, 656–659.

58. Chialvo, D., & Apkarian, A. V. (1993). Modulated noisy biological dynamics: Three examples. *Journal of Statistical Physics, 70*, 375–391.

59. Fisher, G. H. (1967). Measuring ambiguity. *American Journal of Psychology, 80*, 541–547.

60. Michel, S. Geusz, M. E. Zaritsky, J. J., & Block, G. D. (1993). Circadian rhythm in membrane conductance expressed in isolated neurons. *Nature, 259,* 239–241.

61. Kawamoto, A. H., & Anderson, J. A. (1985). A neural network model . . .

62. See also Kohonen, T. (1988). An introduction to neural computing. *Neural Networks, 1,* 3–16.

63. See Ditzinger, T., & Haken, H. (1989 and 1990) On ambiguous patterns. See also Haken, H., & Fuchs, A. (1988). Pattern recognition and pattern formation as dual processes. In: J. A. S. Kelso, A. J. Mandell, & M. F. Shlesinger (Eds.). *Dynamic Patterns in Complex Systems.* Singapore: World Scientific.

64. This is an image inspired by C. H. Waddington's epigenetic landscape, and in particular, artist Yolanda Sonnabend's rendition of it that I was privileged to see at the fourth Waddington conference "Significance and Form in Nature and Art," Perugia, Italy, May 1993.

65. Kawamoto, A. H., & Anderson, J. A. (1985). A neural network model . . .

66. Reported in Kelso, J. A. S., et al. (1994). Multistability and metastability in perceptual and brain dynamics.

67. Borsellino, A., DeMarco, A., Allazetta, A., Rinesi, S., & Bartolini, B. (1972). Reversal time distribution in the perception of visual ambiguous stimuli. *Kybernetik, 10,* 139–144.

68. For example, Crick, F., & Koch, C. (1990). Towards a neurobiological theory of consciousness. *Seminars in the Neurosciences, 2,* 263–275; and (1992). The problem of consciousness. *Scientific American, 267,* 152–160.

69. Pomeau, Y., & Manneville, P. (1980). Intermittent transitions to turbulence in dissipative dynamical systems. *Communications in Mathematical Physics, 74,* 189.

70. Epstein, W. (1988). Has the time come . . .; and (1993). The representation framework in perceptual theory. *Perception and Psychophysics, 53,* 704–709.

71. Marr, D. (1982). *Vision.* New York: Freeman.

72. Rock, I., & Palmer, S. (1990). The legacy of Gestalt psychology; Stadler, M., & Kruse, P. (1986). Gestalttheorie und Theorie der Selbstorganisation; and Rumelhart, D. E., & McClelland, J. L. (Eds.). *Parallel Distributed Processing.* Cambridge: MIT Press.

CHAPTER 8

1. Bunge, M. (1980). From neuron to behavior and mentation: An exercise in levelmanship. In: H. M. Pinsker & W. D. Willis, Jr. (Eds.). *Information Processing in the Nervous System.* New York: Raven Press.

2. Rose, S. P. R. (1980). Can the neurosciences explain the mind? *Trends in Neurosciences, 3,* 1–4.

3. Edelman, G. M. (1978). Group selection and phasic re-entrant signaling: A theory of higher brain function. In: G. M. Edelman & V. B. Mountcastle (Eds.). *The Mindful Brain,* Cambridge: MIT Press.

4. Kelso, J. A. S., & Tuller, B. (1984). A dynamical basis for action systems. In: M. S. Gazzaniga (Ed.). *Handbook of Cognitive Neuroscience.* New York: Plenum Press. This work was first published in *Haskins Laboratories Status Report,* 1981.

5. Reeke, G. N., Finkel, L. H., Sporns, O., & Edelman, G. M. (1989). Synthetic neural modelling: A multilevel approach to the analysis of brain complexity. In: G. M. Edelman, W. E. Gall, & W. M. Cowan (Eds.). *Signal and Sense: Local and Global Order in Perceptual Maps.* New York: Wiley.

6. Reeke, G. N. et al. (1989). Synthetic neural modelling, p. 37.

7. Consider, for example, a recent approach called synthetic neural modeling by Edelman and colleagues. (1989). *Signal and Sense*. In at least two of the main modules of this model (organization of cortical maps and their integration) "all anatomical connections are generated at the start of simulations and once generated they are never changed" (p. 12; see also p. 20 where the anatomical connections are fixed). Thus, the essential ability of the CNS to flexibly recruit or disengage different anatomical components when the environment, task, or phenotype changes is ignored. In numerous places the modelers recognize that alternatives to the mechanisms they propose are possible and that a vast amount of simplification is necessary. Yet little or no rationale or motivation is provided for the choices behind these simplifications. The authors remark that one of their automatons displays a primitive form of "attention," but state further that "there is no implication that this system simulates the kind of attention seen in conscious animals" (p. 62). Similarly, in their module for reaching behavior they note "that unlike the case of real animals, the motor cortex is entirely autonomous, that is, unregulated by other regions" (p. 65). Furthermore, although "the components resemble those in real brains, this is not an explicit model of the cerebellum" (p. 66). The claimed biological reality, it seems, gives way to the constraints of simulation. None of these or other criticisms of synthetic neural modeling are particularly damning in themselves. It is only against ambitious claims of a new more complete approach to the *entire nervous system* that they must be judged.

8. Kelso, J. A. S., & Tuller, B. (1984). A dynamical basis for action systems.

9. An excellent source that I draw on here is Steward, O. (1989). *Principles of Cellular, Molecular and Developmental Neuroscience*. Berlin: Springer-Verlag.

10. Agnati, L. F., Bjelke, B., & Fuxe, K. (1992). Volume transmission in the brain. *American Scientist, 80*, 362–373.

11. Cajal, R. (1911). *Histologie du Systeme Nerveux de l'Homme et des Vertébres*. Paris: Maloine.

12. This section and the next are based on a trilogy of papers by Larry Liebovitch and colleagues. Liebovitch, L., & Koniarek, J. P. (1992). Ion channel kinetics. *IEEE Engineering in Biology and Medicine*, June, 53–46; Liebovitch L. (1994). Single channels: From Markovian to fractal models. In: D. P. Zipes & J. Jalife (Eds.). *Cardiac Electrophysiology: From Cell to Bedside*, 2nd edn. Philadelphia: W. B. Saunders; and Liebovitch, L., & Tóth, T. I. (1991). A model of ion channel kinetics using deterministic chaotic rather than stochastic processes. *Journal of Theoretical Biology, 148*, 243–267.

13. Cited in Frauenfelder, H., Sligar, S. G., & Wolynes, P. G. (1991). The energy landscapes and motions of proteins. *Science, 254*, 1598–1603.

14. Frauenfelder, H. et al. (1991). The energy landscapes and motions of proteins.

15. Liebovitch, L., & Koniarek, J. (1992). Ion channel kinetics.

16. Mandelbrot, B. B. (1982). *The Fractal Geometry of Nature*. San Francisco: Freeman.

17. Hodgkin, A. L., & Huxley, A. F. (1952). A quantitative description of membrane current and its application to conductance and excitation in nerve. *Journal of Physiology, 117*, 500–544.

18. Liebovitch, L., & Tóth, T. I. (1991). A model of ion channel kinetics using deterministic chaotic rather than stochastic processes.

19. Pellionisz, A. (1990). Neural geometry: Towards a fractal model of neurons. In: R. M. Cotterill (Ed.). *Models of Brain Function*. Cambridge: Cambridge University Press.

20. Abbott, L. F., & LeMasson, G. (1993). Analysis of neuron models with dynamically regulated conductances. *Neural Computation, 5*, 823–842.

21. Strassberg, A. F., & DeFelice, L. J. (1993). Limitations of the Hodgkin-Huxley formalism: Effects of single channel kinetics on transmembrane voltage dynamics. *Neural Computation, 5*, 843–855.

22. Peng, C. K. et al. (1993). Long-range correlations in nucleotide sequences. *Nature, 356,* 168–170.

23. Peng, C. K. et al. (1993). Long-range anticorrelations and non-Gaussian noise in the heart-beat. *Physical Review Letters, 70,* 1343–1346,

24. Crossman, E. R. F. W. (1959). A theory of the acquisition of speed-skill. *Ergonomics, 2,* 153–166.

25. Warren, R. M., & Gregory, R. (1958). An auditory analogue of the visual reversible figure. *American Journal of Psychology, 71,* 612–613.

26. Tuller, B., Kelso, J. A. S., & Ding, M. (1993). *A Note on Fractal Time in the Brain.* Un-published observations; Ding, M., Tuller, B. & Kelso, J. A. S., Characterizing the dynamics of auditory perception. *Chaos: An Interdisciplinary Journal of Nonlinear Science* (in press).

27. Wing, A. M., & Kristofferson, A. B. (1973). Response delays and the timing of discrete motor responses. *Perception and Psychophysics, 14,* 5–12.

28. For an excellent review, see Grillner, S., Wallén, P., & Brodin, L. (1991). Neuronal network generating locomotor behavior in lamprey. Circuitry, transmitters, membrane properties and simulation. *Annual Review of Neuroscience, 14,* 169–199.

29. Wallén, P., & Grillner, S. (1985). The effect of current passage on NMDA induced TTX resistant membrane potential oscillations in lamprey neurons active during locomotion. *Neuroscience Letters, 56,* 87–93.

30. Selverston, A. I., & Moulins, M. (Eds.). (1987). *The Crustacean Stomatogastric System.* Berlin: Springer-Verlag.

31. Moulins, M., & Nagy, F. (1985). Extrinsic inputs and flexibility in the motor output of the lobster pyloric neural network. In: A. I. Selverston (Ed.). *Model Neural Networks and Behavior.* New York: Plenum Press.

32. Hodgkin, A. L. (1977). Chance and design in electrophysiology: An informal account of certain experiments on nerve carried out between 1934 and 1952. In: A. L. Hodgkin et al. *The Pursuit of Nature.* Cambridge: Cambridge University Press.

33. For a good review see Aihara, K., & Matsumoto, G. (1986). Chaotic oscillations and bifurcations in squid giant axons. In: A. V. Holden (Ed.). *Chaos.* Princeton, NJ: Princeton University Press; and Glass, L., & Mackey, M. C. (1988) *From Clocks to Chaos.* Princeton, NJ: Princeton University Press, for many additional references.

34. Hayashi, H., & Ishizuka, S. (1992). Chaotic nature of bursting discharges in the *Onchidium* pacemaker neuron. *Journal of Theoretical Biology, 156,* 269–291.

35. Mpitsos, G. J., Burton, R. M. Creech, H. C., & Soinala, S. O. (1988). Evidence for chaos in spike trains of neurons that generate rhythmic motor patterns. *Brain Research Bulletin, 21,* 529–538.

36. For example, as articulated by Selverston, A., Moulins, M. (1985). Oscillatory neural net-works. *Annual Review of Physiology, 47,* 29–48.

37. See Selverston, A. I. (1988). Switching among functional states by means of neuromodula-tors in the lobster stomatogastric ganglion. *Experientia, 44,* 337; and Pearson, K. G. (1985). Neuronal circuits for patterning motor activity in invertebrates. In: M. J. Cohen & F. Strum-wasser (Eds.). *Comparative Neurobiology: Modes of Communication in the Nervous System* (p. 237). New York: John Wiley.

38. See Katz, P. S., & Harris-Warrick, R. M. (1990). Actions of identified neuromodulatory neurons in a simple motor system. *Trends in Neurosciences, 13,* 367–373; and Marder, E. (1988). Modulating a neuronal network. *Nature, 335,* 296–297.

39. Kelso, J. A. S., Tuller B., Vatikiotis-Bateson, E., & Fowler, C. A. (1984). Functionally-specific articulatory cooperation following jaw perturbations during speech . . .

40. Selverston, A. I. (1988). Switching among functional states . . .

41. Marder, E. (1988). Modulating a neuronal network.

42. Harris-Warwick, R. M. G. Johnson (1989). Motor pattern networks: Flexible foundations for rhythmic pattern production. In T. J. Carew & D. B. Kelley (Eds.) *Perspectives in Neural Systems and Behavior*. New York: Alan R. Liss.

43. Nagashino, H., & Kelso, J. A. S. (1991). Bifurcation of oscillatory solutions in a neural oscillator network model for phase transitions. *Proceedings of the 2nd Symposium on Nonlinear Theory and Its Applications*. Shikanoshima, pp. 119–122; and (1992). Phase transitions in oscillatory neural networks. *Science of Artificial Neural Networks, SPIE, 1710*, 279–287.

44. Cohen, M. A., Grossberg, S., & Pribe, C. A. (1993). Frequency-dependent phase transitions in the coordination of human bimanual tasks. Presented at the World Congress on Neural Networks, Portland, Oregon, July 11–15, 1993.

45. For at least one exception, see Llinas, R. R. (1988). The intrinsic electrophysiological properties of mammalian neurons: Insights into central nervous function. *Science, 242*, 1654–1664.

46. Llinas, R. R. (1988). The intrinsic electrophysiological properties of mammalian neurons.

47. Steriade, M. (1993). Oscillations in interacting thalamic and neocortical neurons. *NIPS, 8*, 111–116; and Steriade, M., McCormick, D. A., & Sejnowski, T. J. (1992). Thalamocortical oscillations in the sleeping and aroused brain. *Science, 262*, 679–685.

48. O'Keefe, J., & Nadel, L. (1978). *The Hippocampus as a Cognitive Map*. Oxford: Clarendon Press.

49. Cooke, J. E. (1991). Correlated activity in the CNS: A role on every timescale. *Trends in Neurosciences, 14*, 397–401.

50. Shatz, C. A. (1992). The developing brain. *Scientific American, 267*, 60–67.

51. Huerta, P. T., & Lisman, J. E. (1993). Heightened synaptic plasticity of hippocampal CA1 neurons during a cholinergically induced rhythmic state. *Nature, 364*, 723–725.

52. Stanton, P. K., & Sejnowski, T. J. (1989). Associative long-term depression in the hippocampus induced by Hebbian covariance. *Nature, 339*, 215–217.

53. Kocsis, B., & Vertes, R. P. (1992). Dorsal raphe neurons: Synchronous discharge with the theta rhythm of the hippocampus in the freely behaving rat. *Journal of Neurophysiology, 68*, 1463–1467.

54. Adey, W. R., Dunlop, C. W., & Hendrix, C. E. (1960). Hippocampal slow waves: Distribution and phase relationships in the course of approach learning. *Archives of Neurology, 3*, 74–90.

55. Gray, C. M., & Singer, W. (1989). Stimulus-specific neuronal oscillations in orientation columns of cat visual cortex. *Proceedings of the National Academy of Science, 86*, 1698–1702.

56. For recent review, see Engel, A. K., König, P., Kreiter, A. K., Schillen, T. B., & Singer, W. (1992). Temporal coding in the visual cortex: New vistas on integration in the nervous system. *Trends in Neurosciences, 15*, 218–226.

57. Schuster, H. G. (1991). Nonlinear dynamics and neuronal oscillations. In: H. G. Schuster (Ed.). *Nonlinear Dynamics and Neuronal Networks*. Weinheim: VCH.

58. Freeman, W. J., & van Dijk, B. W. (1987). Spatial patterns of visual cortical fast EEG during conditioned reflex in a rhesus monkey. *Brain Research, 422*, 267–276.

59. Bressler, S. L., Coppola, R., & Nakamura, R. (1993). Episodic multiregional cortical coherence at multiple frequencies during visual task performance. *Nature, 366,* 153–156.

CHAPTER 9

1. Hawking, S. W.(1992). *A Brief History of Time.* New York: Bantam.

2. Sherrington, C. S. (1940). *Man and His Nature.* Cambridge: Cambridge University Press.

3. Eccles, J. C. (1973). *The Understanding of the Brain.* New York: McGraw-Hill, p. 217.

4. Katchalsky, A. K. Rowland, V., & Blumenthal, R. (1974). Dynamic patterns of brain cell assemblies. *Neuroscience Research Program Bulletin, 12,* p. 152.

5. For a recent review and references, see Freeman, W. J. (1991). The physiology of perception. *Scientific American, 264,* 78–85.

6. Rössler, O. E. (1987). Chaos in coupled optimizers. In: S. H. Koslow, A. J. Mandell, & M. F. Shlesinger (Eds.). Perspectives in biological dynamics and theoretical medicine. *Annals of the New York Academy of Sciences, 504,* 229–240.

7. This work was first reported at a conference held at the Supercomputer Computations Research Institute, Tallahassee, Florida, April 3–5, 1991. Kelso, J. A. S. et al. (1991). Cooperative and critical phenomena in the human brain revealed by multiple SQUIDs. In: D. Duke & W. Pritchard (Eds.). *Measuring Chaos in the Human Brain.* Singapore: World Scientific, pp. 77–112. For more details see Kelso, J. A. S. et al. (1992). A phase transition in human brain and behavior. *Physics Letters A, 169,* 134–144; and Fuchs, A., Kelso, J. A. S., & Haken, H. (1992). Phase transitions in the human brain: Spatial mode dynamics. *International Journal of Bifurcation and Chaos, 2*(4), 917–939.

8. This kind of language shows up when scientists are quoted by science writers, as well as in more serious writings. See Crease, R. P. (1993). Biomedicine in the age of imaging. *Science, 261,* 554–561.

9. Roland, P. E. (1993). *Brain Activation.* New York: Wiley-Liss, p. 418.

10. Leyton, A. S. F., & Sherrington, C. S. (1917). Observations on the excitable cortex of the chimpanzee, orang-utan and gorilla. *Quarterly Journal of Experimental Physiology, 11,* 135–222.

11. Leyton, A. S. F., & Sherrington, C. S. (1917). Observations on the excitable cortex of the chimpanzee, orang-utan and gorilla. p. 137.

12. Fox, J. L. (1984). The brain's dynamic way of keeping in touch. *Science, 225,* 820–821.

13. Fox, J. L. (1984). The brain's dynamic way of keeping in touch. p. 820.

14. Edelman, G. M., & Finkel, L. H. (1984). Neuronal group selection in the cerebral cortex. In: G. M. Edelman, W. E. Gall, & W. M. Cowan (Eds.). *Dynamic Aspects of Neocortical Function.* New York: Wiley, pp. 653–695.

15. Kelso, J. A. S., Stelmach, G. E., & Wanamaker (1974). Behavioral and neurological parameters of the nerve compression block. *Journal of Motor Behavior, 6,* 179–190; and Kelso, J. A. S., Wallace, S. A., Stelmach, G. E., & Weitz, G. (1975). Sensory and motor impairment in the nerve compression block. *Quarterly Journal of Experimental Psychology, 27,* 123–129.

16. For review, see Taub, E., & Berman, A. J. (1968). Movement and learning in the absence of sensory feedback. In: S. J. Freedman (Ed.). *The Neuropsychology of Spatially Oriented Behavior.* Homewood, IL: Dorsey Press.

17. Merzenich, M. M., Kaas, J. H., Wall, J. T., Nelson, R. J., Sur, M., & Felleman, D. J. (1983). Topographic reorganization of somatosensory cortical areas 3b and 1 in adult monkeys following restricted deafferentation. *Neuroscience, 8,* 33–55.

18. Pons, T. et al. (1991). Massive cortical reorganization after sensory deafferentation in adult macaques. *Science, 252,* 1857–1860.

19. Ramachandran, V. S., Rogers-Ramachandran, D., & Stewart, M. (1992). Perceptual correlates of massive cortical reorganization. *Science, 258,* 1159–1160.

20. Lackner, J. R. (1988). Some proprioceptive influences on the perceptual representation of body shape and orientation. *Brain, 111,* 281–297.

21. Edelman, G. M. (1987). *Neural Darwinism.* New York: Basic Books, p. 221.

22. Rose, S. P. R. (1992). Selective attention. *Nature, 360,* 426–427.

23. Rapp, P. E., Bashore, T. R., Martinerie, J. M., Albano, A. M., Zimmerman, I. D., & Mees, A. I. (1989). Dynamics of brain electrical activity. *Brain Topography, 2,* 99–118.

24. For an excellent recent review, see Albano, A. M., & Rapp, P. E. (1993). On the reliability of dynamical measures of EEG signals. In: B. H. Jansen & M. E. Brandt (Eds.). *Nonlinear Dynamical Analysis of the EEG.* Singapore: World Scientific.

25. For example, references 23 and 24. See also Babloyantz, A., Nicolis, C., & Salazar, M. (1985). Evidence of chaotic dynamics of brain activity during the sleep cycle. *Physics Letters A, 111,* 152–156. For a review of the various dimension estimates, see Başar, E. (Ed.). (1990). *Chaos in Brain Function.* Berlin: Springer-Verlag. For more on the various problems surrounding dimension calculations, see Mayer-Kress, G., (Ed.), (1986). *Dimensions and Entropies in Chaotic Systems.* Berlin: Springer-Verlag.

26. Albano, A. M., & Rapp, P. E. (1993). On the reliability of dynamical measures of EEG signals, p. 135.

27. For a recent summary of electroencephalography and magnetoencephalography, see Wikswo, J. P., Gevins, A., & Williamson, S. J. (1992). The future of the EEG and MEG. *Electroencephalography and Clinical Neurophysiology, 87,* 1–9. For review of MRI-based methods, see Tank, D. W., Ogawa, S., & Ugurbil, K. (1992). Mapping the brain with MRI. *Current Biology, 2,* 525–528.

28. See reference 7 for details. For additional discussion and analysis, see Kelso, J. A. S. (1992). Coordination dynamics of human brain and behavior. In: R. Friedrich & A. Wunderlin (Eds.). *Evolution of Dynamical Structures in Complex Systems.* Springer Proceedings in Physics, Vol. 69. Berlin, Heidelberg: Springer-Verlag; Fuchs, A., & Kelso, J. A. S. (1993). Self-organization in brain and behavior: Critical instabilities and dynamics of spatial modes. In: B. H. Jansen & M. E. Brandt (Eds.). *Nonlinear Dynamical Analysis of the EEG.* Singapore: World Scientific, pp. 269–284; and Fuchs, A., & Kelso, J. A. S. (1993). Pattern formation in the human brain during qualitative changes in sensorimotor coordination. *World Congress on Neural Networks Proceedings, 4,* 476–479.

29. I refer here to the independent work of Eckhorn, Gray, and Singer and their colleagues. For a good review see Bressler, S. L. (1990). The gamma wave: A cortical information carrier? *Trends in Neurosciences, 13,* 161–162.

30. Eckhorn, R., Bauer, R., & Reitbock, H. J. (1989). Discontinuities in visual cortex and possible functional implications: Relating cortical structure and function with multielectrode/correlation techniques. In: E. Başar & T. H. Bullock (Eds.). *Brain Dynamics.* Berlin: Spring-Verlag.

31. Fuchs, A., Friedrich, R., Haken, H., & Lehmann, D. (1987). Spatiotemporal analysis of multichannel α-EEG map series. In: H. Haken (Ed.). *Computational Systems—Natural and Artificial.* Berlin: Springer-Verlag; and Friedrich, R., Fuchs, A., & Haken, H. (1991). Synergetic analysis of spatiotemporal EEG patterns. In: A. V. Holden, M. Markus, & H. G. Othmer (Eds.). *Nonlinear Wave Processes in Excitable Media.* New York: Plenum Press.

32. Jirsa, V., Friedrich, R., Haken, H., & Kelso, J. A. S. (1994). A theoretical model of phase transitions in the human brain. *Biological Cybernetics, 71*, 27–35.

33. The Šil'nikov condition for the existence of chaos has been confirmed in EEG time series obtained during petit mal epilepsy seizures. See Friedrich, R., & Uhl, C. (1992). Synergetic analysis of human electroencephalograms: Petit-mal epilepsy. In: R. Friedrich & A. Wunderlin (Eds.). *Evolution of Dynamical Structures in Complex Systems.* Springer Proceedings in Physics, Vol. 69. Berlin, Heidelberg: Springer-Verlag.

34. Kelso, J. A. S., & Fuchs, A. (1995). Self-organizing dynamics of the human brain: Intermittency, antimonotonicity & Šil'nikov chaos. *Chaos*, in press; and Arneodo, A., Coullet, P. H., Spiegel, E. A., & Tresser, C. (1985). Asymptotic chaos. *Physica, 14D*, 327–347.

35. For example, Ding, M., & Kelso, J. A. S. (1991). Controlling chaos: A selection mechanism for neural information processing? In: D. Duke & W. Pritchard (Eds.). *Measuring Chaos in the Human Brain.* Singapore: World Scientific. Control of chaos and chaotic synchronization are now burgeoning topics in the applications of dynamical systems theory. See, for instance, Schiff, S. J. et al. (1994). Controlling chaos in the brain. *Nature, 370*, 615–620.

36. Kelso, J. A. S., & Fuchs, A. (1995). Self-organizing dynamics of the human brain.

Index